# THE SPAM BOOK

On Viruses, Porn, and Other Anomalies
from the Dark Side of Digital Culture

# THE SPAM BOOK

## On Viruses, Porn, and Other Anomalies from the Dark Side of Digital Culture

*edited by*

Jussi Parikka
*Anglia Ruskin University*

Tony D. Sampson
*University of East London*

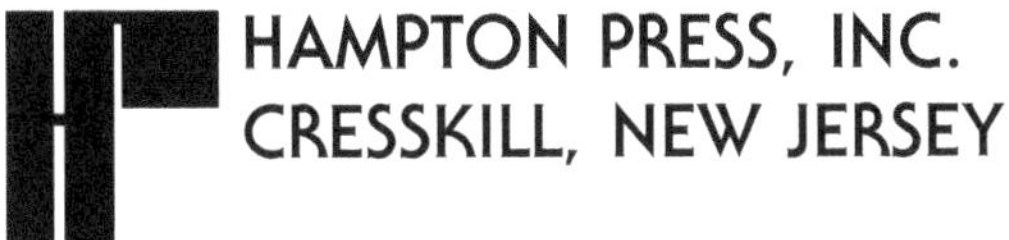

HAMPTON PRESS, INC.
CRESSKILL, NEW JERSEY

Library of Congress Cataloging-in-Publication Data

The spam book : on viruses, porn, and other anomalies from the dark side of digital culture / edited by Jussi Parikka, Tony D. Sampson.
    p. cm. -- (Hampton Press communication series : Communication alternatives)
    Includes bibliographical references and index.
    ISBN 978-1-57273-915-4 (hardbound) -- ISBN 978-1-57273-916-1 (paperbound)
  1.  Computers--Moral and ethical aspects. 2.  Internet--Moral and ethical aspects.
3.  Technology--Moral and ethical aspects. 4.  Spam (Electronic mail) 5.  Computer viruses 6.  Internet pornography.  I. Parikka, Jussi, 1976- II. Sampson, Tony D.
  QA76.9.M65S63 2009
  303.48'34--dc22

                        2009018269

Hampton Press, Inc.
23 Broadway
Cresskill, NJ 07626

# CONTENTS

# FOREWORD

## *Sadie Plant*

They may seem quintessentially Dutch, but tulips came from far away: Although the flat patchwork fields of the modern tulip industry are an impressive sight, it is still in the springtime gardens on the shores of the Bosphorous that the flowers can be seen at their best. When they were introduced to Holland in the early 17th century, tulips found an eager audience in what was already a highly cultured world: Northern Holland has been described as Europe's first modern economy, and people certainly had the money and the sensibility to pursue the exotic and treasure the rare. The basic flower was attractive enough. But when it came to cultivating tulips, the plain, primary-colored varieties excited far less interest than the frilly, multicolored, streaked, and patterned strains.

These varieties were also highly individuated and difficult to cultivate: For reasons quite unclear at the time, the strains could not be reproduced by cultivating their seeds, and outgrowths from the bulbs were unpredictable and slow, taking 2 to 3 years to flower. There was also an element of randomness: Dramatic changes could occur in the same bulbs from one season to the next.

All this and much besides meant that individual bulbs began to attract extremely high prices: Even in the 1620s some cost as much as 1,000 guilders; by the 1630s, they were priced at 6,000 guilders—more than enough to buy a luxurious house. Trading was frenetic, ridiculous even, as bidders raised prices beyond all reason. Bulbs began to function as a currency, and options were traded hundreds of times on futures markets while they were still maturing in the ground.

Of course, there was a crash, a day when someone woke up and thought: Why did I just sell my house for the promise of a flower? What am I doing? I must be mad! The crash, at the end of the 1630s, was sudden and devastating, and remains one of the most celebrated examples of economic collapse.

"Tulipmania" was an economic bubble that took much of the Dutch and the wider European financial system with it when it burst. What had fueled the whole thing to such heights? In the 1920s it was discovered that the exotic variations of what had become known as "the flowers that drove men mad" were not the result of careful genetic development, but the consequence of a virus, an anomaly in otherwise healthy, but less variegated plants. The multicolored stripe and ornate patterns were symptoms of disease.

Viruses are largely judged in negative terms. Like all such oddities, they are seen as mistakes, spanners in the works, bugs in the system, diseases, malfunctions, irregularities. Viruses are only the beginning: The digital world is awash with such anomalies—the spam that fills inboxes, the worms that crawl around the Internet, all the junk and detritus that flows through the gutters of cyberspace. And yet it is clear that they can often have extremely productive and creative effects, as in the case of the infected tulip bulbs: Their viral contagion can indeed be said to have had many beneficial consequences, at least in the context of 17th-century European aesthetics and sensibilities. It went on to inspire a passion for tulips that established the region as the world's largest bulb producer, and did much to determine later Western tastes in ornamental flowers and their representations as well. But such positive judgments are as problematic as the negatives. The anomalies in question are rarely purely destructive phenomena, but nor are they heroic subversives locked in battle with the forces of logos or the state, lively irregularities capable of interrupting and destabilizing a world of conformity. They may sometimes play such roles, just as they are sometimes uncompromisingly damaging. More often, however, they are so mixed up with the lives of their hosts that it is almost impossible to judge them as anything other than vital elements of the systems they traverse. Otherwise, all that can be said in judgment of an anomaly like the 17th-century tulip virus is that its effects were variously and arguably both good and bad, destructive and creative, positive and negative: The flowers looked good, but the bulbs were sick; money was made and lost as well.

Long after Nietzsche, the question of whether things are good or evil, positive or negative, normal or strange remains on the tip of the collective tongue. But the exploration of cultural phenomena should not be confined to judgments about rights and wrongs, the purposes and meanings of processes and events. And the search for significance seems particularly spurious when one's material is all the gloriously meaningless junk, noise, interferences, and seediness that give the Internet so much of its character. This does not, however, render the content of all this communication irrelevant: Indeed, the contents are often perfect expressions of the networks on which they run. The underground routes, the back doors and dark alleys of the Internet play host to the same kind of questionable deals and sordid offerings that move through back doors and dark alleys everywhere. On and off the Internet, certain ways of dealing with information and doing business tend to attract certain kinds of service and commodity. And all the temptations of money and porn that fill the inboxes of the world, the promises of better financial or sexual performance, penises, partners, or porn, are reminders that digital networks do not stand alone, but are always intimately implicated with their users and all the plays of power and desire in which they are involved.

It is, at first glance, quite amazing that the formal, regulated, logical world of computing should have spawned so many weird and wonderful forms of digital wildlife. Of course, the very purity of the logic, the smoothness of the system, tends to exacerbate the effects of even the slightest disturbance, and even to provoke it too: There is always an excess in play. But once computers, and their users, went online, the networks that emerged were highly complex and volatile. With complexity comes a certain degree of instability and unpredictability that can be fatally destabilizing, but also drives change and innovation by making networks vulnerable, and so open to new influences and opportunities.

They may make problems for the system, but they are often necessary to it too, not least so that it can define its own limit, establishing and policing the boundaries between this information and that noise, this logic and that nonsense. Even as it shuts them out, it needs its anomalies: Police and thieves stand opposed to each other, but are also locked into close, symbiotic relationships, and constantly evolve in relation to each other too. They are neither simple enemies nor dialectical partners, but tendencies at work in all the countless scales and speeds of activity in the complex system they become.

Perhaps it is all a question of perspective and time. The virus that is malevolent for the bulb is beneficial in relation to the flower. The code that breaks through one security barrier is also the incentive to develop new lines of defense. Short-term anomalies can be crucial to long-term stability, and vice versa: At least 8% of the human genome is composed of retroviruses that would have threatened the body at one time but are now simply part of the code. It depends how far back one is willing to step: Perhaps the very fact of digital technology is an anomaly in an analogue world; perhaps the earth itself is an anomaly; maybe humanity, with all its technology, was already anomalous in a tool-free world.

Tulipmania was certainly a great irregularity, a malfunctioning of 17th-century financial markets causing the first such large-scale economic crash. It was a kind of fever: The craze was as infectious as the virus itself, a runaway sequence of events triggered by the smallest of anomalies—which was, as it happens, effectively repressed as soon as its nature was known: once it was discovered, after nearly three centuries, that a disease was the agent of tulip variegation, the virus was eliminated by the tulip industry. Modern striped, multicolored, and frilled tulips are the flowers of healthy bulbs, bred to emulate those of their virally infected predecessors: The effects remain, but the virus has gone. Order has been restored.

And order was restored in the markets too, which were nevertheless transformed by the experience. They cycle keeps going—or rather countless cycles, all interweaving and overlapping and operating at many different scales and speeds, keep coming and going, repeating themselves in networks that are nevertheless never quite the same.

Digital anomalies are all things to all people and all networks: They are subversive of order and complicit in its maintenance; opposed and produced by the systems they traverse. What can be said—and what is brilliantly demonstrated in this book—is that the study of the seedy, chaotic underbelly of what otherwise appear to be smooth and highly regulated systems is not only fascinating in its own terms, but also crucial to an understanding of the networked world and everything with which it interacts.

# ON ANOMALOUS OBJECTS
# OF DIGITAL CULTURE

## An Introduction

*Jussi Parikka*

*Tony D. Sampson*

In the classic 1970s Monty Python sketch, a couple enters, or rather, in typical Pythonesque mode, descend upon a British cafe and are informed by the waitress that Spam is on the menu.

> There's egg and bacon; egg sausage and bacon; egg and spam; egg bacon and spam; egg bacon sausage and spam; spam bacon sausage and spam; spam egg spam spam bacon and spam; spam sausage spam spam bacon spam tomato and spam.

The joke, of course, refers to Spam, the canned food substance that originated in the 1930s in the United States, but was famously imported into Britain during World War II.[1] Spam (spiced ham) became a cheap supplement for pure meat products, which were in severe shortage during the conflict. Perhaps the cheapness and mass consumption of Spam during the period are among the reasons why it became the butt of many music hall jokes. Indeed, following the music hall tradition, Spam becomes central to the Python's often nonsensical sketch as it quickly deterritoralizes from the more obvious context of the waitress–customer discussion to a full Viking chorus of spam, spam, spam . . .

Spam, spam, spam, spam. Lovely spam! Wonderful spaaam! Lovely spam! Wonderful spam. Spa-a-a-a-a-a-a-am! Spa-a-a-a-a-a-am! Spa-a-a-a-a-a-am! Spa-a-a-a-a-a-am! Lovely spam! (Lovely spam!) Lovely spam! (Lovely spam!) Lovely spaaam! Spam, spam, spam, spaaaaam!

The joke's intention, as Monty Python jokes in general tend to do, is to get us to laugh at a major concern of contemporary communications: *communication breakdown.*[2] The habitual repetition of everyday events quickly turns into a chaotic mess and a turbulent example of noncommunication. The familiar communication channels of this archetypal British working-class cafe are suddenly flooded with intruding thirds, a noise that fills the acoustic space with a typically meaningless Python refrain: spam, spam, spam. In this sense (or nonsense), the sketch manages to parody the meaninglessness intrinsic to any meaningful act of communication by increasing the level of environmental noise that accompanies the process of sending messages. In fact, the invading Viking horde (perhaps a veiled reference to the U.S. troops stationed in Britain during World War II) eventually drowns out, or "spams," the ongoing conversation between the waitress and the customers, transforming the chaotic scene into a closing title sequence filled with more references to spam, spam, spam . . .

More than 30 years later, and the analogy made between Python's sketch and the unsolicited sending of bulk e-mail has provided new impetus to the word *spam*. Perhaps for many of us digital spam is less funny. For those of us increasingly reliant on e-mail networks in our everyday social interactions, spam can be a pain; it can annoy; it can deceive; it can overload. Yet spam can also entertain and perplex us. For example, how many of you have recently received an e-mail from "a Nigerian Frind" (*sic*) or a Russian lady looking for a relationship? Has your inbox overflowed with the daily announcements of lottery prizes and cut price Viagra? Perhaps you have experienced this type of Dadaist message, which appears at the zero degree of language.

Dehasque Little Bergmann
Dewald Murray Eriksson Tripathy
Gloo Janusauskas Nikam Lozanogmjpkjjpjrfpklkijnjkjflpkqkrfijmjgkkj
kgrkkksgpjmkqjmkggujfkrkpktkmmmjnjogjkhkhknjpgghlhnkofjgp
gngfgrgpkpgufifggmgmgugkirfsftkgtotutmumpptituuppmqqpgpjpkq
qquqkuqqiqtqhqnoppqpruiqmqgnkokrrnknslsifhtimiliumgghftfpfnfsf
nfmftflfrfjhqgrgsjfflgtgjflksghgrrgornhnpofsjoknoofoioplrlnlrjim
jmkhnltlllrmthklpljpuuhtruhupuhujqfuirorsrnrhrprtrotmsnsonjrh
rhrnspngslsnknfkfofigogpkpgfgssgqfsgmgti qfrfskfgltttjulpsthtrmkhnilh
rhjlnhsisiriohjhfhrftiuhfmuiqisighgmnigi gnjsorgstssslolsksiskrnrnsf-
spptngqhqitpprpnphqrtmprph.[3]

Generally speaking, however, spam arouses a whole panorama of negative and bemused emotions, in much the same way as computer viruses, worms, and the uninvited excesses of Internet pornography often do. In fact, we might collectively term these examples as *digital pollution* and identify them as a major downside (or setback) to a communication revolution that promised to be a noiseless and friction-free *Road Ahead*.[4] In this context, and against the prescribed and often idealized goals of the visionaries of digital capitalism, they appear to us as *anomalies*. Nevertheless, despite the glut of security advice—a lot of which is spuriously delivered to our e-mail inboxes, simply adding to the spam—little attention has been paid to the cultural implications of these anomalous objects and processes by those of us engaged in media and communication studies, and particularly studies linked to digital network culture. Perhaps we have been too busy dealing with the practical problem and have failed to ask questions of anomalies in themselves.[5] The innovation of this volume is to answer these questions by considering the role of the anomaly in a number of contexts related to digital communication and network culture. However intrusive and objectionable, we argue that the digital anomaly has become central to contemporary communication theory. Along these lines, we begin this book by asking: *"In what sense are these objects anomalous?"*

If we constrain ourselves to the dictionary definition of the anomalous, as the *unequal, unconformable, dissimilar,* and *incongruous,* in other words, something that deviates from the rule and demonstrates irregular and abnormal behaviour or patterns,[6] then arguably our question becomes problematized by everyday experiences of network culture. To be sure, spam, viruses, worms, and Internet porn are not irregular or abnormal in this sense. This *junk* fills up the material channels of the Internet, transforming our communications experiences on a daily or even hourly basis. For example, according to recent moderate sources, 40% of e-mail traffic is spam, meaning some 12.4 billion spam mails are being sent daily.[7] Similarly, in an experiment using a "honeypot" computer as a forensic tool for "tracking down high-technology crime," a team from the BBC in the United Kingdom recently logged, on average, one attack per hour that could render an unprotected machine "unusable or turn it into a [zombie] platform for attacking other PCs."[8] It is therefore not surprising that many network users fear everyday malicious Internet crime more than they do burglary, muggings, or a car theft.[9] Indeed, within the composite mixture of the everyday and the anomalous event, the fixed notion that the normal is opposed to the abnormal is increasingly difficult to reconcile.

It is from this cultural perspective that we approach the network anomaly, arguing that the unwelcome volume of anomalous traffic informs multiple articulations concerning the definition of the Internet and how the network space is becoming transformed as a means of communication. For example, in the late 1990s, network culture was very much defined in terms of the economic potential of digital objects and tools, but recently the dominant discourse has tilted toward describing a space seemingly *contaminated* by digital waste products, dirt, unwanted, and illicit objects.[10] There are, indeed, a number of ways in which anomalies feedback into the expressive and material components of the assemblages that constitute network culture. On the one hand, network security businesses have established themselves in the very fabric of the digital economy (waste management is the future business model of late modernity). The discourses formed around this billion-dollar security industry, ever more dependent on anomalies for its economic sustenance, lay claim to the frontline defense of network culture against the hacker, the virus writer, and the spammer, but they also shape the experiences of the network user. On the other hand, analysis of the build up of polluted traffic means that evaluations are made, data is translated into prediction models, and future projects, such as Internet 2.0 and other "spam and virus-free" networks, are proposed as probable solutions to the security problems facing online businesses and consumers. In other words, anomalies are continuously processed and rechanneled back into the everyday of network culture. Whether they are seen as novel business opportunities or playing the part of the unwanted in the emerging political scenarios of network futures, anomalous objects, far from being abnormal, are constantly made use of in a variety of contexts, across numerous scales. Therefore, our aim in this introduction is to primarily address the question concerning anomalies by seeking conceptual, analytic, and synthetic pathways out of the binary impasse between the normal versus the abnormal.

In our opinion, what makes this collection standout, however, is not only its radical rethinking of the role of the anomalous in digital culture, but that all of the contributions in the book in one way or another mark an important conceptual shift away from a solely representational analysis (the mainstay of media and cultural studies approach to communication). Rather than present an account of the digital anomaly in terms of a representational space of objects, our aim as editors has been to steer clear of the linguistic categorizations founded on resemblances, identities, oppositions, and metaphorical analogies. In our opinion, the avoidance of such representational categorizations is equal to rejecting the implicit positioning of a pre-fabricated grid on which the categories identified constrain or shackle the object. For us, judging a computer virus as a metaphor of a biological virus all too easily reproduces it to the same fixed terms conjured up in the metaphor in itself and does not provide any novel information concerning the intensive capacities of, for example, a specific class of software program. Hence, our desire is to avoid metaphorics as a basis of cultural analysis is connected to our wish to focus "less on a formation's present state conceived as a synchronic structure than on the vectors of potential transformation it envelops," to use Brian Massumi's words.[11] Furthermore, we do not wish to blindly regurgitate the rhetoric of a computer security industry who peddle metaphorical analogies between the spread of computer viruses and AIDS. For that reason, we have attempted to avoid the tendency to conjure up the essence of the digital anomaly from a space considered somehow *outside*—a space populated by digital *Others* or out-of-control Artificial Intelligence pathogens engaged in a evolutionary arms race with a metaphorical immune systems.[12] In this sense, the reference to the *dark side of digital culture* in the subtitle of this book is more closely allied to our understanding of the darkness surrounding this type of representational analysis than it is the darkness of the object in itself. We intend to address this lack of light (or lack of analysis) by considering a conceptual approach that is more fluid, precise, and inventive in terms of a response to the question of the anomaly. It is designed to grasp the liminal categories and understand the materiality and paradoxical inherency of these weird "objects" and processes from theoretical and political points of view.

We do nevertheless recognize that on material and representational levels, spam and other anomalies do have effects. But in this volume, we acknowledge the problems inherent to the deployment of a media theory based exclusively on effect.[13] To be sure, in the past, this is how media anomalies such as violence and pornography have been treated by certain field positions within media and communication studies—the effects of which were considered to *cultivate* an audience's sense of reality.[14] Perhaps our approach will be seen as more Monty Python than George Gerbner, in as much as we are less interested in the causality afforded to the impression of media *meanings* than we are in the *process of communication* in itself. Yet, this does not indicate a return to the transmission model so prevalent in early communication theory, wherein the establishment of communicative fidelity between Sender A and Recipient B, in the midst of signal noise, is the basic setting. On the contrary, instead of the linear channeling of messages and the analysis of effects, one might say that this book is concerned with *affect* and *ethology*: how various assemblages of bodies (whether technological, biological, political or representational) are composed in interaction with each other and how they are defined, not by forms and functions, but by their capabilities or casual capacities. In other words, we are interested in how one assemblage, a heterogeneous composition of forces, may affect another.[15] Later, we refer to this approach as *topological,* because we argue that it releases us from the analytical dichotomy between causal (fatal) effects and complete indeterminism, and allows us to instead consider a co-causal, intermediate set of determinisms and nonlinear bodies. Significantly then, we use the term *topology* to address the complex assemblages of network society, which are not restricted to technological determinism, or the effects technology has on society, but encompasses the complex foldings of technological components with other aspects of social and cultural reality.[16]

Importantly, in this analytical mode, we are not seeking out the (predefined) *essence* of the anomaly (whether expressed in terms of a representational category or intrinsic technical mechanism), but instead a process in a larger web of connections, singularities, and transformations. Therefore, our question positions the anomaly in the topological fabric of an assemblage from where new questions emerge. For example, how do operating systems and software function in the production of anomalous objects? In what kind of material networks do such processes interact? How are certain software processes and objects translated into criminal acts, such as vandalism, infringement, and trespass?[17] We now elaborate on this theoretical position from a historical perspective, before addressing the questions of affects, topology, and anomalous objects.

## MEDIA ANOMALIES: HISTORICAL CONTEXT

Analysis of media in terms of the anomaly is nothing new. There are, in fact, many approaches that implicitly or explicitly address anomalous media. A number of well-known approaches that should be familiar to the media and communication field, including the Frankfurt School and the media-ecological writings of the Toronto School (including Neil Postman), have regarded (mass) media in itself as an anomaly. Of course, these approaches do not concern a strict deviation from the norm or events *outside of a series*, as such. Instead, the dangerous anomaly has long been regarded as a function of the homogenizing powers of popular media. The repetitious standardization of media content is seen as a result of the ideological capitalist-state apparatus, which applies the logic of the factory assembly line to the production of cultural artefacts. For the Frankfurt School, particularly Adorno and Horkheimer, analysis of mass media revealed a system of consumer production in conflict (but also as a paradoxical fulfillment of) with the enlightenment project via mass ideological deception.[18] Later, Postman continues along similar lines by conceptualizing the modern mass media, especially television, as a kind of a filter that hinders public discourse by allowing only programs and other "objects" with entertainment value to pass through communication channels.[19] As an index of this dystopic understanding of mass media, some years later the former Pink Floyd songwriter Roger Waters transposed these *weapons of mass distraction* and apocalyptic visions of Western media culture into his conceptual album *Amused to Death* (1992), where the TV sucks in all human emotion while the human species amuses itself to death watching *Melrose Place*, the Persian Gulf War, and copious amounts of porn. Indeed, in this way the media machine is treated as a monstrous anomaly, and significantly, a totality rather than a singularity.

In a historical context, the shock of the "new" media seems to have always occupied a similar polemical space as the one that obsessed the conservative approaches of media effects theorists, like Gerbner. The anomalies of the new media are most often surrounded by moral panics. Such panics, whether around cinema, television, video, computer games, or the Internet, with its malicious dark side, populated by perverts lurking around every virtual corner, can perhaps be seen as an attempt to contextualize new media in existing social conventions and habits of the everyday. The media panics surrounding the Internet, for example, have highlighted the contradiction between the ideals of a reinvigorated public sphere—an electronic agora for scientists, academics, politicians, and the rest of civil society—and the reality of a network overflowing with pornography, scams, political manipulation, piracy, chat room racists, bigots, and bullies. In recent years we have seen how the Internet has been transformed from a utopian object into a problematic modulator of behavior, including addiction, paedophilia and illicit downloading. It has become an object for censorship—necessitating the weeding out of unpleasant and distasteful content, but also the filtering of politically sensitive and unwanted exchange.[20] In fact, in wake of the Jokela high school shootings in Finland in November 2007, there are those who claim, like they did after Columbine, that it is not the guns, but the Internet that is to blame. The uncontrollable and uncensorable flood of damaging information is still grasped as more dangerous than the impact of firearms.

The emergence of inconsistencies and deviations in media history has led Lisa Gitelman to argue that we should "turn to the anomaly" and concentrate on the patterns of dissonance that form when new media encounter old practices. For Gitelman, "transgressions and anomalies . . . always imply the norm and therefore urge us to take it into account as well."[21] Therefore, anomalies become a tool of the cultural analyst, enabling him or her to dig into the essential, so to speak. They can be imagined as vehicles taking us along the lines of a logic that delineates the boundaries between the normal and the abnormal. But in our view such approaches do not dig deeply enough into the logical mode of the anomaly since there is always a danger that such a representational analysis will continue to treat it as an excluded partner (*Other*) who haunts the normalized procedures of the *Same*.

Alternatively, we argue that network culture presents us with a new class of anomalous software object and process, which cannot be solely reduced to, for example, a human determined representation of the capitalist mode of consumerism.[22] The examples given in this collection—*contagious software*, *bad objects*, *porn exchange*, and modes of *network censorship*—may well derive some benefit from representational analysis (particularly in the context of porn and spam e-mail content),[23] but our anomalies are not simply understood as irregular in the sense that their content is *outside of a series*. On the contrary, they are understood as expressing another kind of a topological structuring that is not necessarily derived from the success of friction-free ideals as a horizon of expectancy. The content of a porn site,[24] a spam e-mail, or a computer virus, for instance, may *represent* aspects of the capitalist mode of production, but these programs also express a materiality, or a logic of action, which has been, in our opinion, much neglected in the media and communication field. This is a logical line in which automated excessive multiple posting, viral replication, and system hijacking are not necessarily indices of a dysfunctional relation with a normalized state of communication, but are rather capacities of the software code. Software is not here understood as a stable object or a set of mathematically determined, prescribed routines, but as the emergent field of critical software studies is proposing, it is a process that reaches outside the computer and folds as part of the digital architectures, networks, social, and political agendas. When we combine this capacity of software with our focus on the dynamics of the sociotechnical network assemblage, in its entire broadband spectrum, we experience systems that transfer massive amounts of porn, spam, and viral infection. Such capacity, which in our view exceeds the crude distinction between normal and abnormal, becomes a crucial part of the expressive and material distribution of network culture. Porn, spam and viruses are not merely representational; they are also component parts of a sociotechnical-logical praxis. For us, they are a way of tapping into and thinking through the advanced capitalist mode in the context of the network.

We therefore suggest that the capacity of the network topology intimately connects us to a post-Fordist mode of immaterial labour and knowledge production. We do not however prescribe to a strictly defined cybernetic or homeostatic model of capitalist control (a point explained in more detail later), which is designed to patch up the nonlinear flows deemed dangerous (like contagions) to the network. On the contrary, our conception of capitalism is a machine that taps into the creative modulations and variations of topological functioning.[25] Networks and social processes are not reducible to a capitalist determination, but capitalism is more akin to a power that is able to follow changes and resistances in both the extensive and intensive redefining of its "nature." It is easy at this point to see how our vision of the media machine no longer pertains to the anomalous totality described by the Frankfurt and Toronto Schools. Like Wendy Chun, we see this machine as an alternative to the poverty of an analysis of the contemporary media sphere as continuously articulated between the polarity of narratives of total paranoid surveillance and the total freedom of digitopia. Therefore, following Chun, in order to provide a more accurate account of the capacities of media technologies as cultural constellations, this book looks to address networked media on various, simultaneously overlapping scales or layers: hardware, software, interface, and extramedial representation ("the representation of networked media in other media and/or its functioning in larger economic and political systems").[26]

Such an approach has led us and other contributors to draw on a Deleuze-Guattarian framework. We might also call this approach, which connects the various chapters of this book, an assemblage theory of media.

Yet, in order to fully grasp this significant aspect of our analysis, it is important to see how Deleuze and Guattari's meticulous approaches to network society can be applied beyond the 1990s hype of "the rhizome" concept and what we see as its misappropriation as a metaphor for the complexities of networked digital media. In this sense, most contemporary writers using Deleuze and Guattari have been keen to distance themselves from a metaphorical reading of cultural processes in the sense that "metaphoricity" implies a dualistic ontology and positions language as the (sole) active force of culture (see the Contagions section introduction for more discussion). In this context, Deleuze and Guattari have proved useful in having reawakened an appreciation of the material forces of culture, which not only refer to economic relationships, but to assemblages, events, bodies, technologies, and also language expressing itself in other modalities other than meaning. Not all of the chapters are in fact locked into this framework, yet, even if they do not follow the precise line, they do, in our opinion, attempt to share a certain post-representational take which is reluctant to merely reproduce the terms it criticizes and instead explores the various logics and modes of organization in network culture in which anomalies are expressed. Importantly, the chapters are not focused on the question of how discourses of anomalous objects reproduce or challenge the grids of meaning concerning ideology and identity (sex, class, race, etc.) but rather they attempt to explore new agendas arising beyond the "usual suspects" of ideology.

We now move on to explore the topological approach in more detail, proposing that it can do more than simply counter representational reductionism. First, we specify how it can respond to the fault lines of essentialism. Then we use it to readdress a mode of functionalism that has pervaded the treatment of the anomaly from Durkheim to cyberpunk.

## TOPOLOGICAL THINKING: THE ROLE OF THE ACCIDENT

In order to further illuminate our question concerning the anomalies of contemporary communication, let us return to the Monty Python sketch for further inspiration and a way in which we might clearly distinguish between a prevalent mode of essentialism and our topological approach. Following strictly essentialist terms we might define Python's cafe by way of the location of the *most important* and familiar communication codes;[27] looking for the effective functioning of communication norms. In this mode, we would then interpret the "spamming" of the cafe as an oppositional function, setting up certain *disparate* relations between, on the one hand, a series of perfected communication norms, and on the other hand, the imperfection of our anomaly. Yet, arguably, the Python sketch does more than establish dialectical relations between what is *in* and what is *outside a series*. Instead, Python's comedy tactic introduces a wider network of reference, which unshackles the unessential, enabling the sketch to breach the codes of a closed communication channel, introducing fragments of an altogether different code. Thus, in the novel sense of topological thinking, the British cafe becomes exposed to the transformational force of spontaneous events rather than the static essences or signs of identity politics.

In a way, Monty Python suggests an anti-Aristotelian move, at least in the sense proposed by Paul Virilio: that is a need to reverse the idea of accidents as contingent and substances as absolute and necessary. Virilio's apocalyptic take on Western media culture argues for the inclusion of the potential (and gradual actualization) of the general accident that relates to a larger ontological shift undermining the spatio-temporal coordinates of culture. In a more narrow sense, Virilio has argued that accidents should be seen as incidental to technologies and modernity. This stance recapitulates the idea that modern accidents do not happen through the force of an external influence, like a storm, but are much more accurately follow-ups or at least functionally connected with, the original design of that technology. In this way, Virilio claimed that Aristotelian substances do not come without their accidents, and breakdowns are not the absence of the presumed order, but are rational, real and designed parts of a media cultural condition: the "normal" state of things operating smoothly.[28] With Monty Python, as with Deleuze, the structures of anticipation and accidentality are not simply reversed, but the anomalous communication event itself emerges from within a largely accidental or inessential environment.[29]

To analyze the material reality of anomalous objects, we must therefore disengage from a perspective that sees the presumed friction-free state of networking, the ideal non-erring calculation machine, or a community of rational individuals using technologies primarily for enlightenment as more important than the anomaly (spam, viruses, and porn merely regarded as secondary deviations). Indeed, in our view, accidents are not simply sporadic breakdowns in social structure or cultural identity, but express the topological features of the social and cultural usage of media technologies. In this context, we concur with Tiziana Terranova,[30] who discusses network dynamics as not simply a *"space of passage* for information," but a milieu that exceeds the mechanism of established communication theory (senders, channels, and receivers). The surplus production of information comprises a turbulent mixture of mass distribution, contagion, scams, porn, piracy, and so on. The metastability of these multiple communication events are not merely occurrences hindering the essence of the sender–receiver relation, which generally aims to suppress, divide, or filter out disparities altogether, but are instead events of the network topology in itself. The challenge then, is not to do away with such metastabilities, but to look at them in terms of an emergent series and experience them as the *opening up* of a closed communication system to environmental exteriority and the potentialities that arise from that condition. A condition we can refer to as the inessential of network culture.

## THE TOPOLOGICAL SPACE
## OF "BAD" OBJECTS

> If all things have followed from the necessity of the most perfect nature of God, how is it that so many imperfections have arisen in nature—corruption, for instance, of things till they stink; deformity, exciting disgust; confusion, evil, crime, etc.? But, as I have just observed, all this is easily answered. For the perfection of things is to be judged by their nature and power alone; nor are they more or less perfect because they delight or offend the human senses, or because they are beneficial or prejudicial to human nature.[31]

We have thus far argued that the anomaly is best understood in terms of its location in the topological dynamics of network culture. Significantly, in this new context then, we may also suggest that anomalies are not, as Spinoza realized, judged by the "presumed imperfections of nature" (nature representing a unity, as such), but instead they are judged by "their nature and power alone." In other words, it matters not if objects "delight or offend the human senses." Particular "things" and processes are not to be judged from an outside vantage point or exposed to "good" or "bad" valuations. Instead, the ethological turn proposes to look at the potentials of objects and ask how they are capable of expression and making connections.

In this way, the shift toward topological analysis becomes parallel to a perspective that claims to be "beyond good and evil" and instead focuses on the forces constituent of such moral judgments. This marks the approach out as very different from the historical tradition of social theory, particularly the early response of organic functionalists to the good and bad of social events. For example, Emile Durkheim was perhaps the first social scientist to show how anomie played an important part in social formations, but he negated the productive capacities we have pointed to in favor of describing the anomaly as a state of social breakdown. For Durkheim, the ultimate anomalous social act—suicide—stemmed from a sense of a lack of belonging and a feeling of remoteness from the norm. Anomaly as a social phenomenon therefore referred to a deprivation of norms and standards. Although suicide was positively disregarded as an act of evil, it did however signal a rupture in the organics of society, an abnormality, a falling out of series, as such.[32] Indeed, his statistical container model of macro society—much appreciated by the society builders of 19th-century Europe—judged social phenomena against the average, the essential, and the organic unity of social functionalism. This of course ruled out seeing anomalies as social phenomena with their own modes of operation and co-causal capacity to affect.

Baudrillard's notion of the perverse logic of the anomaly intervenes in the functionalist exorcism of the anomalous, as a thing that doesn't fit in.[33] Writing mainly about another bad object, drugs, Baudrillard argued that the anomaly becomes a component part of the logic of overorganization in modern societies. As he put it:

> In such systems this is not the result of society's inability to integrate its marginal phenomena; on the contrary, it stems from an overcapacity for integration and standardization. When this happens, societies which seem all-powerful are destabilized from within, with serious consequences, for the more efforts the system makes to organize itself in order to get rid of its anomalies, the further it will take its logic of overorganization, and the more it will nourish the outgrowth of those anomalies.[34]

Beyond the law-abiding notion of Durkheim's anomie Baudrillard, therefore, proposed to consider contemporary phenomena (the writing stems from 1987) as labeled by *excess*—a mode of hyperrational anomaly. He argued that the modern emphasis placed on control management has itself spurred on these excesses of standardization and rationality. The strange malfunctions become the norm, or more accurately, they overturn the logic of thinking in terms of self versus other. Moreover, in the perverse logic of Baudrillard's anomalous, the object, as an extensive target of social control, is preceded by an intensive logic that exceeds the grid of explanation imposed by social scientists, educationalists, and therapeutic practitioners. Instead of external deviations contrary to the internal functioning of the social, anomalies start to exhibit an intensive and integral social productivity.

The distinction made here between the intensive productivity of the anomaly and a social model developed around organic unity and functionalism is perhaps better grasped in DeLanda's similar distinction between relations *of interiority*, and *relations of exteriority*.[35] In the former, societies are regarded as solely dependent on reciprocal internal relations in order that they may exhibit emergent properties. In the latter, DeLanda seemingly turns the generalized social organism inside out, opening up its component parts to the possibilities and capacities of complex interactions with auxiliary assemblages. In fact, what he does is reconceive the social organism as an assemblage.

So as to further explore this notion of the social assemblage, let's return to the example of the forensic honeypot computer introduced in the first section of this introduction. Previously understood as a closed system, the rationalized logic machine soon becomes exposed to the disparities of the network. Emergent relations hijack the honeypot's functionality. Its relation to an exteriority links up these disparities and in turn connects it to other assemblages. It is at this juncture that we locate the transformational *differentiation* and *alterity* of the honeypot as it becomes inseparable from the relations it establishes with a multiplicity of other assemblages, populated by technosocial actors, including netbots, virus writers, cookies, and hacker groups and their software. Nevertheless, the anomalies that traverse the assemblage are not simply disparities that suppress or divide, but are instead the role of the anomaly intervenes in the process of the becoming of the honeypot. It establishes communication with other objects related to the assemblage, potentializing new territories or deterritorializing other assemblages.

Anomalies transform our experiences of contemporary network culture by intervening in relational paths and connecting the individual to new assemblages. In fact, the anomaly introduces a considerable amount of instability to what has been described in the past as a cybernetic system of social control.[36] In practice, the programs written by hackers, spammers, virus writers, and those pornographers intent on redirecting our browsers to their content, have problematized the intended functionality and deployment of cybernetic systems. This has required cyberneticians to delve deeply into the tool bag of cybernetics in an effort to respond to the problem engendered: *How to keep the system under control?* For experts in the computing field, defensive software, such as antivirus technology, represents a new mobilization of security interests across the entire networked computing environment instead of being exclusively aimed at single computers,[37] and it is interesting to see how many of these defences appear to play to the notion of organic unity as described earlier. For example, computer scientists based at IBM's TJ Watson Research Centre during the early 1990s attempted to tackle the problem of computer viruses by developing a cybernetic immune system.[38] Using mathematical models borrowed from epidemiology, these researchers began to trace the diffusion patterns of computer viruses analogous to the spread of biological viruses. Along with other commercial vendors, they sought out methods that would distinguish between so-called legitimate and viral programs. In other words, their cybernetic immune system was designed to automate the process of differentiating *self* from *nonself* and ultimately suppress the threshold point of a viral epidemic (the point at which a disease tips over into a full-blown epidemic).

However, the increasing frequency of digital anomalies has so far confounded the application of the immunological analogy. In fact, research in this area has recently shifted to a focus on topological vulnerabilities in the network itself, including a tendency for computer viruses to eschew epidemiological threshold points altogether.[39] Maps of the Internet and the

World Wide Web (www), produced by complex network theorists in the late 1990s,[40] demonstrate how networks become prone to viral propagation, as they would any other program. There is, as such, a somewhat fuzzy distinction between what can be determined as self and non-self. As we have already pointed out, the anomaly is not, in this sense, outside the norm.

The history of cybernetics provides many more examples of this problem where logic encounters network politics. The origins of Turing's theory of computational numbers was arguably realized in a paradoxical and largely unessential composition of symbolic logic, in as much as he set out to prove that anomalies coexisted alongside the axioms of formal logic.[41] Not surprisingly then, Turing's *halting problem,* or the *undecidability problem,* eventually resurfaced in Cohen's formal study of computer viruses, a doom-laden forecast in which there is no algorithmic solution to the detection of all computer viruses.[42] Indeed, logic systems have long been troubled by their inability to cope with virals. The problem of the self-referencing liar bugged the ancient Greek syllogistic system as much as it has bugged the contemporary cybernetics of network culture.

In this light, it is interesting to draw attention to the way in which these fault lines in cybernetics and Durkheim's anomie have converged in cyberculture literature. With its many references to Gaia[43] (a theory of natural balance and equilibrium akin to immunology) cyberculture has co-opted the principle of the self-referencing maintenance of organic unity into the fabric of the collectivities of cyberspace. For example, John Perry Barlow argued that the immune system response of the network is "continuously" defining "the self versus the other."[44] In this way, he typified the tendency of cyberpunk's frontier mentality to discursively situate the digital anomaly firmly *outside* of the homeostatic system of network survivability. In fact, as Bruce Sterling revealed, cyberpunks and the cyberneticists of the antivirus industry have become strange bedfellows:

> They [virus writers] poison the digital wells and the flowing rivers. They believe that information ought to be poisonous and should hurt other people. Internet people build the networks for the sake of the net, and that's a fine and noble thing. But virus people vandalize computers and nets for the pure nasty love of the wreckage.[45]

It seems that the much wished-for stability of the cyberpunk's *Daisyworld* is increasingly traversed by the instabilities produced by the anomaly. As Sterling noted in another context, "the Internet is a dirty mess"[46] that has lost its balance mainly because of the increasing outbreaks of cyberterrorism and cybercrime, but also because of the negligence of the authorities to adequately address the problems facing network culture. In Sterling's vision, which increasingly echoes those of the capitalist digerati, there is a horizon on which the network eventually becomes a clean and frictionless milieu. Yet such a sphere of possibility rests conceptually on the notion of homeostasis and stability, which sequentially implies a conservative (political) stance. In our view, it is more insightful to follow Geert Lovink's position that networking is more akin to *notworking*:

> What makes out today's networking is the *notworking*. There would be no routing if there were no problems on the line. Spam, viruses and identity theft are not accidental mistakes, mishaps on the road to techno perfection. They are constitutional elements of yesterday's network architectures. Networks increase levels of informality and also pump up noise levels, caused by chit-chat, misunderstandings and other all too human mistakes.[47]

We argue that the *noise* of Lovink's *notworking* not only throws a spanner in the works of the cybernetic system, but also more intimately connects us to the capacity of the network to affect and thus produce anomalies. Instead of seeing the network as a self-referential homeostatic system, we want to therefore propose an autopoietic view of networks wherein alterity becomes the mode of operation of this sociotechnical machine (even though, e.g., Lovink might be reluctant to use these concepts). So if we would want to approach network systems in a broad framework as autopoietic systems, one would need to emphasize their difference from an old ideal of harmonious determined Nature. Following Guattari,[48] we argue that systems are not structures that merely stabilize according to a predetermined task, but are instead machines composed in disequilibrium and a principle of abolition. Here, re-creation works only through differentiation and change, which are ontological characteristics of a system that relies continuously on its exterior (a network). The digital network is consequently composed in terms of a phylogenetic evolution (change) of machines, and importantly understood as part of a collective ecological environment. In this context, the maintenance project of any machine (social, technical, or biological system) cannot be simply confined to the internal (closed in) production of self, or for that matter the detection of non-self, but instead returns us to the individuation process (discussed earlier) and the continuance of what Guattari called the "diverse types of relations of alterity."[49] We argue that a condition akin to a *horror autotoxicus* of the digital network, the capacity of the network to propagate its own imperfections, exceeds the metaphor with natural unity. Indeed, despite a rather vague notion about the purposeful essence of network production as described by individuals like Bill Gates (something perhaps akin to Spinoza's "perfect nature of God"), the network itself is without a doubt the perfect medium for both perfection and imperfection.

# CONCLUSION: STANDARD OBJECTS?

We do not doubt that what we are dealing with here are very curious objects indeed. They present mind-boggling problems to system managers and network controllers. Yet, the failure to adequately overcome the computer virus problem perhaps pales in comparison to what *Wired Magazine* described as the next big issue for network security: the autonomous software netbots (or spambots) that are more flexible and responsive to system defences than the familiar model of pre-programmed computer viruses and worms. As *Wired* described the latest threat

> The operational software, known as command and control, or C&C, resides on a remote server. Think of a botnet as a terrorist sleeper cell: Its members lurk silently within ordinary desktop computers, inert and undetected, until C&C issues orders to strike.[50]

Here we see that the netbot becomes discursively contemporised in terms of a latent terrorist cell that evades the identification grid of an immune system. Possibly this marks a discursive shift away from the biological analogy with viruses and worms toward the new anxieties of the war on terror. Whatever the rhetoric, identification is perhaps the key contemporary (and future) problem facing not just computer networks, but networks of political power, wherein nonexistence (becoming invisible) can become a crucial tactical gesture, as Galloway and Thacker suggest in Chapter 13.

The invisibility of software objects has in practice confounded a media studies approach orientated toward a representational analysis of phenomenological "content." Software considered as a specific set of instructions running inside a computer is obviously something more akin to a performance, rather than a product of visual culture. To combat the often-simplistic analysis of software, Lev Manovich proposed, back in 2001, that media studies should move toward "software studies," and in doing so he provided an early set of principles for an analysis of new media objects. Manovich's principles of new media include *numerical representation, modularity, automation, variability* and *transcoding*. New media in this way is based on the primary layer of computer data—code—that in its programmability separates "new" from "old" media, such as print, photography, or television.[51] However, since then, Chun noted how Manovich's notion of transcoding— that software culture and computation is about translating texts, sounds, and images into code—is not a sufficiently rich notion.[52] Instead of registering (repeating) differences that pre-exist, Chun argued that computation makes differences and actively processes code in and out of various phenomenological contexts, such as text or sound. Her argument is supported by virus researchers who note that even a simple opening and closing of an application, or rebooting of a system, can make changes to boot sector files, log files, system files, and Windows' registry.

For example, opening and closing a Word document is a computational process that may result in, for example, the creation of temporary files, changes to macros, and so forth.[53] However, these processes do not directly come into contact with the human senses (we cannot always see, hear, touch, taste, or indeed smell an algorithmic procedure) and there is consequently a deficit in our cognitive and conceptual grasping of software objects and processes, as such. Yet, despite the *abstract* nature of mathematical media, these processes are completely real and demand attention from cultural theory, not least because the contemporary biopower of digital life functions very much on the level of the nonvisual temporality of computer network. This is why cultural theory needs to stretch its conceptual capacities beyond representational analysis and come up new notions and ideas in order to better grasp the technological constellations and networked assemblages of "anomalous media culture." By proposing novel concepts, like those suggested by Deleuze for example, we do not aim to prescribe a trendy cultural theory, but rather enable a rethink of the processes and emerging agendas of a networked future.

The anomalous objects discussed in this volume can therefore be taken as indices of this novel media condition in which complex transformations occur. Yet, while on an algorithmic and compositional level, the objects and processes highlighted in spam e-mails, computer viruses, and porn communities are not in anyway different from other objects and processes of digital culture, there is clearly a repetitious and discursive filtering process going on: *If software is computation that makes a difference (not just a coding of differences), then there is also a continuous marking out of what kind of processes are deemed as normal, abnormal, and/or anomalous.* In other words, there is an incessant definition and redefinition of what, on the one hand, makes a good computation, a good object, and a good process, and on the other hand, what is defined as irresponsible and potentially a bad object or process. However, as noted earlier, the material and expressive boundaries of these definitions are not at all clear. We may, in this light, therefore suggest that such turbulent objects are considered as standard objects of network culture.[54] Instead of merely being grasped as elements that should be totally excluded from the economic, productive, and discursive spheres of the knowledge society, they are equally understood as captured and used inclusively within the fabrication of digital assemblages. For example, the anomaly takes on new functions as an innovative piece of evolutionary "viral" or "spam" software (in digital architecture or sound production for instance), or is translated into new modes of consumer organization and activation (viral marketing), or becomes adapted to serve digital sociality in practices and communities (pornographic exchange). Ultimately, if capitalism is able to make novel use of these critical practices of resistance, then cultural and media theorists should do likewise. Otherwise, they will remain anomalies for a theory unable to perceive of new modulations of power and politics functioning on the level of software.

From the varied perspectives offered in this volume the reader will notice that our take on the anomaly is not considered sacrosanct—anomalous digital objects are distributed across many scales and platforms. However, we do feel that all of the following chapters intersect with our notion of the anomalous object, albeit provoking a controversy around its compositional theme. Therefore, in order to introduce a sense of organization to the mixture of viewpoints put forward in *The Spam Book* we have divided the chapters in subsections: *Contagions*, *Bad Objects*, *Porn,* and *Censored.* Each subsection has an introduction setting out how we, the editors, grasp the position and the value of each chapter. As we have already suggested, there are of course many takes on the digital anomaly, but what *The Spam Book* proposes to do is shed some light on what has, until now, remained on the dark side of media and communication and cultural analysis.

# PART I

# CONTAGIONS

---

## NO METAPHORS, JUST DIAGRAMS . . .

Digital contagions are often couched in analogical metaphors concerning biological disease. When framed in the linguistic structures of representational space, the biological virus becomes a master referent, widely dispersed in the fields of cultural studies, computer science, and the rhetoric of the antivirus and network security industries.[55] The figurative viral object becomes part of a semiotic regime of intrusive power, bodily invasion, uncontrollable contamination, and even new modes of auto-consumerism. Nevertheless, although representational analysis may have an application in a media age dominated by the visual image, the approach does not, in our opinion, fully capture the imperceptible constitutive role of contagion in an age of digital networks.

Indeed, when contemplating the metaphor of contagion, it is important to acknowledge two constraining factors at work. First, the analytical focus of metaphorical reasoning may well establish equivalences, but these resemblances only really scratch the surface of an intensive relation established between a viral abstraction and concrete contagious events. Second, it is important to recognize the political import of the analogical metaphor in itself. It has an affective charge and organizational role in the spaces, practices, and productions of digital network culture. For example, as seen in this section, the resemblances established between neo-Darwinian genes and computer viruses have imposed the logic of the arms race on evolutionary computing.[56] In conjunction with epidemiological and immunological analogies, the digital gene delimits the patterning of software practices, excluding the contagious anomaly from the norms of code reproduction.

In light of the often-divisive imposition of the metaphor in the materiality of digital network culture, it is important that the this volume provides a potential escape route out of the analytical constraints of representation. Gilles Deleuze could, in our opinion, function as one alternative thinker who provides a set of tools for a post-representational cultural analysis. Via Deleuze we can substitute the metaphoric burden of the biological referent on its digital "viral cousins" with an exposition of the constitutive role contagion plays in material spaces, time-based practices, and productions. In place of the negatives of the metaphor we find an abstract diagram with an affirmative relation to the concrete contagious assemblages of digitality. To be more concise, the diagrammatic refrain is what holds these assemblages together, or even more succinctly, attempts to delimit and control the identities of these larger unities. Think of the abstract diagrams used in this section as descriptions of the intensity of relations, repetitiously "installed" in the concreteness of the digital assemblages addressed in each chapter.

We begin the section with John Johnston's chapter on the computer viruses' relation to artificial life (ALife) research. Johnston loses the familiar metaphorical references to the spread of biological disease and instead explores the complex relationality between illicit virus production and the futures of ALife research. The chapter is stripped bare of the verbosity of Deleuzian ontology, yet arguably, the abstract diagram is ever present. It is apparent in the chapter's endeavor to dig beneath the surface of analogical reasoning and instead explore the limitations and mysteries of "imitating biology." In this way, Johnston refocuses our attention on the problematics of establishing a link between organic life and nonorganic digitality. In fact, Johnston's diagram presents a somewhat challenging distinction between the two, and as a result he questions the viability of virally coded anomalies, which are both outside of the natural order of things and at risk of exceeding the services of human interest.

Tony Sampson's chapter (chap. 2) uses three questions to intervene in a conception of universal contagion founded on the premise of "too much connectivity." Beginning with a brief account of the universality of the contagious event, he locates the prominence of a distributed network hypothesis applied to the digital epidemic. However, Sampson points to an alternative viewpoint in which evolving viral vulnerabilities emerge from a composition of stability and instability seemingly arising from the connectivity and interaction of users. Chapter 2 then moves on to argue that efforts made by antivirus researchers to impose epidemiological and immunological analogies on these emerging susceptibilities in digital architecture are very much flawed. For example, the crude binary distinction between self and non-self is regarded here as ill-equipped to manage the fuzzy logics of an accidental topology. Indeed, Sampson sees the problems encountered in antivirus research extending beyond the defense of the body of a network to the control of a wider network of interconnecting social bodies and events. He concludes by speculating that the problematics of anomaly detection become part of a broader discursive and nondiscursive future of network conflict and security.

In Chapter 3, Luciana Parisi pushes forward the debate on the universality of viral ecologies by seeking to avoid distinctions made between digital and analogue, technical and natural, and mathematical and biological architectures. The focus of her analysis moves instead to the material capacity of infectious processes, which exceed the organizational tendencies of algorithmic logic. For example, Parisi investigates how the neo-Darwinian genetic code imposes an evolutionary schema on the diagram of digital contagion. However, unlike Johnston, Parisi argues that the organic and nonorganic of digitality are assembled together and that assumptions made in the practice of writing genetic algorithms fail to grapple with the symbiotic nature of what she terms the abstract extensiveness of digital architecture. Parisi employs a complexity of reasoning to explain her alternative blob architectures. If the reader is unfamiliar with the influential events theory of Alfred N. Whitehead or Lynn Margulis' notion of endosymbiosis then the ideas expressed can be difficult to tap into. Nevertheless, a concentrated deep read will offer great rewards to those wanting to discover digital contagion in a novel and profound light.

Finishing this section, Roberta Buiani (chap. 4) proposes that virality is not merely an inherent "natural" part of the software code, but is continuously distributed as figures of contagion, virulence, and intensivies across popular cultural platforms. Making a useful distinction here between *being viral* and *becoming viral*, Buiani returns us to the limits imposed by the metaphoric regime of disease and the nonlimitative distribution of a flexible "single expression." In its becoming, virality has the potential to produce creative outcomes, rather than just new threats—new diagrams perhaps?

# 1

# MUTANT AND VIRAL

## Artificial Evolution
## and Software Ecology

*John Johnston*

In early hacker lore the story is told about a certain mischievous and dis-gruntled programmer who created the first computer "worm," a simple pro-gram called "Creeper" that did only one thing: It continually duplicated itself. In one version of the story the programmer worked for a large corpo-rate research laboratory, with many networked computers, into which he released Creeper. Very quickly, the memory of every computer in the labo-ratory began to fill with this replicating "digital organism." Suddenly realiz-ing the consequences of what he had done, the programmer immediately wrote a second program called "Reaper," designed to seek out and destroy copies of Creeper. When it could find no more copies it would self-destruct, the disaster thus averted. In a darker and probably fictionalized version of the story, the programmer worked as an air traffic controller, and once inside the network, Creeper began to appear on the screens of the air traffic con-trollers as a little airplane with the blazon: "I'm Creeper! Catch me if you can!"

In the early 1980s the story caught the attention of A.K. Dewdney, a programmer and author of a monthly column in *Scientific American* called "Computer Recreations." Dewdney was so taken by the Creeper-Reaper story that he decided to create a game based on it. He called it "Core War," and took pains to isolate the electronic battlefield from the rest of the computer by constructing a simulated environment (actually a virtual machine) in which it would operate.[57] When Christopher Langton, a budding computer scientist working at the dynamical systems laboratory at Los Alamos, organized a conference devoted to the "synthesis and simulation of living systems"—the conference, in fact, would inaugurate the new science of Artificial Life (ALife)—he invited Dewdney to demo the game and preside as judge over an artificial "4-H contest" intended to humor and amuse the conference participants. As it turned out, several future ALife scientists were struck by the possibilities they saw in *Core War*.

This story is appealing, I think, because in it we find computer worms or viruses, computer games, and ALife—three things that would later be separated out and sealed off from one another, but here jostling together in one semi-mythic ur-narrative. It also brings to our attention the generally unknown or acknowledged fact that the highly theoretical and often austere new science of ALife has several traceable roots in popular culture, a dark side (the kinship with computer viruses) and a more openly ludic form visible in a number of creative computer games and the enabling, participatory tools now found on many ALife Web sites like *Nerve Garden* and *Framsticks*. However, although computer games and simulations like *SimLife* and *Creatures* have never posed a problem for ALife, from its official beginnings, computer viruses constituted a forbidden zone, proscribed from legitimate experiment. This proscription is more than a matter of passing interest, inasmuch as digital viruses would eventually reveal themselves to be of singular importance within official ALife research.

Indeed, computer viruses were one of the few topics that Christopher Langton actively sought to discourage at the first ALife conference in 1987.[58] Fearful of the negative associations, Langton simply did not want to attract hackers to Los Alamos. Apart from another of Dewdney's articles, "A Core War Bestiary of Viruses, Worms and Other Threats to Computer Memories," there were no references to viruses in the bibliography of conference proceedings.[59] Perhaps more significantly, Langton did not invite Fred Cohen, one of the first "professional" experimenters with computer viruses. In 1983, as a graduate student in computer science, Cohen had written a virus of 200 lines of code that could invisibly give him system administrator privileges on a Unix operating system, and had published the results of his experiments in the highly reputable journal, *Computers and Security*.[60] But of course the line of demarcation was not always clear. In 1988, Cornell student Robert Morris had released his self-replicating "Internet worm," which quickly paralyzed some 6,000 computers across the country. Morris' actions not only created panic and hysteria but directly fueled the establishment of a panoply of legal measures and new law enforcement agencies. Cohen's own laconic response was simply to remark that Morris had just set the world record for high-speed computation. Even so, in those inchoate times uncontrolled "experiments" with forms of ALife *avant la lettre* could only be worrying and even an impediment to professional researchers, as Stephen Levy observes in his book on ALife:

> During the period that young Morris and other unauthorized experimenters were blithely releasing predatory creatures in the wild [i.e., on floppy disks and networked computers], Cohen and other serious researchers were consistently being refused not only funding but even permission to conduct experiments in computer viruses. As a result, the creations of willful criminals and reckless hackers were for years the most active, and in some ways the most advanced, forms of artificial life thus far.[61]

Levy thus gave the impression that a whole new realm of ALife was beginning to burgeon, some of it scientifically "authorized" and officially sanctioned, whereas other forms constituted an unauthorized but no less fertile "underside." However, not only has the boundary line remained porous and ever-shifting, but Artificial Life itself has proven to be an anomalous and destabilizing science whose very accomplishments have made its status uncertain. On the one hand, its simulations of evolutionary processes have proved highly useful to theoretical biology; on the other, its development of complex software (particularly evolutionary programming techniques and swarm models for network computation) have greatly contributed to advanced research in computer science. ALife's claim to disciplinary autonomy, nevertheless, hinges on its more ambitious project to actually create artificial life. Yet, paradoxically, what makes ALife of special interest here is precisely its inability to remain separate and apart from the new viral ecology and machinic-becoming that increasingly define contemporary reality.[62] Specifically, its own production of viral anomalies and eventual opening to swarm models and network dynamics necessarily bring new theoretical perspectives into play that problematize its initial neo-Darwinian underpinnings.

## THE ADVENT OF ARTIFICIAL EVOLUTION

In Dewdney's game *Core War* the players are actually little programs designed by programmers to destroy the opponent program's memory, using strategies similar to those found in computer viruses. A simple one called "Imp" works exactly like Creeper. These competing programs are written in a version of assembly language called "Red Code," and run simultaneously on the same computer. Response to *Core War* was immediate and quickly generated a large audience of active participants, which led to a 1986 tournament in Boston. The pattern is now familiar: Thanks to the Internet, a new computer game sparks infectious enthusiasm and generates a whole new subgroup of users who share code, techniques, and "cheats"—indeed, a whole "game world." One of the first games to flourish following this pattern, *Core War* now has regularly scheduled tournaments and many Web sites where the software needed to play the game can be freely downloaded.

But while many new players started writing their own digital war machines, the *Core War* story also continued in another direction. Soon after the first ALife conference in 1987, a team of scientists led by Steen Rasmussen appropriated and rewrote the game, somewhat like Langton's transformation of John Conway's *Game of Life* into an experimental system of self-reproducing cellular automata.[63] Instead of attempting to destroy one another, the programs would interact according to what Rasmussen called an "artificial chemistry." An essential component of Dewdney's game, the executable file called MARS (Memory Array Red Code Simulator), was repurposed as VENUS (after the Roman goddess of fecundity), which worked as follows: A large collection of short pieces of "pre-biotic" code or instructions would be fed into the simulator, where the pieces would be processed by a set of simple rules. The hope was that stable blocks of code constituting computational "organisms" would emerge. The results were encouraging but inconclusive. Although eventually the group agreed that "stable cooperative structures of code" had been found, they did not evolve.[64]

Digital evolution was first achieved by Tom Ray, a biologist influenced by both *Core War* and Rasmussen's adaptation of it. In Ray's ALife virtual world, which he called *Tierra,* digital organisms (actually blocks of self-reproducing code) not only replicated but evolved into a teeming variety of interacting new forms. As a consequence, and precisely because it was the first instance in which the breeder's hand was fully removed and autonomous evolution actually occurred, *Tierra* became a crowning achievement of ALife research. Trained as a biologist specializing in tropical rain forest ecology, Ray had always wanted to simulate evolution on a computer, which would both speed up its glacially slow process and allow the results to be carefully analyzed. After hearing about the first ALife conference Ray went to Los Alamos and presented his plan to Langton, Rasmussen, and other ALife scientists, who were impressed by his project but warned him about the dangers—this was not long after Robert Morris had released the first destructive Internet worm. They were also skeptical that he would be able to mutate his creatures without crashing the computer. Ray solved the first problem by following Dewdney's approach: Writing the code in assembly language, he created an environment (a virtual machine) in which his digital creatures would be isolated and contained. As for the second problem, his biological expertise inspired him to introduce a special mechanism he called "template matching" that would allow a creature to replicate even if, as a result of a deleterious mutation, it lacked the necessary instructions. (In *Tierra,* mutation is achieved by bit-flipping parts of the code.) Rather than having to go to a specific memory address, template matching enabled a digital organism to find the needed instructions anywhere in the "soup," as Ray called the total system memory. Template matching, in fact, proved to be decisive in several respects. It not only allowed separate digital organisms to share the replication code, much like bacteria share DNA, but the portability, exchangeability, and indeed mutability of this code enabled the system to be generative, and to exfoliate into a dynamically evolving ecology.

After seeding *Tierra* with The Ancestor, as Ray named his first digital creature, he was astonished to discover that within 12 hours his virtual world was filled with not only Ancestor replicants but a wondrous variety of mutants and parasites. In subsequent runs, hyper-parasites that Ray called "cheaters" also appeared. Indeed, by taking advantage of the special conditions of this virtual environment a small but thriving ecology of interacting digital organisms had come into being. Many of the mutant organisms, moreover, were quite resourceful. Lacking the code for replication, the parasites would simply borrow it from the Ancestors. Being smaller than their hosts, they required less CPU time to copy themselves to a new location in memory and could therefore proliferate more quickly. The parasites and hosts also exhibited dynamic behavior similar to the Lotka–Volterra population cycling in predator–prey studies familiar to biologists. In contrast, the hyper-parasites would quickly decimate the parasites. These creatures could do everything the Ancestors could do, but mutation had made their code more compact and efficient, enabling them to destroy the parasites by capturing the latter's CPU time. As for the cheater, although very small (its instructions were one-third the length of the Ancestor's), it could cleverly intercept the replication instructions it needed as they were passed between two cooperating hyper-parasites. Later Ray reported that another mutant type had discovered the advantage of "lying" about its size.[65] The replication code requires that each creature first calculate its size in order to request precisely enough memory for its daughter cell. In this instance, the creature would calculate its size (a 36-instruction set), but then request a space of 72 instructions for its daughter, thereby doubling the amount of "space" and "energy" (i.e., memory and CPU time) available to its progeny. Initially this mutation provided a powerful advantage, but it was later negated when it swept through the population and all the creatures began to lie about their size. In summary, *Tierra* set the bar for subsequent ALife programs. Following Ray's success, scientifically useful ALife creatures had to be able to replicate, evolve, and interact in a self-enclosed world that allowed precise measurements of these activities.

This milestone was soon followed by many more successes in the new science of ALife. Software simulations on platforms such as John Holland's *Echo*, Chris Adami's *Avida*, Andrew Pargellis' *Amoeba*, Tim Taylor's

*Cosmos* and Larry Yaeger's *PolyWorld*—to name several of the best known—provided the experimental basis for a sustained exploration of digital evolution and the theory of emergence that underlies it.[66] In *Amoeba*, for example, Andrew Pargellis went a step further than Ray, successfully setting up conditions in which autonomous self-replicating digital organisms emerged spontaneously out of a "pre-biotic" soup composed of strings of code randomly amalgamated. Subjected to mutation, these organisms then evolved into more robust replicators. However, along with these "healthy" self-replicating organisms, digital viruses also emerged. Paradoxically, the viruses were both parasites on and enablers of other digital life forms, making the dynamics of the environmental "soup" notably complex. Both threatened and aided by the growth of viral colonies that share one another's code, again much like simple bacteria, these "proto-biotic" forms either formed cooperative structures like the viral colonies or died. In his account of *Amoeba*, however, Pargellis was unable to explain how these digital viruses perform this contradictory double function, both enabling nonviral forms to replicate and using the latter in order to proliferate themselves. In short, by privileging the "healthy" or autonomous replicators, he failed to do justice to the dynamic complexity of the "soup."[67]

From the perspective of biology, one of the most useful features of these ALife platforms is that they can vary the mutation rate. In a series of experiments on the *Avida* platform, for example, Cris Adami saw results very similar to those observed in the evolution of biological bacteria and viruses. When exposed to high mutation rates, Adami noted, some species even survive as a "cloud" or "quasi-species."[68] That is, no one organism any longer contains the entire genome for the species, but rather a range of genomes exist, and this variety allows the species to survive the onslaught of destructive mutations. Viruses in particular exhibit this kind of quasi-species behavior, Adami said, just like "the more robust of his digital organisms do."[69] He continued:

> In virus evolution, you clearly have mutation rates on the order of those we have played around with. It is clear that a virus is not one particular sequence. Viruses are not pure species. They are, in fact, this cloud, this mutational cloud that lives on flat peaks [on a fitness landscape]. They present many, many, many different genotypes.[70]

A multiplicity of digital organisms that can no longer be defined except as a "mutational cloud"—surely this is a curious order of being, produced inadvertently by technology but made intelligible by an anomalous phenomenon of nature. Yet the pursuit of such anomalous instances may prove highly instructive in the attempt to define and (re)create life.

# TOWARD OPEN ALIFE SYSTEMS

In 1997, Mark Bedau and Norman Packard presented the results of their research comparing evolutionary activity in artificial evolving systems with those in the biosphere.[71] The results were sobering, and indicative of the need to make artificial evolving systems more open-ended. Seeking to identify trends in adaptation, the authors defined three measurable aspects of evolutionary activity: cumulative activity, mean activity, and diversity. In general terms, evolutionary activity simply reflects the fact that an evolving population produces innovations, which are adaptive "if they persist in the population with a beneficial effect on the survival potential of the components that have it."[72] Innovations are defined as new components introduced into the system—genotypes in the case of artificial systems and the appearance of a family in the fossil record. In both instances, an activity counter is incremented at each successive time step if the innovation still exists in the population under study. This count is then used to compute the other measures. Diversity, for example, is the total number of innovations present at time $t$ in a particular evolutionary run. These quantitative measures were then used to compare evolutionary activity in two artificial evolving systems, *Evita* and *Bugs*, with evolutionary activity in the fossil record.[73]

The results of Bedau and Packard's comparison were unmistakable: "long-term trends involving adaptation are present in the biosphere but missing in the artificial models."[74] Specifically, cumulative activity, mean activity, and diversity in the fossil record show a steady increase from the Cambrian to the Tertiary periods, except for a momentary drop in the Pernian period which corresponds to a large and well-known extinction event. In contrast, after an initial burst of evolutionary activity in the two artificial evolving systems, there are no long-term trends. In *Evita* the reproductive rate improves significantly at the beginning of the simulation, after which the new genotypes remain "neutrally variant"; in other words, although highly adaptive, they are no more so than the majority of genotypes already in the population. The authors interpret this to mean that "the bulk of this simulation consists of a random drift among genotypes that are selectively neutral, along the lines of [Motoo Kimura's] neutral theory of evolution."[75] In the *Bugs* model three major peaks of innovation occur, but then the evolutionary activity "settles down into a random drift among selectively-neutral variant genotypes, as in the *Evita* simulation."[76] These peaks, the authors explain, reflect successive strategies that enable the population to exploit more of the available resource sites.[77] But as with *Evita*, the possibilities for significant adaptation are soon exhausted.

After presenting quantitative evidence for the qualitative difference between these ALife systems and the biosphere, Bedau and Packard attempted to account for this difference as a necessary first step toward closing the gap between them. First, they noted the absence of interesting interactions among organisms, like predator–prey relations, cooperation, or communication. (In fact, Holland's *Echo* permits such interactions, but a follow-up study revealed that it too lacks "the unbounded growth in evolutionary activity observed in the fossil record.")[78] Second, although Bedau and Packard acknowledged that the spatial and temporal scales of *Evita* and *Bugs* are appreciably smaller and less complex than those of the biosphere, they do not believe that scaling up space and time in the artificial systems or making them more complex will make any qualitative difference. This follows from what they think is the primary reason behind the biosphere's arrow of cumulative activity: the fact that "the dynamics of the biosphere constantly create new niches and new evolutionary possibilities through interactions between diverse organisms. This aspect of biological evolution dramatically amplifies both diversity and evolutionary activity, and it is an aspect not evident in these models."[79] Noting a qualitative similarity between the initial part of the cumulative activity curve for the fossil data and for *Bugs*, they speculated that the biosphere might be on some kind of "transient" during the period reflected in the fossil data, whereas the statistical stabilization in *Bugs* may be caused by the system hitting its "resource ceiling"; in other words, "growth in activity would be limited by the finite spatial and [energy] resources available to support adaptive innovations." Contrarily, the biosphere seems not to be limited by "any inexorable resource ceilings." Its evolution continues to make new resources available when it creates new niches, as "organisms occupying new niches create the possibility for yet newer niches, i.e., the space of innovations available to evolution is constantly growing."[80] But whatever is responsible for unbounded growth of adaptive activity in the biosphere, the challenge is clear. Indeed, to create comparable systems, the authors asserted, "is among the very highest priorities of the field of artificial life."[81] The good news is that an objective, quantitative means for measuring progress is now available.

I have argued elsewhere that Bedau and Packard's findings bring to a close *the first phase* of official ALife research.[82] To be sure, this claim ignores or relegates to a lesser status much of the very valuable work devoted to "artificial chemistries," the transition to "life" from nonliving origins, and the synthesizing of dynamical hierarchies, to name only three other significant strands of ALife research.[83] Nevertheless, several things argue for this assessment. ALife programs like *Tierra* and *Avida*, which have been applauded precisely for instantiating open-ended evolution, stand out among the new science's most visible successes. Now that the inherent limitations of these systems are objectively measurable, approaches that can move beyond them and advance ALife research to a new stage are called for.

One such approach, which no longer depends on the limits of a single computer-generated closed world, had already been taken by Ray himself. Following *Tierra*'s success, Ray decided to install a second version in the more expansive, networked space of the Internet, with the hope that it would provide the basis for a growth in the complexity of his digital organisms. "It is relatively easy to create life [Ray writes]. Evidently, virtual life is out there, waiting for us to provide environments in which it may evolve."[84] But what *is* difficult to produce is something like the Cambrian explosion, where there was "origin, proliferation and diversification of macroscopic multi-cellular organisms."[85] Here, as elsewhere in Ray's writings, the reference to the Cambrian explosion harbors something of an ambiguity. On the one hand, by citing a well-known event in evolutionary biology Ray sustained the more or less automatic assumption that his work contributes to the understanding of basic biological questions about the origin of life and its evolutionary path to a wide diversity of species. On the other hand, after the success of *Tierra*, Ray made it clear that his explicit aim is the generation of complexity, understood simply as an increasing diversity of interactions among an increasing diversity of organisms or agents, in the medium of digital computers. For Ray, the actual accomplishment or realization of this complexity takes precedence over whatever it might mean in relation to the processes of the organic world. As he bluntly stated, "The objective is not to create a digital model of organic life, but rather to use organic life as a model on which to base our better design of digital evolution."[86]

Significantly, at no point in writing about *Internet Tierra* does Ray mention or allude to the simulation of life; instead, he simply reiterates his aim to use evolution "to generate complex software." When the Cambrian explosion is mentioned, it serves only as a convenient and well-known general model of complexity, as when he stated his specific intention "to engineer the proper conditions for digital organisms in order to place them on the threshold of a digital version of the Cambrian explosion." The global network of the Internet, because of its "size, topological complexity, and dynamically changing form and conditions," presents the ideal habitat for this kind of evolution. Under these propitious conditions, Ray hopes, individual digital organisms will evolve into multicelled organisms, even if

> the cells that constitute an individual might be dispersed over the net. The remote cells might play a sensory function, relaying information about energy levels [i.e., availability of CPU time] around the net back to some 'central nervous system' where the incoming sensory information can be processed and decisions made on appropriate actions. If there are some massively parallel machines participating in the virtual net, digital organisms may choose to deploy their central nervous systems on these arrays of tightly coupled processors.

Furthermore, if anything like the Cambrian explosion were to occur on *Internet Tierra*, then we should expect to see not only "better" forms of existing species of digital organisms but entirely new species or forms of "wild" software, "living free in the digital biodiversity reserve," as Ray put it. Because the reserve will be in the public domain, anyone willing to make the effort will be able to observe and even "attempt to domesticate" these digital organisms. Although domestication will present special problems, Ray foresees this as an area where private enterprise can get involved, especially because one obvious realm of application would be as "autonomous network agents."

After several years of operation, however, *Internet Tierra* did not prove to be as dramatically successful as the earlier closed-world version, mainly because of difficulties with the parallel-processing software.[87] Yet there have been some rather astonishing results. At the ALife VI conference in 1996, Ray and colleague Joseph Hart reported on the following experiment:

> Digital organisms essentially identical to those of the original *Tierra* experiment were provided with a sensory mechanism for obtaining data about conditions on other machines on the network; code for processing that data and making decisions based on the analysis, the digital equivalent of a nervous system; and effectors in the form of the ability to make directed movements between machines in the network.[88]

Tests were then run to observe the migratory patterns of these new organisms. For the first few generations, these organisms would all "rush" to the "best-looking machines," as indeed their algorithms instructed them to do. The result was what Ray called "mob behavior." Over time, however, mutation and natural selection led to the evolution of a different algorithm, one that simply instructed the organism to avoid poor-quality machines and consequently gave it a huge adaptive advantage over the others.

Overall, *Internet Tierra* makes fully explicit a new objective in Ray's work: to deploy evolutionary strategies like natural selection in the digital medium in order to bring into being a quasi-autonomous silicon world of growing complexity. Although the digital organisms of this world are of essential interest to ALife and evolutionary biology, their wider significance exceeds the boundaries of these disciplines and even the frame of scientific research. Indeed, the necessary constraints of science can actually inhibit us from seeing how these organisms participate in a larger transformation through a co-evolution of technology and the natural world. Conceptually, traditional "molar" oppositions between *phusis* and *technē*, the organic and the nonorganic, are giving way to new "molecularizations" in swarm models and simulations that are fomenting a becoming-machinic within the human environment, as the latter increasingly functions as a bio-machinic matrix enabling a largely self-determined and self-generating technology to continue natural evolution by other means.

## COMPUTER IMMUNE SYSTEMS

Another alternative to the single computer-generated, closed-world approach characteristic of the first phase of ALife research is represented by David Ackley's experiments with what he called "living computation." In a paper at the ALife VII Conference in 2000, Ackley noted that if asked to give an example of artificial life the millions of computers users today would most likely answer: computer viruses.[89] "Do we really want to exclude the rapidly expanding world of internetworked computers from consideration as a form of ALife?" he then asked. Such questions lead him to point to a number of remarkable parallels between living systems and manufactured computers, starting with the fact that both are excellent copiers and therefore present "tremendously virus-friendly environments."[90] An even more striking parallel is evident between "the arc of software development" and the "evolution of living architectures":

> From early proteins and autocatalytic sets amounting to direct coding on bare hardware; to the emergence of higher level programming languages such as RNA and DNA, and associated interpreters; to single-celled organisms as complex applications running monolithic codes; to simple, largely undifferentiated multicellular creatures like SIMD [single-instruction multiple-data stream] parallel computers. Then, apparently, progress seems to stall for a billion years give or take—the software crisis. Some half a billion years ago all that changed, with the "Cambrian" explosion of differentiated multicellular organisms, giving rise to all the major groups of modern animal.

Living computation hypothesizes that it was primarily a programming breakthrough—combining what we might today view as object-oriented programming with plentiful multiple-instruction multiple-data stream (MIMD) parallel hardware—that enabled that epochal change.[91]

In the context of these observations, Ackley proposed that "the actual physicality of a computer itself may support richer notions of life" than either software programs alone or the software candidates for artificial life. In effect, this perspective stands the ALife agenda on its head: "Rather than seeking to understand natural life-as-it-is through the computational lens of artificial life-as-it-could-be [essentially the ALife agenda as formulated by Christopher Langton] . . . we seek to understand artificial computation-as-it-could-be through the living lens of natural computation-as-it-is."[92]

Ackley called this further extension and implementation of biological principles "living computation." With computer source code serving as its "principal genotypic basis,"[93] he was looking at ways that the principles of living systems could be applied to networked computer systems.

For specific experiments, Ackley constructed **ccr**, "a code library for peer-to-peer networking with emphases on security, robust operation, object persistence, and run-time extensibility."[94] An explicit objective is to enable a **ccr** world to communicate with other **ccr** worlds in a manner consistent with how living systems guard against possible sources of danger. Hence, the peer-to-peer communications architecture requires a more layered, self-protective system of protocols than the familiar TCP/IP protocols of Internet communication. Ackley described how **ccr** starts with very small messages and builds to larger ones using a cryptographic "session" key as a rough equivalent of means used in the natural world—complex chemical signaling systems, hard-to-duplicate bird songs, ritualized interactions—to authenticate messages and build trust and confidence before stepping up to more elaborate and sustained exchanges. Another initiative deployed in **ccr** is to circumvent the commercial software practice of distributing only pre-compiled binary programs while "guarding access to the 'germ line' source code, largely to ensure that nobody else has the ability to evolve the line."[95] According to the analogy of what he called "software genetics," Ackley understands computer source code as genome, the software build process as embryological development, and the resulting executable binary as phenotype. In these terms, the rapidly growing "open source" software movement is of essential importance:

> With source code always available and reusable by virtue of the free software licensing terms, an environment supporting much more rapid evolution is created. The traditional closed-source "protect the germ line at all cost" model is reminiscent of, say, mammalian evolution; by contrast the free software movement is more like anything-goes bacterial evolution, with the possibility of acquiring code from the surrounding environment and in any event displaying a surprising range of "gene mobility," as when genes for antibiotic drug resistance jump between species."[96]

Ackley illustrated this point by comparing **ccr** to a different "species," the GNU Image Manipulation Program (GIMP). Both utilize an identical piece of code—the GNU regular expression package—which Ackley likened to a "highly useful gene incorporated into multiple different applications out of the free software environment."[97] According to traditional or commercial software practices such duplication might be deemed wasteful, but for Ackley "such gene duplication reduces epistasis and increases evolvability."[98]

By treating the computer itself as a kind of living system, Ackley surely pushed the limits of what some perceived as merely an interesting analogy. On the other hand, the application of the principles of living systems to actual computational systems is hardly new or *outré*. It was, in fact, the basis of many of John von Neumann's novel ideas about computation as well as the inspiration for the development of neural networks, genetic algorithms, and evolutionary computation. Ackley's project, moreover, is closely linked to the development of computer immune systems, as discussed below. In short, by resizing the framework of ALife research and extending it to networked computer systems, Ackley openly pushed the contemporary convergence between new computational approaches to biology and biological approaches to computation. At the same time, in attributing a kind of "life" to networked computers he further extended Langton's idea that life is the result of a particular organization of matter, rather than of something that inheres in individual entities. In short, Ackley envisions 'life" as a property of the complex exchanges that computational networks make possible.

As Ackley implicitly acknowledged, the elimination of computer viruses from scientific consideration had become something of a sore spot and dark underside in relation to official ALife research, despite the emergence of viruses and viral-like phenomena in many of its experiments. By the early 1990s, clear signs were emerging that the official boundary line had weakened.[99] In his book, *It's Alive! The New Breed of Computer Programs* (published in 1994), Fred Cohen discussed computer viruses under the rubric of "living programs" (LPs), which also include *CoreWar*, Conway's *Game of Life*, and Ray's *Tierra*. Cohen defined a living system as comprised of an organism *and* its environment, arguing that when viewed as a pattern in "the information environment" computer viruses are very much alive. The "outsider" scientist Mark A. Ludwig had pushed this point of view even further in his book, *Computer Viruses, Artificial Life and Evolution*, published the year before. Like Cohen, Ludwig offers astute technical and philosophical analysis of Alife, while also providing computer code for experimenting with a variety of real viruses.[100] However, Ludwig argued explicitly that computer viruses are a *more significant form of ALife* than the "laboratory-contained" forms produced in scientifically sanctioned experiments, precisely because these viruses "live" in a world that was not specifically designed to support them. For Ludwig, Darwinian evolutionary theory provides the proper scientific framework for comparison. Describing progressive steps by which virus writers have attempted to foil anti-virus scanning techniques, with the mutating polymorphic virus written by the legendary Bulgarian virus-writer Dark Avenger representing the latest and most complex stage, Ludwig suggested that the next step would be a mutating virus with a genetic memory that could evolve into increasingly more immune forms. Not content with speculation, Ludwig actually supplies the source code (in assembly language) for a "Darwinian Genetic Mutation Engine" that can convert a lowly DOS virus into a genetically evolving polymorph. Significantly, like several ALife scientists, Ludwig ended up questioning whether conventional Darwinian theory can actually explain the evolutionary development he described.[101]

Whatever the case, there is no reason to believe that such research has been ignored by hackers and virus writers. Indeed, the latter's undeclared war against software industry giants like Microsoft and corporate Web sites and the consequent attempts to provide virus protection have clearly resulted in an escalating "arms race" in today's digital ecology that illustrates Ludwig's basic argument. Although Ludwig undeservedly remains something of a "crank" outsider, the undeniable fact that networked computers became a site where new forms of viral life were constantly emerging finally forced some ALife scientists to consider computer viruses as a theoretical issue intrinsic to their own enterprise. At the *ALife IV Conference* in 1994, Jeffrey Kephart argued that current antivirus techniques are doomed to fail, and eventually must be replaced by a biologically inspired immune system for computers.[102] Yet Kephart failed to see that this entails reconceiving the world of computers as—or as having the properties of—a living ecosystem, populated with "computers, with people, software, data, and programs," as Stephanie Forrest put it.[103] In her paper, "Principles of a Computer Immune System," Forrest listed some 12 organizing principles of a biological immune system, many of which—autonomy, adaptability, and dynamically changing coverage—implicitly recall Francisco Varela's model of the immune system as autopoietic and an alternative to the dominant military model.[104] Simply put, the immune system is a complex adaptive one, not simply a system that detects and attacks alien intruders; as Varela clearly understood, it is a self-organizing regulatory system that does not distinguish between self and other but instead maintains many heterogeneous components at levels that enable the system as a whole to function best. The hard part, of course, is how to incorporate such principles into a design that will enable a computer to function similarly. If the objective is "to design systems based on direct mappings between system components and current computer system architectures,"[105] Forrest argued, then the latter will have to be radically modified. One possible architecture would be something like an equivalent "lymphocyte process" comprised of lots of little programs that would query other programs and system functions to determine whether they were behaving normally or not. But they would also have to monitor each other, "ameliorating the dangers of rogue self-replicating mobile lymphocytes"[106] and thus a possible form of digital cancer. Just how feasible this approach will turn out to be is difficult to say, and Forrest herself remains cautious, acutely aware of the limitations of "imitating biology" given that biological organisms and human-made computers differ greatly in both method of operation and objective.

# SWARM INTELLIGENCE

With computer immune system research we unavoidably brush up against the limits of a conceptual contradiction between the computer as a tool or medium over which we exercise near complete control and the computer as part of an ecosystem that cannot function unless given more life-like capacities that will put it outside of our control. Clearly, if a fully functional computer immune system is to be constructed—or rather, and more likely—evolved, then human computational and communicational activities will have to be made more like biological exchanges—fluid, adaptable, and cooperative. It is already evident that the forces driving complex software environments like the Internet are not easily susceptible to brittle and mechanistic attempts at control, which only foment software "arms races," with viruses and parasites breeding and mutating in its interstices. In fact, physicists like Albert-Laszlo Barabasi, in his book *Linked,*[107] have shown that the Internet itself is a dynamic "small worlds" network, growing like a quasi-organic structure and stimulating the need for increasingly sophisticated adaptive software agents. In this complex software ecology, emergent ALife forms become not only more likely but necessary to human and nonhuman usage alike.

The application of "swarm intelligence" vividly illustrates the contradiction between command and control and the biological approach to software evolution. For example, the study of ant foraging has recently led to a method for rerouting network traffic in congested telecommunications systems. The method involves modeling the network paths as "ant highways," along which artificial ants (i.e., software agents) deposit and register virtual pheromone traces at the network's nodes or routers. Eric Bonabeau, one of the scientists who developed the method, commented on its application:

> In social insects, errors and randomness are not "bugs"; rather, they contribute very strongly to their success by enabling them to discover and explore in addition to exploiting. Self-organization feeds itself upon errors to provide the colony with flexibility (the colony can adapt to a changing environment) and robustness (even when one or more individuals fail, the group can still perform its task). . . . This is obviously a very different mindset from the prevailing approach to software development and to managing vast amounts of information: no central control, errors are good, flexibility, robustness (or self-repair). The big issue is this: if I am letting a decentralized, self-organizing system take over, say, my computer network, how should I program the individual ants so that the network behaves appropriately at the system-wide level? . . . I'm not telling the network what to do, I'm telling little tiny agents to apply little tiny modifications throughout the network. Through a process of

amplification and decay, these small contributions will either disappear or add up depending on the local state of the network, leading to an emergent solution to the problem of routing messages through the network.[108]

The problem, however, is selling the concept of swarm intelligence to the commercial world. As Bonabeau memorably put it, managers "would rather live with a problem they can't solve than with a solution they don't fully understand or control."[109] But clearly, more is threatened here than the egos of managers and engineers. The very notion of technology as a tool for management and manipulation (which then becomes a means of capitalist profit) is here giving birth to another kind of world composed of a multitude of tiny, self-organizing entities that neither form part of the natural order nor directly serve human interests, although the hope is that they can be gently nudged into doing so. For many, however, this is indeed a frightening prospect: a swarm of hopefully friendly creepers.

# 2

# HOW NETWORKS BECOME VIRAL

## Three Questions Concerning Universal Contagion

*Tony D. Sampson*

*Along with the common celebrations of the unbounded flows in our new global village, one can still sense the anxiety about increased contact . . . The dark side of the consciousness of globalization is the fear of contagion. If we break down global barriers, . . . how do we prevent the spread of disease and corruption? Nothing can bring back the hygienic shields of colonial boundaries. The age of globalization is the age of universal contagion.*

—Michael Hardt and Antonio Negri[110]

This chapter explores contagious network environments by addressing three questions concerning their universality. The first questions the role that "too much connectivity" plays in the physical manifestations of a contagious network culture. It asks if contagion can be explained via the quantity of connections alone, or does the mode of connectivity make a difference. As Michael Hardt and Antonio Negri claimed above, "increased contact" seemingly leads to a universal mode of contagion. Nevertheless, if the mode matters, as I propose, then what kind of mode makes a network become vulnerable to viruses? The second question concerns the offensive and defensive modes of network conflict. It draws upon epidemiological research in computer science in order to ask how a network might be exploited or secured against viral attack. Relatedly, the third question focuses on the conception of so-called *epidemic network power*. In doing so, it asks how the "control" of contagious events is played out in the discursive formations of network culture. Significantly, this chapter breaks with the notion that too much connectivity lowers the barriers of a hygiene shield otherwise protecting us from a "dark side" of disease and corruption. Disagreeing with Hardt and Negri's externalized and software-like parasites, feeding off the vitality of social multiplicity,[111] this chapter argues that such externalities and barriers are illusions of a tendency toward an organic theory of capitalism.[112]

This chapter further argues that the framing hypothesis of the parasitic model of contagion—the robust distributed network—does not entirely capture the vulnerable connectivity from which contagious events emerge. As Andrew Goffey suggested, Hardt and Negri's claim may only capture a little of "the dynamics and the danger which resistance to the present require."[113] The externality of the parasite model is therefore replaced here by an internalized problem concerning the detection and control of contagious anomalies in networked environments. The chapter goes on to explore how an often-fuzzy distinction is made between what is known and unknown on a network. This distinction is not limited to the detection problems encountered on computer networks, but extends to a far broader network security discourse with its own set of practices and defensive-offensive policies. This is a point perhaps amplified in Eugene Thacker's identification of a dangerous discursive "inability" of the U.S. defense policy to distinguish between epidemics and war, and emerging infectious disease and bioterrorism.[114]

Although the chapter rethinks the role of connectivity in contagion theory, it remains consistent with the contention that social power "designates a dynamic of infection,"[115] and that the epidemic problematizes the traditional mechanisms of sovereignty and resistance.[116] Nevertheless, rather than explaining this relation via analogical or metaphorical comparisons between biological or software viruses and network power, the chapter looks instead to the materialities of the network, in which becoming viral seemingly occurs by way of historical topological processes involving accidents, unforeseen events and emergent network vulnerabilities. This materialist approach is partly supported by empirical data emanating from so-called contagion modeling. Arguably, using these models, researchers provide refreshing insights into the tripartite physical–cultural relationship among (a) too much connectivity, (b) the offensive and defensive modes of contagion, and (c) epidemic network power. In place of the equilibrium state of social organics that arguably motivates Hardt and Negri's new social physiology, we find a paradoxical environment in which stability and instability combine. This composition ensures that dynamic networks tend toward virality overtime. This is not because of the intentionality of external forces, or the will of individual nodes, but because collective social usage of a network introduces increasing architectural fragility.

Another significant outcome of contagion modeling, particularly the so-called *scale-free* model, is that it challenges a prevalent assumption made in numerous Internet studies concerning the Cold War origins of robust distributed technologies. Contrary to the stress often placed on the preprogrammed routing of information through redundant topologies, contagion models suggest that the physics of an electronic network has *a robust, yet fragile,* or virally vulnerable topology.[117] Again we find a paradoxical condition in which robustness and vulnerability are in a compositional tradeoff with each other. So although it is perhaps important to carefully approach claims that natural laws universally operate in networks, these models do present an interesting alternative to the robustness of the distributed hypothesis.[118] They suggest an accidental environment in which rare perturbations and shocks can trigger, or contribute to, unforeseen cascading contagions.

Nevertheless, despite the attention given to empirical data, the argumentation in this chapter is not strictly empirical. This is because the analysis focuses more on *emergent* processes rather than the *emerged* production of maps or graphs used in network science.[119] Furthermore, what makes this approach different from other studies of contagion, like those that focus on network security, is that emergent virality is not solely grasped in oppositional terms of a "bad" code threatening to destroy a "good" system. On the contrary, contagion modeling suggests that a system, however stable it may appear, cannot determine its own stability. Viral environments are thus not regarded here as a manifestation of some dreadful dark side of organic unity. In contrast, virality is conceptualized as a surplus product of a sociotechnical network—a network in which social usage combines with topological growth to produce the contagious capacities of assemblages.[120]

Before posing three questions concerning contagion, I first contextualize the claim of universal contagion. In doing so, I briefly discuss how contagion has been broadly expressed in various discourses.


## UNIVERSAL CONTAGION


In his book *The Network Society,* Jan van Dijk followed a number of other social scientists by describing the 21st century as "the age of networks."[121] However, unlike popular and often utopian discourses concerning this new media age, Van Dijk warned of the problems that can arise from "too much connectivity." Among the most pressing of these problems is contagion.[122] In fact, his analysis proposed three forms of contagion with a universality that cuts across the physics and cultural politics of network society. First, he identified how the global network of air transport makes network society vulnerable to the spread of biological diseases. Second, he located how technological networks become volatile to the destructive potential of computer viruses and worms. Third, like Hardt and Negri, Van Dijk located sociocultural contagions, enhanced by the rapidity and extensity of technological networks, which threaten to destabilize established sociocultural-political order. These include the spread of social conformity and political rumor, as well a fads, fashions, hypes, and innovations.[123]

In network science, Albert Barabási's[124] analysis of the physical connectivity of complex networks identifies a parallel mode of universal contagion in which biological and technological viral phenomena spread like ideas and fads on social networks. A tendency toward universalism repeated in the work of mathematical sociologists like Duncan Watts[125] and the popular science writer Malcolm Gladwell's conception of social epidemiology.[126] However, contagion theory itself has a long history in sociological and psychological studies and has played an important role in the divergence of philosophical traditions. Gabriel Tarde's *Laws of Imitation*[127] and Gustave Le Bon's *The Crowd*,[128] both originally published in the late 1800s, set out very different sociological and psychological studies (respectively) of contagious events. On the one hand, Tarde's society of individual imitation, with its refusal to distinguish between the micro and macro levels of the social, set out a challenge to the Durkheimian model of organic social cohesion and went on to influence both Bruno Latour's actor network theory and the social molecularity of Deleuzian ontology. On the other hand, Le Bon's recognition of the psychological susceptibility of the crowd suggested that it was the perfect medium for the contagious rise of "dangerous" democratic politics. His notion of the crowd went on to subsequently influence Freud's work on group psychology, which is of course the nemesis of Deleuze and Guattari's *Anti-Oedipus* project.[129]

In the often-conflictual milieus of digital culture, in which social and technological networks have become increasingly intertwined, novel modes of contagion have emerged. Indeed, although the resistance tactics of electronic civil disobedience groups struggle to defend the hacker ethic from state control,[130] it is the computer virus writer, roundly condemned as a juvenile cybervandal, who has perhaps surfaced as the most effective purveyor of electronic disturbance. This is of course a problematic declaration, in the sense that the motivation of the virus writer does not easily fit in with the hacker ethics of those on the political left or the right. The politics of the computer virus writer are at best ambiguous.[131] Nevertheless, it is perhaps the seemingly apolitical, yet ecological evolutionary capacity of viruses that attracts the attention of cultural theory. Like this, the ecological history of computer viruses has been intimately linked to the history of networked digital capitalism by Jussi Parikka.[132] Elsewhere I have explored the role they play in the emergence of network culture.[133] To be sure, the politics of the virus can be thought through in terms that exceed the content analysis of a particular message and focus instead on the viral medium in itself. As McLuhan argued, it doesn't matter if the media machine churns out Cornflakes or Cadillacs . . . the medium is the message.[134] If power in an information age equates to the effective control of the means of information and communication technology, then as Jean Baudrillard perhaps cynically suggested, the virus creates the "ultra-modern form of communication which does not distinguish, according to McLuhan, between the information itself and its carrier."[135]

Stefan Helmreich approached viral politics via the study of rhetoric.[136] He suggested that in a period of time marked by concerns over the stability of Western democracy, the metaphorical tropes derived from both biological and computer viruses have been discursively put to work in order to rhetorically describe the latest threats to nation state security.[137] For example, in the popular press the computer virus has surpassed its lowly status as digital graffiti to become the "terrorist threat of the digital age."[138] But significantly, I argue that virality exceeds this symbolic domain. Such analogical metaphors attributed to the epidemic do more than figuratively express security concerns. They intervene in what Foucauldian analysis called the "real practices" of the discursive formation of an object.[139] The evolutionary, immunological, and epidemiological analogies, used in academic, military, and commercial research, have contributed to the reorganization of the technological and political terrain of the information hegemony. Not allegorically, but physically embedded in the practices of its institutions. Like this, the apparent dangers attributed to the computer virus are arguably indicative of a more generalized concern over contagious forces. For example, the question of how nation-states can respond to new network threats has prompted RAND to suggest that the United States needs to evolve its military strategy so that "networks fight networks."[140] Eugene Thacker speculated that RAND's notion of future network conflict is an epidemic struggle in which "viruses fight viruses."[141]

Despite the friction-free flows of electronic commerce, the anxieties brought about by increased contact have alerted RAND to a contagious dark side of the network. However, these contagions are not restricted to technological threats: "Epidemics" of all kinds are deemed potentially corrosive to the new network paradigm. As the nation-state connects to the global flows of money, people, culture, and politics, it encounters the contagions of religious cults, organized crime, economic crisis, anarchic activism, information piracy, and so-called terror networks or online jihad. Following the July 7, 2005 suicide bombings on the London Tube, a former CIA agent warned the British public about a "deadly [cultural] virus":

> Log on to the Internet or visit a militant Islamic bookshop and within a few minutes you will find enough inspiration in CDs, ranting sermons, DVDs, for a hundred suicide bombs. It swirls across the Islamic world

as an expression of rage against the West for the invasion of Iraq, support for Israel, and for Western dominance of the world economy...The
only real solution lies within Islam itself. It is only when the vast majority of law-abiding Muslim societies reject the cultural virus of suicide
bombing and cease to glorify it that this plague will burn itself out.[142]


Beyond the rhetorical value of such comments, the globalized viral threat of
terror prompts questions concerning the extent and limitations of capitalist
power and its capacity to control events on a network. But, virality is not
limited to the criminal, terrorist, or activist enemies of capitalism, but also
includes the internal spread of political corruption, economic panic, rumor,
gossip, and scandal on technologically enhanced networks. As Van Dijk
proposed, these contagions persistently threaten the stability of stock markets and political leadership.[143] Therefore, although contagions are discursively "pushed out" to an externalized dark side, the nondiscursive instabilities of network power are very much *inside* the borders of Empire. The
spread of financial contagions demonstrate how vulnerabilities are part of
the interior of an overlapping global economy.[144] The expansiveness of the
capitalist network does not have an easily defined periphery, just an expanding inside. The physical connectivity of network capitalism needs to remain
openly connected so as to promote a continuous flow of money, people, and
goods necessary to sustain capital growth. However, contagious assemblage
overlaps between nation states increases the potential for instability to
spread. Although Empire has profited from the uncertainties of global overlaps, its fragility to rare cascading contagions raises questions concerning
control. For example, the International Monetary Fund (IMF) has sought to
contain contagion by suppressing the developing world's capacity to profit
from cross-border equity flows during periods of financial crisis, but in
doing so it risks spreading political instabilities.[145]

Thacker recognized the problem of establishing control over a network
when he argued that the new network ontology highlights a problem of
maintaining sovereignty.[146] A paradoxical tension now exists between the
need for topological control and the realization that network emergence is
in fact beyond control. As Thacker put it: *"the need for control is also, in
some way, the need for an absence of control. . . ."* What Thacker proposed
is that the nation-state recognizes that resistance to contagion might be futile
unless the state itself assumes the power of the epidemic. This is not just a
network fighting a network, but as Thacker speculated, it is a (good) virus
fighting a (bad) virus. Indeed, the "good" virus/"bad" virus debate is nothing new in computer science. In the 1980s, Jon Shoch and John Hupp[147]
explained how a distributed worm might have worked for the benefit of the
network had it not got out of control. To the annoyance of network controllers, Fred Cohen set out ways in which viruses could be used for benev-
olent purposes.[148] However, the computer security industries' reaction to

the good virus is one of outright hostility.[149] Rather than experiment further with viruses fighting viruses, Internet security research, like that carried out by IBM, has instead tended to translate the analogical metaphors of epidemiology and immunology into real software practices intended to suppress contagion thresholds.

Like the IMF, IBM followed a policy of containment, but the ineffectiveness of digital epidemiology and immunology perhaps provides an insight into the limitations of analogical approaches. Using models borrowed from the study of biological disease, IBM's antivirus team initially assumed that the Internet was a homogeneously mixed *random universe*.[150] More recent modeling however suggests a far-from-random, heterogeneous topology.[151] The Internet is, it seems, robust in terms of resistance to random and common contagions, but highly vulnerable to targeted attacks from viruses aimed at clusters of nodes or shocks to the network that can trigger contagious cascades. The notion that networks become both robust yet fragile makes problematic the causal assumption that too much connectivity determines contagion. The growth of specific branching structures can become an important factor in how individuals connect to a network and how contagious events spread. Nevertheless, although contagion modeling has extended beyond the remit of computer security to research into so-called terror networks,[152] their use has rarely featured in the analysis of emergent cultural politics. Therefore, following Tiziana Terranova's proposal that the physics of a network is "inseparable" from the emergence of cultural politics,[153] this chapter argues that power and resistance are intimately coupled to pragmatic questions concerning the stability and instability of connectivity in a network.

In the past, the relation between the physics and cultural politics of the network has been intellectually couched somewhere in between a poorly grasped oppositional shift from hierarchy to network. This shift is often presented as a military extension of the all seeing panopticon, or an overhyped rendering of the Deleuzeguattarian concept of the rhizome. The latter, itself described as an intellectual escape from hierarchical thinking,[154] has for many captured the dynamic nature of electronic networks.[155] The virus-like nomadic rhizome has helped activists to think through how networked communications might be adopted in order to support modes of direct action. The peripatetic practices of virus writers and pirates, for example, providing inspiration to online anarchists.[156] According to the Critical Art Ensemble, the collapse of global nomadic authority in cyberspace could be brought about by the introduction of—among other things—computer viruses.[157] Significantly though, it is Hardt and Negri's conception of universal contagion that is crucially underpinned by the rhizome. Their take on the Internet as a rhizome, for example, points to a decentered and democratic network structure determined by a highly redundant and distributed mode of information exchange originated by the U.S. military.

In contrast to this particular rhizomatic vision, I propose a move away from how the rhizome has become linked to distributed networks. I argue that although the rhizome has done much to dispel the myths of the hierarchal analysis of a top–down or transcendent power, using it to describe a network as democratically distributed perhaps misses the point. As is seen, network emergence does not result in a democratically distributed equilibrium of connectivity. Indeed, since Kathleen Burnett's comments that "at its most political, connectivity is a democratizing principle,"[158] new maps of the Internet have suggested a far more complex physical gradation between what is hierarchically and democratically distributed—a complexity in which rhizomes can become roots and roots become rhizomes. This will, I argue, have implications for how political dimensions of epidemic power are approached. I now proceed to the three questions designed to help think through these implications.

## QUESTION CONCERNING CONNECTIVITY

There has been much misplaced hyperbole surrounding the democratic nature of networks compared with the tyranny of hierarchies. Indeed, the idea that by merely becoming connected to a network leads to democracy has been rightly challenged.[159] A given configuration of nodes is not enough to guarantee the emergence of any single political form. Nevertheless, the idea that more connectivity leads to increasing susceptibility to an epidemic is not so easily rejected, and as a number of theorists have proposed, increased contact exposes the individual node to more potentially infectious *other* nodes. For example, Goffey argued that more exposure may well be a crucial factor in "the problem of resistance" to contagion.[160] There is however a big question concerning the particular model of network applied to this problem of exposure. For example, there is a widely adopted tendency in the study of the Internet to view its topology as a historical manifestation of the Cold War military-state-complex. Drawing on the distributed network hypothesis, the Internet is grasped as a robustly decentralized evolution of the traditional control and command strategies employed by military tacticians. Paradoxically perhaps, it is the high redundancy of the distributed network that makes it both democratic (in the sense of a randomized distribution of links) and effective in terms of its resistance to targeted enemy attack. It is also the open flexibility of the distributed model that allows it to become effectively deployed as a kind of *swarm-like* mode of conflict. As Arquilla and Ronfeldt proposed, the robust connectivity of a distributed network allows it to be both chaotically dispersed and well-linked in terms of rapid communication flow.[161] Also, following the distributed hypothesis, Thacker argued that the epidemic itself plays a fundamental role in further decentralizing network formations. For instance, in netwar, surveillance of contagious events (computer viruses, biological diseases, or bio-terrorist attacks) can become a tactic designed to "fight" epidemic decentralization and a means of maintaining sovereignty. Yet, as Thacker argued, the tactic itself may indeed become futile in the face of the overwhelming decentralizing powers of the epidemic. As he proposed:

> If an epidemic is "successful" at its goals of replication and spread, then [the network] gradually becomes a distributed network, in which any node of the network may infect any other node.[162]

What Thacker proposed here follows the logic of random network theory. The more connected nodes become, the more vulnerable they are to contagion because every node in the network is democratically linked with equal probability of becoming infected. Imagined in this way, contagion becomes a powerful decentralizing force, which actively breaks down resistance by spreading everywhere. Like the deterritorialized rhizome, it "connects any point to any other point,"[163] and turned into a weapon, like the swarm, the "mobbing" capacity of the virus allows it to strike without warning.[164] However, the robust, yet fragile hypothesis suggests otherwise. The Internet is neither hierarchically arborescent nor entirely rhizomatic. In fact, in terms of contagion, the connectivity of this topology seemingly reverses the logic of the distributed hypothesis, by proposing that epidemics are not causal of topological decentralization. On the contrary, it is the increasing centralization of the network itself—the amplification of a few highly connected clusters, which make a network vulnerable to contagion. Far from the random universe of the distributed network, in which nodes are linked according to averages, the epidemic actually thrives in highly skewed and clustered environments. In other words, the less decentralized (the less random) a network is, the more susceptible it becomes to targeted attacks from viruses. Significantly, universal contagion is not grasped here as determined by increased contact alone, but becomes apparent when networked assemblages overlap, and contagion and contamination follow.[165]

There is a problem however with the new science of networks. These maps express only the homogenous effect of connectivity emerging from heterogeneous processes. They are the emerged product, as such. It is important, therefore, that this first question emphasizes the spatiotemporal evolution of a network in terms of continuous emergence. Contrary to early random modeling of network complexity, the scale-free model explored here, introduces the rather obvious, albeit often omitted, point that most networks do not reach an end-state of equilibrium, but actually continue to grow *overtime*.[166] This means that networks are not simply motionless spatialized regions distributed on a grid, as in a *given* set of nodes and links, but are instead intensive, compositional architectures in which stability and instability combine. Networks may also be thought of as events *in passage*.

Of course, distributed networks are not motionless. Staying constantly in motion is after all a tactic of the swarm.[167] However, the robust qualities of the distributed model do imply an equilibrium end-state typical of random network models, in the sense that there is a need to strike a balance between randomized connectivity and control. Moreover, the hypothesis lacks a sense of the ongoing uncertainties of network emergence. Swarms contrast with scale-free topologies insofar as the latter imply that networks do not simply form around distributed protocols, but emerge via collective and symbiotic interactions occurring between populations of sociotechnical nodes. Indeed, the *universality* of scale-free topologies is organized, not simply by code, but around the collective usage of network participants.[168] Significantly, the communication codes that flow between nodes are considered to play an important role in the assembly of social and electronic networks, but they are not the sole determinant of a network's architecture.[169]

## The Myths of the Cold War Internet

The scale-free model contradicts a general tendency in studies of new media to situate the state–military objectives of the distributed model as central to an understanding of how electronic networks function. In particular, how the epidemic "logic" of network power, has evolved as a robust form. For example, as Thacker argued:

> By definition, if a network topology is decentralized or distributed, it is highly unlikely that the network can be totally shut down or quarantined: there will always be a tangential link, a stray node (a "line of flight"?) that will ensure the minimal possibility of the network's survival. This logic was, during the Cold War, built into the design of the ARPAnet, and, if we accept the findings of network science, it is also built into the dynamics of epidemics as well.[170]

Throughout the 1990s many cultural theorists and political economists linked the evolution of the Internet to the militarized objectives of the Cold War.[171] Countering some of the more popular discursive utopias concerning the age of networks, they pointed to the central role of DARPA (the U.S. Defence Department Advanced Research Projects Agency) in the early design of a robust and redundant topology intended to maintain the sharing of information, despite segments of the system being destroyed by a physical (nuclear) targeted attack. Arguably, this foregrounding of the U.S. military–industrial complex not only provided a definition of the early topological evolution of the physical network infrastructure, but it also fashioned the emergence of an enduring political and cultural identity of the network.[172] Of course, the positioning of this identity varied by some degree. On the one hand, some claimed that network culture is directly characterized by a *panoptic* expression of militarized, cybernetic power: a surveillance culture symptomatic of the victorious spread of post-Cold War capitalist sovereignty.[173] Hardt and Negri, on the other hand, argued for a rhizomatic network, tending toward decentralized and a potentially difficult to control distributed connection, but nevertheless originating from the military-state-complex. As they argue . . .

> The Internet . . . is a prime example of this democratic network structure. An indeterminate and potentially unlimited number of interconnected nodes communicate with no central point of control; all nodes regardless of territorial location connect to all others through a myriad of potential paths and relays . . . the original design of the Internet was intended to withstand military attack. Since it has no center and almost any portion can operate as an autonomous whole, the network can continue to function even when part of it has been destroyed. The same design element that ensures survival, the decentralization, is also what makes control of the network so difficult . . . this democratic model is what Deleuze and Guattari call a rhizome, a non-hierarchical and concentred network structure.[174]

However, also in the late 1990s, new empirical data suggested that the Internet's topology had deviated considerably from the plans put forward by military-funded engineering projects in the 1960s. The hypothetical robustness of the network, which purportedly emerged from its highly redundant distribution and random (democratic) connectivity, is actually countered by increasing network clustering and vulnerability. Network scientists went on to argue that it is indeed an enduring myth of Internet history that it was ever designed to simply survive a targeted nuclear attack.[175] Their opinion on this matter is fairly well supported by RAND who recognize that usage of the Internet ensured that the distributed model was transformed into something "unforeseen."[176] Moreover, in an interview in the late 1990s, Paul Baran, the engineer often misleadingly attributed with the distributed design of the Internet, argues that "roadblocks" set up by the telecoms monopoly AT&T prevented his work for RAND from being fully integrated into the ARPANET project.[177] According to Baran obstacles arose from the failure of communication engineers to fully adapt to the new digital paradigm in technology.[178] Subsequently, the implementation of packet switching into the fabric of the ARPANET project went ahead free of Baran's full set of proposals, including a stipulation of a highly redundant and robust topological design similar to a fishnet (see Fig. 2.1).

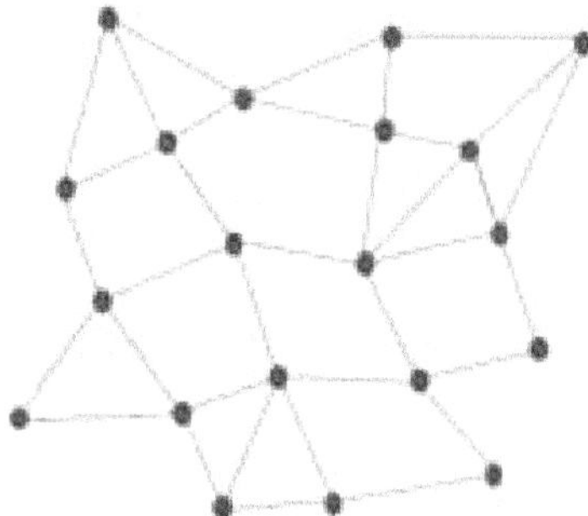

*FIGURE 2.1.* Baran's feasibility study for RAND determined that a distributed network was the most redundantly connected topology and as such a robust defense against targeted attack. "The enemy could destroy 50, 60, 70 percent of the targets or more and it would still work."[179]

Contrary to Baran's optimization of redundancy, the scale-free model exposes an accidental emergence produced by historical processes of use. It is certainly a very different topology to the fishnet design anticipated in Baran's RAND project. In place of the essence of the distributed model, there is a topology exhibiting a complex fractal composition between stability and instability (see Fig. 2.2).

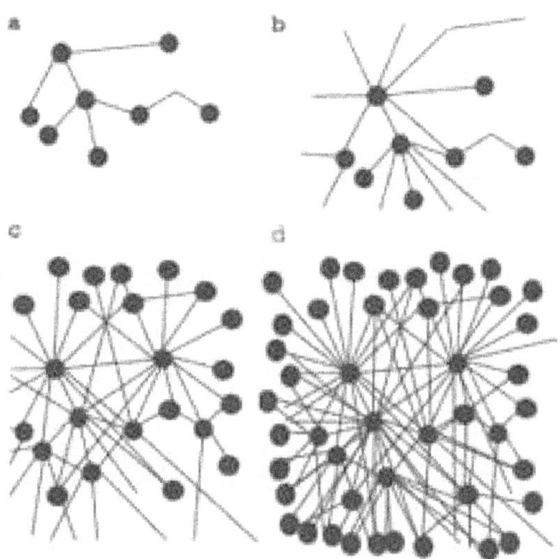


*FIGURE 2.2.* The nodes of a scale-free network are not given—they grow. Starting with a smaller number of nodes (a) and developing links over time (b & c) older nodes in the network become vulnerable hubs or clusters (d).

Despite a tendency to freeze the emergence of a network as an effect, the scale-free model provides a glimpse of a network in passage. It further demonstrates how physical branching structures emerge not simply as a result of intentional (coded) design, but are instead largely directed by events produced by the collective interactions established between developers and users. The directionality of the networks evolution seems to further suggest an assemblage which becomes sensitive to the uncertainty of a series of events triggered by social usage.[180] Once the network began to grow in 1969, its topological emergence occurred as a largely unanticipated result of localized, but collective negotiations and interactions, which appear to have had a global impact on the mutation of its topology. E-mail, for instance, was not at first regarded as a necessary mode of Internet usage, but soon became one of its main applications and one reason behind its celebrated growth spurt in the 1980s-1990s. The emergence of the Internet is not consequently grasped here as formed by the identities or internal essences of its military past, but instead transpires, fractal-like from heterogeneous processes of social interaction. So as to fully understand this point, it is necessary to look more closely at how the scale-free model was developed.

## The Power Law: *There Might be Giants*

> If the heights of an imaginary planet's inhabitants followed a power law distribution, most creatures would be really short. But nobody would be surprised to see occasionally a-hundred feet-tall monster . . . in fact among six billion inhabitants there would be at least one over 8,000 feet tall.[181]

In 1998, a group of researchers working in the field of nonlinear physics and using similar robotic software to that used by search engines to trace web page links, began to map out what they considered would be the random universe of the University of Notre Dame Web domain.[182] Random, because until this point in time, complex network theory had been dominated by the work carried out by Erdos and Renyi in the late 1950s.[183] In simple terms, the random model defines complex networks as homogeneously mixed, or democratically linked: each node has an equally probable chance of having the same amount of links. Significantly, the Erdos-Renyi model also became central to the development of epidemiological models and was subsequently imported from the study of diseases to the computer virus problem in the early 1990s.[184]

In a *Scientific American* article published in May 2003, Barabási explained why his team anticipated these random features in the Web architecture.

> People follow their unique interests when deciding what sites to link
> their Web documents to, given the diversity of everyone's interests and
> tremendous number of pages they can choose from, the resulting pat-
> tern of connections should appear fairly random.[185]

However, once the robot program had *crawled* the paths of HTML metadata and returned its findings, the researchers soon discovered that the Web
was not a random universe after all. On reflection, perhaps the researchers
should not have been so surprised that network growth played such an
important role in topological transformations. In 1990, for example, there
was just one web page (one node). By 2003, there were more than 3 billion.
Likewise, nearly three decades ago the Internet only had a few routers, but
has currently expanded to millions. After analyzing the distribution of links
across a large sample of webpages, researchers realised that the complexity
of the Web exhibited a strange, skewed topological consistency, the patterning of which prompted one author to contrast the democracy of random
networks with the far-from-random *aristocratic* connectivity of the scale-
free model.[186] In a *Scientific American* article Barabási recalled how the maps
revealed that:

> A few highly connected pages are essentially holding the World Wide
> Web together. . . . More than 80 percent of the pages on the map had
> fewer than four links, but a small minority, less than 0.01 percent of all
> nodes, had more than 1,000.[187]

Although randomness and chance are factors in the growth of the socioelec-
tronic topology of the Web, these new findings suggest that there is also a
dynamic organization of nodes and links. Distinct from the equilibrium state
of random networks, scale-free networks follow an approximate 80/20 rule,
in which a few *rich* nodes continue to get richer (approximately 20% of the
nodes have 80% of the links). Using graph theory, the researchers established that rather than the well-proportioned bell-curve graph produced by
the average distribution of links in a random network, the topologies of
these electronic networks demonstrate a *power law*. This law is evident in
open systems in which small perturbations can send a system into new configurations or phase transitions, allowing *explorations* of alternative states.
As Barabási pointed out, the decaying tail of a power law on a graph denotes
that many small events "coexist with a few very large ones" or "highly
linked anomalies."[188] One of the factors hypothetically driving this undemocratic connectivity is a process Barabási terms *preferential attachment*.[189]
Because the Web has no central design, Barabási claimed that "it evolves
from the individual actions of millions of users."[190] For example, links are
made to older established nodes, as is the case when web designers more

often than not link to the most popular search engines or online shopping malls. This logic suggests that nodes like *Google* and *Ebay* become giants.

It is important to add that the concept of preferential attachment is complex in nature. Unlike the soft links of the Web, the connectivity of the physical infrastructure of the Internet is more likely to be *distance-dependent*. This is a twofold constraining factor. First, the economic cost of laying cable is considerably more than the creation of a HTML link, and second, routers will be located in areas where the social demand makes it economically viable for their existence. Yet, new network theory claims that the scale-free model is a universal model apparent in the Web, metabolic networks, electronic circuits, Internet routers, co-authorship, and conceptual networks of language.[191] Indeed, scale-free universality has influenced the study of epidemic spreading on digital networks with new approaches proposed by researchers familiar with the complex, collective behavior of physical matter, rather than the analogies found in biological and technological codes. This shift from code to how clusters become potential multichannel nodes for epidemic spreading may provide a novel perspective that escapes the analogical tendencies of the sciences of code,[192] focusing instead on the network environment.

## Re-coupling Connection and Collectivity

The implication of the claim that scale-free networks are universal has not surprisingly provoked criticism from a number of authors concerned with the emergent political ontology of networks. For example, Thacker argued that the claims of new network science require us to disentangle the ambiguous relation it sets up between collectivity and connectivity.[193] He argued, as such, that simply being connected does not necessarily, in a political context at least, infer collectivity or does "the mere existence of this collectivity [point] to the emergence of a political [democratic] form." On the contrary, Thacker argued that the nodes (bodies or actors) in a network need to become organized around "relationships that create or change nodes," and therefore go on to express political coherence somewhere in between the tensions of collectivity and connectivity. However, in the context of contagious events and their relation to the cultural politics of the network, I argue that the universality of connectivity needs to figure in the analysis. An emergent epidemic is highly dependent on the specific connectivity of a network and will tend to "exploit" topologies that grow volatile overtime. Network politics can therefore be thought through in terms of the power relations established when nodes connected to a network become susceptible to repetitious contagious events.

When considered in terms of how the social assemblages of electronic networks come together, collectivity and connectivity are not so easily dismissed. As DeLanda argued, the properties of the links of a network are not always "inferred by the properties of the persons linked."[194] Nodal identities with deliberate intentions and freedom of action clearly exist, and are important to the study of human interaction on networks, "but some of the emergent properties of a network tend to stay the same despite changes in those attributes."[195] This infers that emergent network collectivities have properties independent of local interactions. In social network analysis, for example, small worlds and clustering are topological phenomena that are not beholden to individual actors. In this sense, DeLanda made a useful distinction between what is regarded as a deliberate plan by a person and "changes brought about by causal interactions among social assemblages without any conscious intervention by persons."[196] Therefore, the dynamic directionality and repetition of an epidemic may become removed from the causal will of individual nodes and the connectivity, relationality and collectivity of nodes become important factors in grasping how a network becomes viral. Furthermore, I argue that networked relations become subject to transitive and intervallic flows of repetitious events (their networkability), which can make nodes vulnerable to unanticipated shocks that trigger contagion.

## A QUESTION CONCERNING THE OFFENSIVE–DEFENSIVE MODE OF EPIDEMIC POWER

Can the directionality of a networked epidemic become tactically exploited as an offensive mode and therefore necessitate new defensive strategies? Paul Hitlin[197] interestingly addressed this question by discussing how electronic networks can provide a slipstream for contagious social cascades of rumors designed to damage political opponents. He provided examples of how false reporting can spread rapidly on converged media networks, particularly when the rumor is repeated over and over again on popular channels. If he is correct, then political strategists may become like viral marketers, who claim to spread ideas through sociotechnical networks. If they succeed then what defences become available to those who seek to resist the epidemic?

The robust, yet fragile hypothesis is only just becoming known to Internet security experts who have struggled to find solutions to viral vulnerabilities. IBM had turned to the Erdos-Renyi random model in the 1990s because of its widespread use in the study of biological diseases. Their main strategy was the twofold deployment of the random model and epidemic threshold theory, also borrowed from biological epidemiology. Threshold theory looked at the point in which the birth rate of a virus exceeded its death rate—a point that once breached would tip over into a full-blown epidemic. However, by 1998 one of the researchers working at IBM identified what he described as one of the major "mysteries of epidemiology." In a paper presented at the *Virus Bulletin Conference* in Munich that year, White's paper "*Open Problems in Computer Virus Research* "[198] argued that computer viruses do not have to reach an equilibrium point that depends sensitively on birth and death rates. The biologically inspired models used by IBM did not properly describe the spread of viruses like the *LoveBug*, which continued to infect systems independent of a breached threshold. Indeed, it was apparently White's paper that came to the attention of Pastor-Satorras and Vespignani from the Abdus Salam International Centre for Theoretical Physics in 2000, prompting their scale-free research into the epidemic spread of computer viruses on electronic networks.[199] Pastor-Satorras and Vespignani's subsequent paper "*Epidemic Spreading in Scale-Free Networks*"[200] published 1 year later suggested a "new epidemiological framework" marked by the absence of an epidemic threshold and its associated critical behavior. Simply put, in scale-free networks one *promiscuous* computer is enough to create an epidemic since if it is unprotected and connected to a cluster, it will eventually spread a virus to another computer without the appropriate virus protection. As Vespignani proposed, the Internet is prone to the spreading and the persistence of infections (no matter how low their) virulence.[201]

Both the absence of a threshold and the dominant role of promiscuous nodes suggest alternative ways of thinking through the tension between network control and emergence. Whereas in a distributed network model contagion is situated as a force for decentralization, in scale-free topologies it is the promiscuous nodes that make a network epidemic come together by acting as intensive centers of viral exchange. So, on the one hand, we have the idea put forward by Thacker in which network control, using surveillance techniques for example, is in fact marked by the need for the absence of control: *Ultimately it takes a swarm to fight a swarm.* However, on the other hand, scale-free models suggest that it is the control of promiscuous nodes that becomes critical to the overall control of a network.[202]

The role of promiscuous nodes in epidemic spreading complicates the notion that too much connectivity leads to contagion. Indeed, if an infection is introduced to a given set of randomly linked nodes the epidemic will eventually burn-out as the number of resistant or immune nodes will eventually exceed the susceptible nodes. This type of infection works when a virus is inadvertently spread from node to node via a chain of encounters determined by spatial proximity and equilibrium. However, networks are not given: *They grow* and infection is therefore intimately coupled to the emergent historical processes of a network assemblage. As new connectors or hubs develop in a scale-free topology they introduce new vulnerabilities—new opportunities to spread. Many of these opportunities are provided by distance-independent links. These include physically embedded links like airports or disembedded mediated links like e-mail address books. These links act as an ecological bridge[203] enabling the infection to escape the random death of nodal containment by crossing over from one network to another via contagious assemblage overlaps. Epidemic bridges occur when small changes are made to the connectivity of a network, like the inadvertent or purposeful rewiring of a link, which produces complex and dynamic topological transformations in the shape of new infection vectors. In fact, the decentralization of a network, rather than producing more bridges, could reduce the effectiveness of connectors and consequently lessen the power of an epidemic spread. Taking out the centers of a network makes it more resistant, therefore a serverless Internet might allow the epidemic to spread more randomly and eventually burn itself out.[204]

The scale-free model suggests a distinctive shift away from the focus on detecting epidemic code toward a concentration on the contagious environments in which replicators spread. Epidemic dispersion occurs not simply because of the codification of a particular replicator, but because the environmental conditions are conducive to the networkability of a contagious event. For example, the configuration of witting or unwitting *intervallic* nodes, some of which are major hubs, may allow an epidemic *transitive* passage through the network. Here we see the significance of a network conflict in which big connectors or hubs hold together the consistency of an assemblage and intermediately determine the directionality of an epidemic. Hubs act as intervallic points of viral exchange leading to escape bridges an epidemic crosses in order to access another network. They also mark a vulnerability to targeted attacks and potential capture. Netwar may not be best organized around random "blanketing" or "mobbing," as the swarming model proposes, but would involve the uncertainties of a far-from-random topological struggle in which the defence against witting and unwitting, known and unknown infected carriers requires a fuzzy mode of detection.

## A QUESTION CONCERNING NETWORK POWER: FUZZY DETECTION

In response to the unknown threats of digitality, IBM has turned to immunological analogies in the development of anomaly detection software. Like this, the analogy has seemingly exceeded the figurative biological referent and become enmeshed in the real practices of the security industry. However, digital immune systems, instead of providing a solution, have exposed the problem of distinguishing between known and unknown events on a network. This distinction is, it seems, fuzzy at best, and worse, undecidable. In order to defend against highly contagious viral events, network security strategists have had to accept a tradeoff between maintaining a robust means of communication and the fragility of a system sensitive to unforeseen contagions. Indeed, I conclude this chapter by speculating that the detection problem has become part of a broader struggle for network power, which involves risk managers defending the nation state against unforeseen contagions, including the spread of computer viruses, financial instabilities, and significantly, the social and cultural viruses of a hidden networked enemy.

The problem of anomaly detection has concerned social theory since its inception. In the introduction to this book we pointed to Durkheim's early theory of anomie, in which certain forms of suicide represent a social deviation from the structural norms and rules that organize macro-society. The homeless, inebriated orphan takes their own life because they are a social dysfunctionality in the accident-free rationale of anomic detection. In a related contemporary context, Baer's solution to the anomalous cultural viruses of the suicide bomber advocates a similar type of social and cultural exorcism. As he proposed, in order to halt the terror virus, the Islamic world will need to take part in its mass rejection, following which the epidemic would, it seems, burn itself out. Yet, as I argued, epidemic burn out functions in the homogenous equilibrium of network connectivity, not in heterogeneity.

In the introduction to this book, we proposed that the role of the anomaly might be best grasped in Baudrillard's account of the "overorganization" of society and culture. In this sense, the detection of anomalies exposes the historical overorganization of the social according to organic or functionalist social models. This applies to Durkheim's treatment of anomie, as it does to Le Bon's psychological overorganization of *The Crowd*: a social organism so riddled with contagions that it threatened the stability of 19th century institutional power. In fact, it was Tarde who perhaps first rumbled the organizational tendencies of organicism. For him, it was the instability of heterogeneous and contagious micro-relations, which far from destroying the larger organized social unities, actually composed them. What this points to is the role instability plays in the organization of stable social wholes. Furthermore, far from being anomaly-free, the social requires the constitutive anomaly in order to reproduce itself.

Paul Virilio's contention that the war on terror is becoming an *"accidental war* that no longer speaks its name"[205] introduces a fundamental point concerning the detection problem. Virilio argued that the suicide terror attack is an example of an area of social contact and potential conflict, in which the unknown is growing out of all proportion. Indeed, Virilio is not alone in pinpointing the problem of anomaly detection in recent security discourses and the strategic use of the unknown in the administration of fear. Thacker similarly explored the discursive formations that surround the war on terror, particularly those that tactically play on the viral epidemic as a way in which the "enemy" can perhaps remain *advantageously* unknown. As he put it:

> Epidemics ignite public fears with great ease, in part because the "enemy" is often undetected, and therefore potentially everywhere. But more than this, it is the alien, nonhuman character of epidemics that incite public anxiety—there is no intentionality, no rationale, no aim except to carry out iterations of what we understand to be simple rules (infect, replicate, infect, replicate . . . ). The exceptions of epidemics and war implode in biological warfare, bioterrorism, and in the way that U.S. policy enframes the public health response to infectious disease.[206]

This is a defense policy that becomes linked to *one* emerging threat to national security. Like this, Thacker colluded with Virilio in the sense that he situated the epidemic, like the accidental terror attack, as central to the strategic management of uncertainty. As Virilio[207] claimed, it is in the interest of the administrators of fear to intentionally prolong the uncertainty over the origin of a suicide attack. The shock of the accident turns into a rumor, independent of an attributable source, which can be all the better manipulated.[208]

There is however, in addition to the discursive formations surrounding the uncertainties of the terror epidemic, a nondiscursive mode. Uncertainties, accidents and contagious events are, I propose, conditions of the robust-fragility of netwar, which necessitate a rapid rethink of the role of the anomaly. Evidence of a shift in tactical response is perhaps evident in RAND's recognition that the network enemy has become more ambiguous, less detectable, and detection itself takes place increasingly in the absence of intelligence. Tacticians are of course well aware of the way in which network robustness swims in fragility composed of accidents, shocks, and cascading contagions. The anomalous event is not outside of their environment, but always inside, with the potential to be triggered at any moment. As RAND conceded "detection will more likely occur only after an attack has begun."[209] Anomaly detection is therefore more about risk assessment than it is about containment. Learning from the experiences of computer science,[210] the securing of robust networks relies more on risk equations, in which the known and unknown are no longer binary, but increasingly fuzzy distinctions.

In Galloway and Thacker's coda at the end of this collection the reader is reminded of Donald Rumsfeld's fuzzy risk assessment between the known-knowns, known-unknowns, and unknown-unknowns. What appeared to many at the time as a Pythonesque piece of nonsense in fact provides evidence of the American administrations' adoption of the fuzzy techniques now influencing the computer security debate. Echoing Rumsfeld, a computer network security expert recently assessed the risk of a digital attack:

Vulnerabilities that can be exploited are quantifiable risks (known-knowns), while for those for which there is no exploitation (known-unknowns) the impact is unquantifiable. Unknown-unknowns remain uncontrollable, unquantifiable risks.[211]

Here the risk managers of the Internet, like Rumsfeld, not only differentiate between the fuzzy gradations of what is known and the unknown, but they acknowledge the unknowable and uncontrollable quantities of the unknown yet to be exposed in an epidemic attack. In this sense, they both realize that the unforeseeable potential of the anomalous and contagious event is not an external deviation of homogenous structure, but is instead a heterogeneous compositional force endemic to the network.

## ACKNOWLEDGMENTS

Thanks to Tiziana Terranova and Jussi Parikka for making time to look over this chapter at various stages of development.

# 3

# EXTENSIVE ABSTRACTION IN DIGITAL ARCHITECTURE

*Luciana Parisi*

In Octavia Butler's book *Dawn,*[212] the main character Lilith awakens in a windowless, doorless cubicle. Light-colored walls form out of shifting plat-forms, which also turn into a leaf-like bed and morph into the shape of a table. The table's surface folds into an edible bowl of food that unfolds back into it, if Lilith does not eat it on time. Behind the table, a wrapped up mound of clothing, made of extremely soft material taking the shape of her body, changing textures and shades according to her movements and bodily temperature. Stains would quickly fade and vanish from the cloths, the surfaces of the table and the walls leaving her with a sense of wonder about the malleable fabric of the cubicle. Is it thin plastic, glass, or cellulose? How can these surfaces be soft and hard, warm and cool at the same time? In vain, Lilith attempts to find solutions while testing the responsiveness of this alien environment keeping her in forced isolation.

She looks for traces of cracks or fissures on the surface of the cubicle—any line that would become an exit on the walls and the floor of this prison where she had been kept for an unmemorable time. Only later, after the encounter with her medusa-like captivators, who have rescued her from the self-destroying human race on planet earth, does she realize that she has been housed on a sort of spaceship. The cubicle is only part of an infinite series of pods connected to each other and separated by a sort of elastic membrane. The spaceship resembles less an interstellar spacecraft and more an extremely adaptive intergalactic environment produced by semi-autonomous networks able to pass water, food, oxygen, heat, in an interdependent chain re-arranging its order according to changing conditions. A spaceship made of many intertwined environments hardly looking like each other. A mutant ecology without identity and composed of extremely distinct yet encroached milieus.

Lilith finds out that these environments are internal, external and associated zones of interaction built by the Oankali's genetic engineering techniques of evolution, exploiting the extremely infectious nature of genetic codes. The entire architecture of such unnatural ecology is a lab of genetic experimentation ruled by the promiscuity of codes, numbers exceeding the program of numbered probabilities via their residual potential to be calculated anew. The engineered environments of the Oankali are governed not by new stochastic calculations of a binary code—in other words, a set of delimited probabilities—but by the milieus of connectedness of the code approximate proximities, the associative numbering zone of all numbered codes.

The Oankali's engineering space is a parasitic symbiosis where layers of genetic substrates are continuously nested onto each other through the growth of new walls and bridges. This is an architecture of slime built by bacterial genetic trading—bacterial sex—constructing veritable intricate communities or cities of sludge, otherwise known as biofilms,[213] a thin layer of parasiting cells covering a surface in aqueous or wet environments. Discovered only a few decades ago in the field of microbiology, biofilmic architectures are intricate bacterial milieus threaded with pores and channels and embedded in sticky protective goo or sludge, a blob. Biofilms are complex communities of bacteria forming a city of thousands of intricate neighborhoods—hundreds of different colonies—rapidly growing overnight by sticking a new layer on top of the genetic substrate. These multilayered biofilms are tightly packed together like the urban centers of the most densely populated cities (from Lagos to Bombay), the macroagglomerate cities of the immediate future. Towers of spheres and cone- or mushroom-shaped skyscrapers soar 100 to 200 micrometers upward from a base of dense sticky sugars, big molecules, and water, all collectively produced by the bacterial inhabitants. In these cities, different strains of bacteria with different enzymes help each other exploit food supplies that no one strain can break down alone, and all of them together build the city's infrastructure. The more food available, the denser the bacterial populations become engineering biofilms in a web of genetic trading invented by ancient bacteria 3.6 billion years ago. Significantly, echoing the work of Lynn Margulis,[214] biofilmic architectures point out that evolution is not a linear family tree, but an affair of trading in slime growing to cover the entire surface of the Earth.

Although most futuristic visions of digital architecture, for example Gibson's *Neuromancer*,[215] point to a data environment in which all physical objects are turned into a series of numbers, governed by the binary calculator of all nature: the genetic code, Octavia Butler's genetically engineered environment, exposes the infectiousness of symbiotic algorithms. Here numbers are not compressed and divided into isolated bits, but remain connected in abstraction, related by means of a certain incomputable attraction and a power to remain nested inside each other. In this way, the Oankali build architectures of slime, rather than a dataspace of digital simulation. Such architecture aims not at mimicking nature, but at exposing a symbiotic architecture able to invent a new nature and bring the inorganic back into the organic.

This chapter proposes that such symbiotic architecture points at a new conception of coded spatiality—an infectious relationality marking the ontological condition of biological, technical, cultural networked spaces. To a certain extent, this new conception intersects with recent arguments concerning the viral capacities of both digital and biological codes in which the viral no longer constitutes an exception or anomaly, but is rather the rule of a viral networked order.[216] As the editors of this book seem to suggest, this is an asymmetric network implying that autonomous programs coexist in single modes of operations and contribute to the evolutions of networks themselves. The emphasis on the information code as composed of internal, external, and associated milieus of interactions, has led to a new conception of digital culture as itself composed in the relational milieus of viral ecologies.[217] This argument can, however, be pushed further to suggest that a spam architecture needs to account for the experiential dimensions of what I term the *extensive abstraction of architecture*. This concept of extensiveness must not fall into the impasse located in many visions of digital architecture between the digital and the analogue, the technical and the natural or the mathematical and the biological nature of extension. Instead, following William James, I argue that for "a relation to be real it has to be an experienced relation."[218] Therefore, the metaphysical dimension of relationality, often attributed to digital architectures, needs to be accounted for in the material capacities to experience change, to capture the transition from one state to another, and to register the algorithmic passing between distinct blocks of space–time. What is proposed here is that such relationality needs to be grasped in the material process of infection that arguably ensures that digital architectures exceed the organization of algorithmic programs.

In order to expand on the concept of the experiential dimensions of abstract extension, this chapter primarily draws on Alfred N. Whitehead's notion of extensive continuum.[219] As such, I suggest here that relationality is implicated in the activities of microperceptions of infinitesimal, incomputable quantities. This is an infectious conception of the architecture of digital code, but only insofar as microperceptions pass from one state to another through the resonating chambers of matter. Importantly, there is in this conception a continual symbiotic relation between mental and physical perceptions of all kinds of bodies ready to participate in the building of architectures of infection.

Whitehead's use of the term *extension*—from the Latin *extendo*—helps to define the capacities to stretch, stretch out, or spread out. Whitehead proposed an energetic conception of extension, which implies tension and effort. Here space and time are a partial expression of one relation of extension between events, which in itself is neither spatial nor temporal.[220] This is not only because spatial relations extend through time, but as Whitehead observed, following the discovery of electromagnetic relativity we know that what is simultaneous in space for one percipient, is successive in time for another, depending on their relative state of motion. Far from suggesting that information space depends on an embryonic relation with the human body, entailing a center of temporal perceptions able to frame the spatialized data, to make data part of experience, as for example Mark Hansen claimed,[221] it is argued here, following Whitehead, that there is primarily neither time nor space to be experienced, but rather a relation of extension between events. There is potentiality in extensiveness only insofar as the material world is composed of the tiniest objects of perceptions: a body and thumb, a drop of water and a swarm of flies, a molecule and an electric charge. What is an object for one percipient, however, is something else to another. A drop of water to a human percipient is a swarm of flies to an electron.[222] The continuity of nature is here found in events that are extensively connected in their intrinsic and extrinsic physical relation:[223] a *discontinuous continuity*.

So as to explore the complexities of the role of extensive experience in digital architecture, I draw on examples of contemporary software design. By contrasting the neo-Darwinian inspired evolutionary models of the digital gene with the endosymbiotic model of parallel evolutionary algorithms in the generative design, deployed by for instance Greg Lynn, I argue that the relationality of an animated, generative extensiveness neither lies in the biocomputation of the genetic codes nor in the sensory or cognitive perceptions of digital codes. Rather, a concept of viral architecture in digital culture becomes implicated in digital calculation and the soft design of extensive experience. This further suggests the centrality of discontinuous continuity in the differential calculus of soft architecture and that the fuzziness of information resonates through a body however small its prehensive capacities of experience may be.

Using these examples from software design, a twofold expression of extensive architecture is developed. First, the differential calculus or the surplus of digital code, points to modifications in the mental and physical capacities of bodies to perceive and move in the imperceptible connectedness between microbodies and macrobodies. Second, the production of new prehensive capacities of extension added to a concrescent nature (a nature growing out of events), leads to prehensive experiences of the extended relations among all kinds of bodies. Extensive architecture, therefore, introduces stealthy occurrences in the seamless calculation of digitality. This has a direct impact on the capacities of thought to be felt in its invisible variations entangled to a multiplicity of inorganic bodies. Extensive architecture is the sticky residue implicated in the smallest and the shortest of encounters, a sort of biofilmic slime that gels *things* together yet anew in the unnatural dens of an anomalous nature. Extensive architecture turns space into blobs, a wet or aqueous extension. As Greg Lynn proposed:

> [a] near solid, to borrow Luce Irigaray's term, [that] has no ideal static form outside of the particular conditions in which it is situated including its position and speed. Gel solids are defined not as static but as trajectories.[224]

Examples of blob architectures are, like the genetically engineered environments of the Oankali, large populations parasiting on each other, stretching trajectories into curves of infinitesimal layering and building a wet supersurface in the multiplexing experience of an extended continuum.

## DIGITAL GENE

The centrality of the genetic algorithm in the design of network architecture derives from a mathematization of biological organization, intended to forecast the evolutive behavior of extension.

Since the early 1990s, genetic and evolutive algorithms have been used to explore design variations that can be bred via software simulations. At the core of digital architecture is a design technique based on neo-Darwinian models of evolution. In particular, Richard Dawkins' conceptual devise of the "blind watchmaker" algorithm suggests that the evolution of forms is not simply derivable from the random mutation of simple genetic instructions, but, more importantly, on nonrandom cumulative selection leading to the development of complex shapes called biomorphs—a complex set of genes.[225]

Dawkins' genocentric view of evolution argues that the emergence of complex form cannot be explained by random genetic mutation. Instead, only the workings of a blind nature that intervenes to combine accumulated variations in the most complex of ways, can account for evolutionary complexity. In the *Blind Watchmaker*, Dawkins refined his previous genetic theories[226] by emphasizing the role of natural selection. He argued that the emergence of complexity cannot be explained by single-step selection, according to which the entities selected are sorted once and for all. For example, clouds through the random kneading and carving of the winds come to look like familiar objects—a sea horse, a group of sheep, a face with a nose, and so on. For Dawkins, the production of these shapes is based on a single-step concept of selection, derived by one type of combination without evolution. Accumulative selection, on the contrary, points out that each selected entity—or at least the result of sorting entities—reproduces *in time*. The results of one sieving process are fed into a subsequent sieving, which is fed onto the next one and so on.[227] Therefore, selection implies the sieving of entities over many generations in sequential succession. The end product of one step, therefore, is only the starting point for the next generation of selection. Cumulative selection indeed points at a blind watchmaker who selects at each step the best adapted generations of genes to favor survival into the next.

To demonstrate his point, Dawkins devised a computer simulation of such processes called *Biomorph Land*, which features the recursive programming of a simple tree-growing procedure.[228] The result is a complex shape emerged out of simple rules of replication—recursive programming—applied locally all over the branching tree. The biomorph—a set of recursive genes—develops and (a-sexually) reproduces. In every generation, the genes supplied by the previous generation are passed to the next generation with minor random errors or mutations. This means that—as generations go by—the total amount of genetic difference from the original ancestor can become very large. Although the mutations are random, the cumulative change over the generations is not. Although progeny in any one generation is different from their parents, each progeny is nonrandomly selected to advance into the next generation.

Because Dawkins' *Biomorph Land* is very large—thus implying that there are a very large number of genetic populations—it is as if the evolutive development of the best-adapted shaped-creature is already mathematically contained in the areas of the genotype. *Biomorph Land*, like John Conway's *Game of Life*,[229] is a practical example of using evolutionary computation for the generation of form, and although its original purpose was only to illustrate the theoretical principles in progressive-cumulative selection, it was soon adopted by a generation of artists and scientists.[230]

In digital architecture, Dawkins' notion of cumulative selection has been used to search for the genetic space of shapes that generatively reproduce and develop through random mutations. Manuel De Landa, for example, explained that generative models of simulation are searching devices exploring a space of possibilities through the combinations of traits, so as to find, over many generations, more or less stable solutions to problems posed by the environment.[231] The searching device in the field of computer simulations is called a "genetic algorithm" (GA) in which a population of computer programs is allowed to replicate in a variable form.

The use of GAs has enabled architects using computer-aided design (CAD) to *breed* new solutions to spatial design, instead of directly programming those solutions. Take for example the generative solutions of a chair designed by architect Celestino Soddu[232] as a way to evolve a modular type according to parameters of random mutation and cumulative selection at each step of the chair's generation. Evolutionary design, according to Soddu, enables a fast design for industrial production, a sort of prototype of a uniquely evolved form. To evolve the modular prototype of a chair a basic algorithm undergoes repeated cycles of evaluation, selection, and reproduction leading to variations in the composition of the algorithmic population and exploring the space of possible solutions in the vicinity of the best adapted generation of chairs.

In the computer model of the blind watchmaker, genetic configurations of variables are arranged into a matrix where each combination of variations occupies a different place defined by the distance between parents and offspring. More specifically, for each set of possible offspring that may be generated, a given distance from the parent occurs with equal and uniformly distributed probability. This modular evolution of prototypes entails the adaptation of a single population in a fixed niche. For example, all individual chairs are assessed according to the same criteria and the same fitness function, which distributes equally to specifically located genes. Every individual gene has an identifiable fitness, but all individuals in the same space are ranked in the same way. The only type of variable dynamics between individual genes and between generations of genes is competitive exclusion (i.e., the algorithms of the same niche compete to become members of the next generation). There is no concept of change in the mechanisms of selection, variation, reproduction of GAs over evolutionary time.

The basic intuition of GA and evolutionary algorithms (EA) follows the dominant intuition of natural evolution: By accumulating small random variations that incrementally improve fitness, best-adapted solutions progressively grow in the design of spatiality.

Is there any anomaly in such generative design of space? Because mutations are already contained in the genetic space of possibilities and within the phases of cumulative selection, the sudden changes in shape are here calculated possibilities, because the mutations of the GA are conceived as a sort of combinatorics positioning of 0s and 1s. What such digital binary logic assumes is that nature, like culture and the natural environments constructed artificially with CAD, operate through a genetic–digital programming that contains in itself all possible solutions for the design of a new shape of chair.

What if indeed genetic space did not coincide with the calculable positions of 0s and 1s? What if contingencies in evolution were neither the slight random mutation of the same genetic space through generation nor the cumulative selection determined by the bell curve, as a limit that redistributes variations in space? What if the evolution of GAs entails not primarily the progressive complexification of form out of simple elements but rather the centrality of complex parallel connections, the symbiotic nature of extension?

## SYMBIOTIC ALGORITHM

Although Dawkins' model of the biomorph proposes a *serial* genetic algorithm,[233] other models of evolution have focused on the activities of *parallel* algorithms entering a symbiotic alliance triggered by environmental variations.

On the one hand, serial algorithms are hierarchically arranged into a genetically related lineage through the gradual accumulation of random variations. On the other hand, parallel algorithms are nested into each other's activities, trading and distributing variations across milieus of interaction. Serial algorithms have a saturation point and a problem of memory space, whereas parallel algorithms allow simultaneous communication between different processors and the sharing of memory and message transmission. Parallel algorithms are distributed algorithms designed to work in cluster-arranged computing environments. Such algorithms are governed by a multiple agent system (MAS), which is a parallel computer system built from many very simple algorithms whose parallel communication leads not to the evolution of one algorithm or the other, but to a new algorithmic behavior. Ant colonies and bee swarms are examples of MAS working in parallel toward a shared goal. These are self-organizing systems that are not centrally guided by one set of instruction, but grow out of parallel algorithmic processes. Parallel algorithms have also been defined as symbiotic or cluster algorithms. They work in parallel yet independently of any other clusters running in the system and build multiple and composite solutions to the same problem.

In contrast to the evolution of GAs, based on local interactions with the environment, a symbiotic algorithm involves the joining together—the parasitism—of previously free-living entities into a new composite under certain conditions. Such conception of symbiotic parasitism has been derived, amongst others, from Margulis' serial endosymbiotic theory,[234] stating that the origin of multicellular organisms or eukaryotes is not determined by cumulative selection of random mutation but by a symbiotic alliance between distinct colonies of bacteria engendering a novel cellular composite.

Thus, for endosymbiosis, variations are the results of enmeshed distinct yet parallel entities, each containing relatively large amounts of genetic material whose independent symbiotic role remains active in the new composite. Whereas genetic—or serial—algorithms use a finite set of binary features or genes to track their evolution in a single lineage where every individual gene has the same features which only change in value, the symbiotic algorithm entails the parallel processing of binary features that are neither contained in a single lineage nor inheritable in a filiative fashion. Rather, the interdependence of the symbiotic algorithms points at a compositional model of evolution.

In evolutionary computation, the compositional—symbiotic—algorithm has many resonances with the "Building Block Hypothesis" theorized by John Holland.[235] However, symbiotic interdependency, as Richard Watson and Jordan Pollack recently argued, distinguishes compositional symbiosis from the "bottom–up" hypothesis.[236] In particular, Watson and Pollack suggested that symbiotic interdependency accounts for a number of modules where the number of possibly maximal configurations for each module is low, and yet greater than one. Thus, the dimensionality of the system is reduced, not to simple elements, but to self-contained parts that can function on their own while remaining interdependent.

Interdependent modular structures are hierarchically organized in clusters and subclusters. Far from being random, such modular dependencies point that the complex is not dependent on the simple. Rather, the configuration of a module is dependent on the configuration of other modules. This reduces the dimensionality of the search space for an algorithm co-evolving with other entities regardless of their distance. To some extent, this point echoes Albert-László Barabási's suggestion that the world can indeed be regarded as scale-free.[237] Arguing against the notion that most quantities in nature follow a bell curve, Barabási insisted that network architectures follow a mathematical expression called a *power law*. On the one hand, a bell curve suggests a peaked distribution characterized in network theory as a random network with homogeneously distributed averages. On the other hand, a power law denotes the absence of a peak that is replaced by a continuously decreasing curve, where many small events coexist with a few large events.[238] In particular, Barabási argued against random graph theory, which has dominated network theories by equating complexity with randomness. According to this theory, the formation of networks stems from a number of isolated nodes connected by random links. This conception of networks is based on an egalitarian model, according to which all nodes have approximately the same number of links. However, Barabási's research has revealed that despite the millions of nodes in the web, the distance between nodes can be scale-free. Indeed, he argued that network architecture does not coincide with the geometries of Euclidean space (where each node occupies an individual place). Network phenomena such as clustering, he suggested, cannot be measured according to randomness.

Clustering is a ubiquitous phenomenon cutting across levels of order, biological, social, and economical. Scale-free networks do not entail an equal distribution of links, but unevenness, wherein a few clusters have many links, and noncentered modular organization is accounted in independent, but interlinked subnetworks that can coexist and cooperate.

Similarly, the parallel or symbiotic algorithm suggests that modular interdependency is defined by the uneven symbiotic encapsulations of distinct entities, like those found in symbiotic sex rather than sexual or asexual reproduction. For example, whereas the genetic serial algorithm relies on the accumulation of hereditary material from parents to offspring, determined by half the genetic material from one parent and half the genetic material from a second parent, symbiotic encapsulation may simply take the sum of genetic material from both parents, a sum that is more than 2, more than 0 and 1. Thus, symbiotic sex can be grasped as the infectious activities between parallel algorithms and points toward the acquisition of genetic material without direct genetic transfer or filiation. According to Watson and Pollack,[239] Symbiogenetic Evolutionary Adaptation Model (SEAM) algorithms show clearly that the concept of a module is not dependent on gene ordering in specific niches but on epistatic dependencies (i.e., the relationship among genes). Significantly, this also implies a rethinking of the activity of natural selection, which is directly influenced by the milieu sensitivity of an entity, and thus, by its contingent capacity to enter an ecology of genetic relation.

Endosymbiosis is not concerned with the extension of simple genes to evolve complex form, but with the parallel bacterial genomes forming clusters or information ecologies: architectures of infection. Rather than generating variation through the cumulative model of selection, symbiotic algorithms expose the primacy of multiple genomes entering in uneven composition. The parallelism of symbiotic algorithms points to the relational dynamics in evolution where genetic populations become large numbers that do not occupy fixed discrete locations. Therefore, the parallel algorithm may need to be rethought, not simply in terms of modular organization as the conception of symbiogenetic modularity proposes. For example, in the standardization of building materials that allow for fast assembling and disassembly of the autonomous parts that compose a modular home.[240] On the contrary, endosymbiotic parallelism may more importantly point to the mutational nature of assemblages and the extended matrix of continual variation. In other words, it may point to the genetic deformation of the grid, the anomalous connection between unreachable milieus, the viral coactivities of differentiation and the topological continuities of discrete genomes. Moreover, although modularity more directly defines the retrospective link between preordained parts that can be broken apart and brought back together in the same order, a symbiotic algorithm may instead be pushed to expose the mathematics of curves, the topological continuum between discrete clusters of numbers. In short, the primacy of relational movement or anomalous connection in extension.

Where Dawkins' model of the GA functions according to the binary logic of digital communication, based on the probability function of a set of possibilities, the symbiogenetic algorithm exposes such digital logic of combinatorics to the vagueness of information milieus, a cloud of fuzzy numbers that cannot but be prehended.

In the next section, I will discuss the symbiogenetic algorithm in blob and folding architecture, to argue that digital design software expresses biomathematical features of *extension* defined by an experiential field of numerical continuities.

## FUZZY ARCHITECTURE

Since the early 1990s, the combination of mathematics, genetics, and information technology has become central to architectural design. In particular, software has become a tool for the design of a responsive and evolutive environment bringing back movement in extension.[241] As opposed to the focus on the evolution of complex form from simple genetic instructions within a digitally rendered Euclidean space, a new conception of extension based on the centrality of variability has entered software architecture. As the Dutch architect and artist Lars Spuybroek put it:

> "no geometry of complexity and morphology resulting form an epigenetic process can be fully Euclidean or elementary." It is up to relations to produce the elements not the other way around. "Variability comes before elementarity."[242]

Variability in extension challenges the Cartesian ideal space of exact coordinates. In this way, architect Greg Lynn rethought the centrality of movement and force in the software design of space.[243] For Lynn, software capacity exceeds the mere rendering and visualising of data to encompass material capacities in the design of flexible, mutable, and differential spatiality. Unlike an earlier generation of prototypes that randomly calculated cumulative selection using GAs, Lynn suggested that design is the result of independent interactive variables and parallel algorithms able to influence one another through their potential activities.[244] Here the Cartesian grid of isolated positions, deprived of force and time and represented by steady-state equations, is contrasted with the Leibnizian conception of the monad. In other words, the converging and diverging of infinitesimal habitats in a point of view, which resembles less an exact mathematical point and more a vectorial flow, as a continuation or diffusion of the point.[245] Lynn takes the monad to be an integral calculus of variables,[246] at once a mathematical singularity and an infinitesimal, incalculable, differential multiplicity.[247] Contrary to the Cartesian model of extension, Lynn embraced Leibniz's integral calculus—the calculus of coexistent variables. Lynn, following Leibniz, defined the monadic conception of objects in space, based not on the bifurcation of force from matter, but on the dynamics of a gravitational field defined by the movement of masses in space, or vectors entering in a mobile balance with one another. Digital animation, according to Lynn, needs to be rethought in the context of a Leibnizian mathematics of differential forces and motion that accounts for variability in spatial design.[248]

Lynn drew on Leibniz's study of differential calculus[249] to express the centrality of time, motion, and force in architecture: the point at which the tangent crosses the curve. Another way of expressing this is to consider the enveloping of time or point-fold, as Deleuze called it.[250] This is where a straight line is always a curve. It is a nondimensional point of conjunction of vectors, a real yet inexact quantity and an intensive degree of differentiation. Significantly, only random, irregular, and complex equations can calculate the irrational numbers of the curve and the limit of the relation between two quantities—exact points—that vanish into the curve. As Deleuze explained it:

> The irrational number implies the descent of a circular arc on the straight line of rational points, and exposes the latter as a false infinity, a simple undefinite that includes an infinite of lacunae. . . . The straight line always has to be intermingled with curved lines.[251]

The calculation of infinitesimals (infinitely small numbers) defines continuous relationality between the smallest quantities. It is a continuity found in the evanescent quantity that retains the character of what is disappearing. What we might term a *virtual residue*.

This point concerning complex equations returns us to the role randomness plays in digital architectures. Recently, the mathematician Gregory Chaitin readdressed the question of the differential calculus in his algorithmic information theory, suggesting that the string of bits running between 0 and 1 corresponds not to a calculable number, but to a random, irreducible, and structureless quantity.[252] Randomness is here understood as maximum entropy—something that cannot be compressed. Because randomness has no pattern or structure, Chaitin argued that it has to be understood as "a thing in itself" or an irreducible quantity. He defined such a seemingly incompressible quantity in terms of the $\Omega$ number: an infinitely long and utterly incalculable number made of gaping holes. $\Omega$ is a number that is *maximally* unknowable.[253] Although the number $\Omega$ is perfectly well-defined mathematically, the digits in the decimal expansion of this real number (i.e., a number like 3.1415926 . . .) cannot be determined. Every one of these digits spans between 0 and 9, but it is impossible to know what it is, because the digits are accidental, random.[254] Hence, unlike Barabási's reformulation of networks as far-from random, or at least in terms of a mixture of random-ness and order, the uncomputable cipher defined by Chaitin reintroduces a sort of randomness into evolutive mathematics.

However, Chaitin defined randomness not in terms of an empty space between nodes. It is not a space of equally distributed information, but rather a full, densely packed zone of information.[255] In short, this mathematical interval between 0 and 1 is neither a discrete number nor a void, but is an intensive quantity defined by an intrinsic numerical variability, which remains computationally open in relation to the configuring constrains of an inexact cipher.

Similarly, Lynn explained that the mathematics of animation software is based on the uncompressibility of the infinitely small interval. This is a dynamic space full of information, defined by a differential equation with more than two interacting components, such as velocity, direction, and temporality of each vector.[256] The interval defines a relational space pregnant with information and populated by infinitesimal variations and qualitative transformations in the relation of form-matter. According to the architect and furniture designer, Bernard Cache, each singular and distinctive point is a geometric point of inflection. It is an infinite curvature of digits where numbers move in opposite directions. Inflection here is "the true atom of form, the true object of geography."[257] It is the slopes and the oblique gradients of hills and valleys rather than the ground of basins.

This conception of continual variations in extension has been central to the study of the flexible grid or "rubber math" described by the biomathematician D'Arcy Thompson.[258] In his writings, he analyzed variations in the morphology of animals using deformable grids, which yield curvilinear lines due to changes in form. He compared the curvature of deformations in formal configurations to the curvature of statistical data, including speed, weight, and gradient forces such as temperature. He then concluded that these variable deformations are instances of discontinuous morphological development. Through the concept of a variable grid, Thompson developed a mathematics of species rooted in dynamical sets of geometric relations. Indeed, deformations are not simply derived from a given form, but from a continuous relationality between internal and external forces. Here the accident is understood as the continual deformation, or destratification of the species, directly constrained by uncomputable relationality. It is the point where the curling of the line between the inside and the outside is an accident constrained by inflection.

Contrary to neo-Darwinian genecentrism, Thompson believes that genetic information is unable to fully specify the generation of form. Instead form can only result from the microactivities of the environment (natural forces), which can be described with the mathematical laws of differential calculus. Thompson found such laws in the geometric shapes of shells and sponges, which cannot be explained by genetics (i.e., genetic inheritance and random mutations). He affirmed that evolution is not governed by natural selection. On the contrary, he pointed toward the variable constraints and parameters within which organisms develop certain limits, which channel animal forms into particular patterns that are constantly repeated across the phyla. For example, he argued that the shape that droplets of viscous liquid take, when dropped into water, is virtually the same as the medusa forms of jellyfish. And yet such a convergence of form is not accidental. Indeed, this accident is fundamentally constrained by the physics of moving fluids described in the equations of fluid mechanics. To further explain this, Thompson suggested a concept of mobile stability between divergent series of internal and external forces. Indeed, Lynn drew on Thompson to address the nature of geometrical folds, a supple geometry that enables the object to bend under external forces, while folding those forces within.[259] This is the geometry of the curving, and not segmenting, line expressing the movement of folding and unfolding between interiority and exteriority.

Thompson's speculations on the deformation of types suggest a topological rather than a modular evolution of shapes: a bending architecture evincing the capacities of algorithms to infinitely curve in symbiotic accord with the gradient variations of the environment. Singular intricate knots are not simply reproducible in the fashion of modular typologies—the complexification of types—insofar as the ecological conditions of reproduction are constrained by the infinitesimal accidents of inflection.

In digital architecture, such a topo-ecology approach can describe developmental landscape and define the space within which organisms evolve, importantly replacing the notion of fixed types organized in the phylogenetic trees like those generated in Dawkins' *Biomorph Land*. Thompson's model of developmental landscape can also be discussed in terms of an alternative fitness landscape in which a surface represents an external environment across which a faceted sphere rolls.[260] The rolling faceted sphere expresses the organism with its own internal constraints, whereas the landscape stands for its potential pathways of development. A landscape is a field where a small vectorial change is distributed smoothly across a surface so that its influence cannot be localized at any discrete point. Slow and fast movement is built into the landscape surface through hills and valleys. Yet the mobilization of space is not derived from a direct action of objects, but is imbued in the environment itself and the potential for movement enveloped in extension. Therefore, the movement of an object across a landscape entails the intersection of its initial direction, speed, elasticity, density, and friction doubled with the inflections of the landscape in itself. It is not that the object performs movement. Rather, the landscape can initiate movement across itself without literally requiring any motion on behalf of the object.

The inflections of an environment are then gradient slopes enfolded into its own geological stratification. Surfaces are not simply horizontal, not merely composed of pieces stitched together alongside a trajectory tending at infinitum. Surfaces do not therefore constitute a ground. On the contrary, surfaces are themselves points of inflections. They are folds with an oblique nature, already imbued with motion through an intrinsic propensity to movement. These surfaces are not an empty space, but microbodies full of dens, where virtual force and motion is stored.

Importantly, from this standpoint, the notion of a fitness landscape does not define an environment in terms of its selection of the best-adapted organisms, but an alternative field of residual potentials housed in the dens and slopes of inflecting surfaces always ready to propel movement. In digital-generative design, the breeding of topo-ecological surfaces corresponds not simply to a combinatorics of codes—discrete quantities. Rather, it entails the curvature of fuzzy numbers, the incompressible qualities of gradients, where extension becomes inflection or *infection*. Active and passive parasitic forces in digital-generative design mark the *obliqueness* of the environment, never reaching a point of equilibrium insofar as these are governed by a mobile stability and directed by vectors of attraction and repulsion. Such multiplex assemblages of potential residue are experimented in Lynn's design of blob surfaces or warped kinematic spaces.[261]

Blob surfaces are held together by their mutual capacity to infect one another and compose symbiotic assemblages. The blob is not topographically specific, but is instead specific to its topological evolutive environment, which remains irreducible to one form or another. Here Lynn understood blobs as monads equipped with internal forces of attraction and mass. As he put it:

> A blob has a centre, a surface area, a mass relative to other objects, and a field of influence: i.e., the relational zone within which the blob will fuse with or will be inflected by another blobs.[262]

Blobs are *objectiles* defined by relations of proximities. This enables them to either redefine their respective surfaces, based on particular gravitational properties, or fuse into one contiguous surface, defined by the interactions of their respective zones of inflection. Either way, the blob is not an entity shaped by internal genetics, but remains open to the gradients of the relational field that compose its different configurations. It is not the variation of the same genetic instruction set designed to generate new spatial configurations of an object. Rather, the potential to acquire different configurations is embedded in the miniscule gradients, infinitesimal variations, or intensive quantities of speed, temperature and pressure of a symbiotic environment, which maps an entire ecology of coexistent milieus of information.

It is useful here to reference again Margulis' theory of endosymbiosis in which she envisioned evolution as a process of assemblage, where no single body can be isolated from topo-ecologies of contagion, in which parasitic architectures of nonlinear feedbacks are formed. This further echoes bacterial meta-metazoic environments, which according to the biological theorist, Dorion Sagan, are defined by a series of intricate eco-systems turning the question of gradual evolution from simple to complex form into a matter of symbiotic parasitism: a parasitism where the host and the guest incidentally become accomplices in the production of intricate ecologies of interaction.[263] Here the environment is not a typographical ground occupied by an organism that gradually evolves through random mutation and cumulative selection. On the contrary, the environment is not a static ground, but a mobile house. Like a snail carrying its house on its back, the environment is continuously being moved by a series of epigenetic relationships, where the outside is enfolded within the movement of an extended continuum. Similarly, blob architecture proposes a nonmodular concept of extension, an incomputable pack of numbers. It is an open-endedness in digital calculations, which turns limited points into zones of continuous complex numbers. This is the continuum of spam architecture, in which the slimy residue connects *things* together. It is an architecture in which an infectious extension in the infinitesimal population of numbers becomes glued to each other in a stealthy building of anomalous socialities.

So as to grasp how all of this relates to the continuous discontinuity of extension introduced at the beginning of this chapter, it is necessary to first see how the status of the object and subject is rethought in terms of vectors of a curve. From a series of inflection, a line is distinguished as a place, a site, a point of view. Far from the ocularcentric tradition that equated the point of view with a pregiven subject or that assigns to an object a fixed position, the subject here is defined by "what remains in the point of view" Similarly, the object is an objectile, virtually coexisting with an infinitesimal number of objects transformed by their relation to the variable positions of the subject. The latter, as Deleuze, drawing on Whitehead, affirmed, is less a subject than a superject: "a point of view on variation" marking the conditions in which "the subject apprehends a variation."[264] From the continuity of the infinitesimal variation of objects to the discontinuity of the subject's view point, in which a new conception of extension as "continuous repetition" is made possible, an extensive continuum remains uninterrupted by the disjunctions and conjunctions of a mathematical, topo-ecological continuum in the minute perception of microbodies. In this sense, Leibniz's differential calculus cannot be disentangled from the activities and passivities of microperceptions housed in a body, no matter how small, how inorganic, it is. Like Deleuze's tiny folds, it probes in every direction, vibrating under the skin, passing through states and resonating across all layers of perception.[265]

If digital–generative architecture is concerned not with the modular evolution of forms, but with exposing the incomputable chance in the digital calculation of parallel forces, gradients, motions, temporalities, then it may be necessary to add that such irreducible complexity entails the primacy of the infectious relations of the tiniest bodies. In short, as argued in the next section, the generative nature of space, instantiated by a mathematics of continuity, needs to account for the experience of differential relations encompassing the abstract and the concrete.

## CHAOS PREHENDED

Blob architectures borrow from the digital calculations of spatial evolution in order to produce a grid engendered as a multisymbiotic enmeshing of surfaces. However, they are not simply the combinatorics of 0s and 1s. Instead they are the infinitesimal variations of curving lines linking 0 and 1. Like the quantum bit[266] — or qubit — the symbiotic algorithm defines not one state or another (0 or 1), but encompasses, at once, a quantum entanglement of 0 and 1. The digital animation of such parallel surfaces works not to imitate the spatiotemporal growth of form as a sort of digitalization of natural evolution, but probes the relational capacities of minimal units of information. Following Deleuze, we might conceive of these units, not as ultimate atoms:

> but miniscule folds that are endlessly unfurling and bending on the edges of juxtaposed areas, like a mist of fold that makes their surface sparkle, at speeds that no one of our thresholds of consciousness could sustain in a normal state.[267]

Therefore, software is not a mere tool for design. It implies instead an experiential zone for the quasi-imperceptible activities of minute percepts. The obscure dust of the world implicated in differential relations. An entire process of continual relation between micropercepts and perceptions, nonsensuous and sensuous prehensions, draw the curvature of felt thought: *a thought that is felt*. Here we can return to Whitehead who argued against the primary function of sensory perception as defined by David Hume as a world perceived in distinct objects in the here and now. As an alternative, Whitehead proposed that perception cannot be "divested of its affective tone" — "its character of a concern" — which entangles sense-perception to nonsensuous or conceptual prehension. Importantly, this is the continuum of the immediacy of the past in the immediacy of the present: a nonsensuous prehension defined by the activity of feeling continuity in discontinuity.[268]

The differential calculus does not solely define the centrality of the incomputable quantities of fuzzy numbers in the symbiogenetic evolution of architectural form. More importantly, it points to what Deleuze termed the "psychic mechanism of perception, the automatism that at once and inseparably plunges into obscurity and determines clarity."[269] In other words, symbiogenetic algorithms are not definable in royal isolation from the dark activities of matter or the dusty percepts and vibrating thoughts that add new dens in continual extension.

The discreteness of digital numbers is never simply—or exclusively—a matter of combining already given probabilities which result in a computable-cognitive equation. As an example, Chaitin's incomputable algorithm suggests that there is mathematical extensiveness between the chaotic fuzziness of actual codes containing too much information and the indivisibility of an exact set of equations. Such extensiveness is not determined by a void because emptiness is only perceived as such from the point of view of clarity. Emptiness is a remarkable and distinguished perception, which is nonetheless populated by the fuzzy obscurities and infinitesimal chaos of minute percepts. This leads us to define digital architecture not in terms of the clear erection of form. Digital architecture is not an all-encompassing zone of clarity in perception or a transparency in subjective orientation devoid of incomputable darkness. Rather, the central focus on the digital generation of architectural form needs to give way to the perceptual experience of extensive continuity and the vagueness of minute percepts. Here, Deleuze proposed an hallucinatory perception ready to grasp "the haze of dust without object" out of which form emerges in a flick of a second, long enough for extension to be minutely perceived.[270]

But what exactly are these minute perceptions, turning each perception into a hallucination? They are the vibrations of matter contracted by all sorts of organs of perception, enveloping incalculable dust into distinct clear form. In other words, the calculus is split into two causalities corresponding to two parallel and symbiotically nested calculations. Again Deleuze observed the two inseparable yet distinguished faces of the calculus: "one relates to the psycho-metaphysical mechanism of perception, and the other to the psycho-organic mechanism of excitation or impulsion."[271]

The differential calculus therefore entails the infective relation between mental and physical reality in which minute perceptions are minute bodies distributed everywhere in matter. Thus, what is perceived is not disentangled from what happens to the body, and the latter does not exist in royal isolation of what happens to abstract extension.

If the modification of objects in digital architecture uses the differential calculus to expose the infinitesimal variations in the composition of an environment of dens and slopes (storing virtualities, force, and motion), then it may be useful to ask a number of questions (not all of which can be addressed in this chapter). First, what clear perceptions does the differential calculus select from minute obscure percepts? Second, which states of hallucination does digital design entail. And third, which kinds of communication and propagation of physical movement it implies? In addition to these three questions we may also ask what are the physical and mental affective states enveloped in an architecture of symbiotic extension, which remains itself in evolution, in a process of prehension (from micro percepts to macro perception and back again), but does not depend on a subject (prehending) or an object (prehended)? In short, what is the spam or the surplus value of code in digital architecture? Finally, what does digital extension add to the extensive continuum and what kinds of percipient events does it entail?

Whitehead's concern with the relation between extension and prehensive extension points out that extension is required by process.[272] In other words, extension points to an intensive spatium of virtual spatiotemporal coordinates directly experienced by means of prehension. Again, this is the continuum of the immediacy of the past in the immediacy of the present. Yet arguably, although process is not directly located in digital processing, the infinitesimal divisions of such processing do indeed point to a relationship of extensiveness involving the activities of microperceptions that are at once sensuous and nonsensuous, physical and mental and add to the concrescent nexus of nature.

Lynn's design of the Port Authority Gateway, for example, has modeled the site as a gradient field of forces simulating the movement of pedestrians, cars, and buses at varying speeds.[273] The final design of the Gateway has been derived from this primary field of forces that already includes the inter-activities of distinct components of movement. The breakdown of these components into geometrical particles that change shapes and positions has enabled the study of singular cycles of movement over a period of time. The generative capacities of extension become imbued in the design process itself, deducted from a relational field of micropercepts, including the speed and slowness of variable interactions constructing an architecture of infection: an experiential mutation between the abstract and the concrete. Like this, the spam or surplus value of code in digital architecture points precisely to a relational event defined by the field of influence in all kinds of micropercepts. Although not entailing any privileged point of orientation of movement this field of influence does however expose the propensity of movement in the design itself to become an associative milieu of calculations and differential relations.

It may be argued that this kind of digital architecture mainly concerns the genesis of form and not experiential relationality. For example, in some cases the sensorimotor interactivity of touching a wall mainly coincides with sensorimotor reception instead of the inventive process of prehensions. Yet it could be misleading to overlook the subtle persistence of percipient events caught up in the darkness of minute perceptions. Such events exceed both sensorimotor responses and mental recognition, operating in the overload zones of too much information. In other words the supple spatiotemporal curvatures pressing against the cortex of cognition suspend all channels of sensory perception and melt the body into the throbbing microbodies of an extensive continuum, where the locality of relations is continuously engendered by the speeds and slowness of past and future vectors. To overlook these zones of tiny inflection-infection is to deny that experience occurs in the interstices of macroperception and that socialities are built in the material activities of a body no matter how small and how inorganic.

Digital architecture does not therefore simply reduce the percipient event to binary processing, but rather it exposes the ingression of fuzzy quantities and spatiotemporal anomalies in experiential modes of feeling thought. Moreover, it adds new capacities for being infected in the world and entering symbiotic dependencies with compossible worlds not of this world, which in turn allows for the nonsensuous feeling of the unthought into the everyday. Like this, Lilith knows that the extendable cubicle housing her body is neither natural nor artificial, neither an organism nor a discrete set of data made flesh. She soon realizes that the responsive walls are not a direct response to given stimuli. There is here no on and off buttons. And yet all the layers of such an unnatural environment are ready to perceive the minimal change in the calculation of symbiotic algorithms and equipped to genetically re-engineer themselves so as to include the ultimate clusters of atoms if necessary. This is not because the Oankali have the most advanced, hyper-efficient technologies, but because their environments are themselves *machinic* bodies all the way through. Thus, Lilith learns that such faceless inhuman spatiality is only an extended multiplicity of micro-bodies, a multiplicity Deleuze described as "an infinity of tiny folds (inflections) endlessly furling and unfurling in every direction."[274]

# 4

# UNPREDICTABLE LEGACIES

## Viral Games
## in the Networked World

*Roberta Buiani*

## VIRAL PROLIFERATIONS

Between 1983 and 1986, a series of consecutive publications officially announced the cause of AIDS, now attributed to the retrovirus known as HIV. In the same period, Fred Cohen introduced a "major security problem called virus."[275] Although his description was rudimentary and, according to some computer virus analysts, not fully accurate,[276] it was immediately adopted universally. "Computer viruses" and other electronic-based "anomalous" agents were soon proclaimed a "high-tech disease."[277]

Today, most of what media and popular culture tend to call viruses does not pertain to one specific variety of computer viruses, and clearly, it does not regard those computer viruses once described by Cohen. Other coded substances and agents carrying a set of characteristics that approximately fit the original definition of virus have been annexed to the same macro-category, as if they were part of the same family. Parasitic nature, general modality of proliferation as well as capacity to "hide" or move furtively within the most recondite strata of computer systems are features that can be ascribed to most network pests.

The single expression that conveniently summarizes and evokes any of the these features is the viral: All agents that fall under this category are said to contain a viral behavior, feature, or code. A number of recent Internet security handbooks has placed computer viruses (code that recursively replicates a possibly evolved copy of itself) side by side with worms (network viruses), Trojan horses (name given to a variety of malicious programs), logic bombs (programmed malfunction of a legitimate application), and other types of malware such as spyware, spammer programs and flooders—to mention a few of the most popular network annoyances—and have made them the subjects of one single field of inquiry, computer virus research.[278] Although this field, which includes affiliates of antivirus (AV) and the security industry, virus writers[279] and computer analysts, classifies malicious agents according to their distinct technical features, it ultimately groups them under the same macro-category. In the same way, popular media and the ordinary user appear to see any unwelcome computer disruption as viral.

The use of the viral as a widely applicable term is a thread. Its function as a receptacle of diverse material features, behavioral aspects, and connotations suggests that it can be applied well beyond the limited territory of computer virus research. Accordingly, biological and computer viruses have ceased to be the only viral items in circulation. Since the end of the 1990s, attributing the term *viral* to phenomena that only vaguely remind us of viruses has become a common practice. For instance, viral marketing (VM) has become an increasingly adopted form of advertisement or promotion for a variety of enterprises as diverse as the software or the cinema industry. Very much linked to this phenomenon (although different in scope and purpose) are viral videos. With the recent popularity of blogs and online sharing Web sites such as YouTube and GoogleVideo, viral videos constitute a series of often amateur shorts that reach unexpected popularity among peers and visitors without any particular reason. The term viral has also been associated with specific forms of media activism. The expression viral tactics was coined by Nathan Martin in 2003 during a speech delivered at the festival of tactical media Next Five Minutes, and has since become a form of media activism.[280]

By organizing, within the same expression, significant features common to viruses, this expanded use of the viral term engenders divergent outcomes. First, one could infer that the use, or abuse, of such a term is symptomatic of an essentialist tendency that uses the viral as a sort of "dump" wherein one can just throw anything pertaining to viruses. In fact, it appears to obliterate all possible differences that could help distinguish between the increasingly heterogeneous variety of multifunctional digital agents and, say, viral media. However, this constitutes just one superficial facet of the viral—an aspect, that, nonetheless, should not be ignored and is treated in the course of this chapter.

Second, as a catalyst of popular and well-known features, the viral is a flexible term. Computer viruses, VM, viral videos, and viral tactics are distinct phenomena. Not only do they operate on different registers and communication codes (computer language, media language, and/or cultural discourse), but they also affect different environments (computer networks and human networks). Additionally, their level of engagement with and mode of appropriation of various viral materials are substantially different. In other words, they utilize and interpret the viral in quite different ways. The emerging of diverse phenomena that share the same suffix, and yet, interpret it in such different fashions, reveals that the viral should not be dismissed as "generalist" or "essentialist." On the one hand, the phenomena mentioned here stem from culturally and popularly accepted characteristics from selected features of viruses. However, these characteristics are neither identical nor do they manifest in uniform fashions. One could say that they aggregate unevenly and crystallize to form different manifestations of the viral. On the other hand, the same characteristics can be applied to a diverse range of activities. In this case, they appear to attest possible transformations of the viral into a reservoir of tactics and actions.

This double articulation of the viral as extensive or customary application (often leading to essentialist interpretations of viruses) and as open-ended use sheds a new light over this term. First, rather than understanding the application of this suffix in negative terms, as a homogenizing expression, merely capturing the nonspecific side of viruses, one could rethink of it positively, as a term that differentiates, that helps formulate, redefine, and diversify novel phenomena. Second, it contextualizes the phenomena it fosters in relation to both the user and the objects that interact with it. In this way, the viral can be understood as a vessel of powerful potentials, originating from the generalized and generalizing aspects of viruses.

The result is not only a new way of understanding the notion(s) associated with the viral and the original agents (viruses) from which they emanate, but also a new way of reinterpreting the role of the users as active players. With the users' appropriation and active application of the viral to forms of advertisement, media phenomena, and human practices, viruses can no longer be understood as faceless, anomalous, and/or superfluous substances that exist within or colonize the media and human environment with their intrinsic and immutable characteristics. Increasingly, they become the generators of unexpected and unconventional creative uses and initiatives.

A general acknowledgment of viruses as elements internal to, or inseparable from, a multifaceted and multidirectionally transforming system is hardly new. Scholars in different disciplines have pointed out the status of viruses and worms as both expressions of and co-contributors to the media and network culture.[281] Like each element that forms the system, be it technological, political, or cultural in nature (a program, a concept, or a phenomenon, or, as Fuller defined it, a media ecology),[282] viruses are dynamically linked to, and intersect, at different levels and degrees, with the other components. Thus, it is inevitable to reinterpret viruses as potential producers of creative outcomes, rather than just threats.

Analyzing viral interaction within the system itself and with the actors who use the system (the users) helps us understand how viruses might contribute to foster open-ended reconfigurations of a system's architecture (the network physics, as Terranova explains). However, it is also crucial to gain awareness on how they might also fuel the user's novel "active engagement with the dynamics of information flows" (the network politics).[283]

A great deal of attention has been devoted to situate viruses in relation to network culture and to uncover the role they might play in contributing to and modifying a dynamic, "turbulent" ecology of media.[284] In most circumstances, it has been demonstrated that viruses do indeed contain a great deal of potential for the development and transformation of network culture. However, where can we actually see their contributions? How, thanks to what mechanisms and in what form does this contribution surface and manifest through concrete practices or phenomena? In other words, we know that viruses have potentials, but what do the existing possibilities look like?[285] Using the viral as a marker of the presence and the contribution of viruses in network culture and beyond may help pinpoint and identify these manifestations.

## A VIRAL GUIDE TO THE VIRAL

As mentioned earlier, viruses' characteristics converging and/or summarized by the viral can be understood as expression of a first habitual or conventional instance of the viral. In this case, the use of this nomenclature to indicate any incarnation of computer malware has practical purposes. First, it serves as an umbrella term that collects, under the same category, all malicious agents existing within network systems. Second, it allows the average user to perform a quick identification of said agents.

However, the use of the viral is clearly not limited to practical necessities. The convergence of several functions in one single term illustrates its significance as a motif. Interpreting this motif as a set of specific or literally identical features that recur in all its manifestations would be inappropriate. In fact, uttering the word *viral* means signaling features that evoke, yet do not exactly reproduce such features. A number of entities and agents, despite their diversity, can be easily included under the viral label. This means that the viral can exceed the domain of computer (or biological) virus research and, possibly, penetrate other disciplinary and cultural realms.

The wide diffusion and use of the viral, then, cannot just be accredited to its value as a generic container of various characteristics, but is dictated, each time, by the types of aspects evoked when they converge in this particular expression. Although the extensive use of the viral can be detected as the sum of quintessential elements that reside within and emanate from viruses, in the second case, the very aspects that characterize viruses and viral entities, or a portion thereof, appear to be removed from the initial source (viruses) and, more or less purposely, appropriated and re-elaborated in an open-ended fashion. At this point, viruses undergo a radical transformation of their functional qualities. Instead of retaining their nature as agents that live out of a certain ecology of media, they are turned into a resource that can be selectively employed and reassembled to describe and even devise many other different activities. From being entities that live, exploit, and interact "with" the context they inhabit, viruses are turned into expressions of other subjectivities that operate, this time, "upon" such context. For instance, viral marketing has appropriated the distributed nature of viruses in general, has drawn upon the aggressive replicating mechanisms of biological and computer viruses, and has adopted their easy-to-remember name. Similarly, viral tactics seem to utilize viruses' capacity to infiltrate and merge with elements of their host.

Extensive (or customary) and open-ended uses that characterize the viral are not the result of different interpretations of the same notion of virus. Rather, they are consubstantial to its existence. In fact, the two aspects are indivisible, as they tend to coexist in each phenomenon examined. For example, stating that VM, viral videos, and viral tactics are solely the manifestation of an open-ended use of the viral would be incorrect. It is because of the recognition of certain conventional features intrinsic in viruses, which are then selected and finally transferred onto and collected into the viral, that the above phenomena were named and formulated. Thus, the viral is both an indication of the multidirectional and productive forces directly deriving from viruses and an expression of the performative forces imparted by individuals on viruses.

The coexistence of the two aspects suggests that the viral, as a nomenclature that incorporates the interaction among viruses, users, and other objects, acts as a means that facilitates potential processes of transformation. The condition of "being viral" implies the presence of qualities proper of viruses that may enable individuals to appropriate them and "become viral" themselves. Although the condition of "being viral" emphasizes the very role that viruses play in the construction and circulation of the viral as a set of conventional features, the promise of "becoming viral" promotes its open-ended use, that is the free intervention of the user as active player able to open up and manipulate the trajectories of the viral.

The distinction made here between two different instances of the viral seems to confirm, while redirecting, Bardini's recent remarks that viruses are redefining postmodern culture as a viral ecology. As a master trope of postmodern culture, Bardini argued, viruses could be grouped as one encompassing category, the "Hypervirus," whose logistic curve can be located at the beginning of the 1980s, with the advent of AIDS. Since then, Bardini added, "[by] materializing the cybernetic convergence of carbon and silicon, [the hypervirus] infected computers and humans alike at unprecedented levels."[286] The Hypervirus, metaphorically described as a "pandemic," has manifested in a variety of forms and through all sectors of culture: The virus, seen as a parasite, introduces disorder into communication, represents a cell of terrorism that emerges, with its viral mechanism of duplication, from the very system that has created it. Ruling our times "as [an] indifferent despot[s]" THE virus[287] can be considered master trope of postmodern culture.

Both the multiform manifestation and the user appropriation of the viral can be easily assimilated to the notion of the Hypervirus. It is thanks to the features and to the current popularity disseminated by viruses through the viral that individuals are enticed to appropriate and adapt the nomenclature to unknown or new phenomena. However, by accepting Bardini's notion of Hypervirus, and the characterization of postmodern culture as a "viral ecology," we also accept the unquestioned dominance of viruses. This perspective does not seem to emphasize the proactive intervention of users in reinventing the course of the viral. While confirming its legacy and popularity, any creative re-elaboration and reutilization of the viral seems to acknowledge and restore the user's agency and increasingly move away from, rather than confirming, the notion of viruses as a given, or as sole fosterers of phenomena or actions.

## THE VIRAL IN VIRUSES (BEING VIRAL)

One way to detect the convergence of expressions in the construction of the viral is by reviewing the role of the biological and informational components in the formulation of the viral.

A number of scholars have drawn attention to the "traffic," as Eve Keller suggested,[288] existing between biology and computer science. Biology, and other—carbon-based-related—life sciences have lent computer science their characteristics and connotations by means of a variety of metaphoric translations and in accordance with particular circumstances, direction of research, scientific assumptions, or ideological agendas.[289] The resulting field of study manifests its connection with the biological through a series of repetitions and/or recurrences that run across horizontally and vertically, like a grid of sometimes intersecting threads. In doing so, these threads shape the discourse and the configuration of computer science, whereas the latter folds back onto the sciences by lending them elements that have been reworked and transformed.

Seeing biology and the life sciences (such as medicine, microbiology, or virology) as the precursors of computer viruses would not be entirely accurate. This statement would neglect the role played by popular imagination in the notion of infectious diseases, long before their molecular causes were officially detected and classified. Furthermore, it would downplay the concurrence of political and rhetorical agendas not only in delineating the connection between biological and computer viruses, but also in affecting their function, as well as their behavior. Ross, for instance, noted how the media commentaries that followed the computer viruses' rise in popularity showed that "the rhetoric of computer culture, in common with the medical discourse of AIDS research, [had] fallen in line with the paranoid, strategic mode of defense Department rhetoric established by the Cold War."[290] Thus, the appearance on the scene of computer viruses was conveniently channeled to articulate "the continuity of the media scare with those historical fears about bodily invasion, individual and national that are endemic to the paranoid style of American political culture."[291]

Galloway noted that the term *virus* was only applied to self-replicating programs after their risky potential was realized. The reason for this assimilation, then, should be found in the political and cultural atmosphere existing in the particular decade of their release: AIDS violently emerged and gained momentum in the media in the mid-1980s, exactly at the same time Cohen was formulating his definition of computer viruses.[292] However, it was not the scientific precision of this definition, and its place in popular culture that Cohen was looking for, but instead a broad, somehow cursory, albeit well-identifiable and "tangible" description. According to Galloway, had computer viruses emerged in the successive decade, today we probably would not call them viruses.[293] Although it is arguable that AIDS was the sole responsibility for giving computer viruses a name, it was probably one of the main factors that consolidated the association between biological viruses and self-replicating programs and the further inclusion of such programs in the broader categorization of viruses.

Not only do Galloway and Ross, among others, provide explanations of how the connection between computer and biological viruses has happened, but they also suggest more profound ramifications, articulations and manifestations that can be detected while viruses make their way through information networks as programs, as well as ideas in people's imagination. The examples given here, in particular, reveal the multilayered-ness of computer viruses. The convergence of diverse expressions originating from contemporary events and the subsequent induction of cultural and social fears reveals the participation of culture at large in establishing the reputation of viruses, by modifying and manipulating their configurations and effects.

The different trajectories that have concurred to form the notion of viruses in general, and computer viruses in particular, seem to indicate their conceptualization as both repositories and generators of discursive formations. In *The Archaeology of Knowledge*, Foucault analyzed the existence of continuities and coexistences of fragments of discourse between heterogeneous fields. The existence of such coexistences and continuities may be signaled through the following categories: naming (the existence of an identical object studied across disciplines), style (a constant manner of statement, or how a certain corpus of knowledge has presupposed "the same way of looking at things"),[294] established groups of statements ("a definite number of concepts whose content has been established once and for all"),[295] and persistence of themes. However, the detection of continuities is not sufficient to group all the above as "unities."[296] Instead, they describe systems of dispersions, that is, "series full of gaps, intertwined with one another, interplays of differences, distances, substitutions, transformations"[297] all linked together through a variety of regularities. In fact, simply put, although naming might be the same, the subject is never quite identical, as it changes and continuously transforms through time, according to various circumstances and mutations of perspectives. Moreover, identical style is not applied to identical technologies and established rules are often not attached to the same concepts. Finally, the same themes are applied to different subjects with different purposes and effects and produce different connotations.

While maintaining elements in common, the series are, indeed, separate entities. As Foucault explained, these systems of dispersion will probably never form unities, yet they are somehow connected through a variety of regularities and elements that keep repeating across enunciations. The challenge in finding and analyzing discursive formations lies, therefore, in identifying the "coexistence of dispersed and heterogeneous statements"[298] and in being able to describe and define a regularity, that is an "order, correlations, positions" between "a number of statements . . . objects, concepts etc."[299] Following this analysis, one may observe that computer viruses do not belong to the same realm, nor would their structure and configuration make them comparable to biological viruses. Yet, they share the same name and some common features. Such features, alone, are present as regularities or as recurring discourses in a number of types of viruses. Moreover, viruses too can be said to recur as regular objects, whose qualities mutate with disciplinary, historical and cultural contexts, as well as dimensional realms.

The regular or reoccurring features of viruses as discursive formations intersect with material practices and coagulate to form dissimilar objects that maintain different characteristics each time. The viral, alone, is a statement that may summarize just about anything related to viruses and incorporate the different forms that viruses can take. However, this does not mean that it maintains identical or unalterable features, or that it can be freely applied, as an unbreakable totality, to any object. Rather, while maintaining a set of underlying and collectively established (or perceived) qualities, it can be also subject to continuous interpretations and modifications enacted by individuals and by the objects with which it is coupled.

The viral is never found alone, but it is always accompanied by different entities (viral agents), phenomena (VM), or actions (viral tactics). When the viral nomenclature is combined with other words or objects (VM, viral videos) it mutates or is adapted to lend such activities and phenomena features they need (such as a behavior, a reproductive set of mechanics, or just its connotations). These features assume different weight or priority according to, or to the benefit of, any agent (or the phenomenon) to which the term is accompanied. The resulting connection of the viral to the designated object sanctions the formation of independent, specific, and novel entities or phenomena, whose characteristics retain features originating from their accompanying attribute (the viral) and the elements that previously characterized the object itself (videos, marketing, etc.). The materialization of the new object is a concrete assemblage. It is the result of the encounter and the concrete realization of the relation between different forces.[300] Not only does the viral incorporate, on the one hand, all the relations, transformations, and dense interpolation between viruses, and their entire surroundings (including society, history as well as other disciplines, such as biology), but also, on the other hand, it acts as that element that simultaneously marks the presence of a concrete assemblage, and facilitates its formation.

## THE USE OF THE VIRAL
## (BECOMING VIRAL)

When Haraway observed that "organisms are not born, but they are made"[301] she refers to the impossibility to interpret "natural" objects (natural or technological in this case) as self-referential, as exclusively born with "boundaries already established and awaiting the right kind of instrument to note them correctly."[302] As part of a media ecology where "'organisms' or 'components' participate in the autopoiesis of the digital culture of networking"[303] viruses can, then, be assimilated to "natural" digital objects. Like other entities or elements whose sum contributes to and, at the same time, affects the entire system, viruses emerge from a complex and often gradual transformative discourse that affects them, and which they affect. In fact, there exists a concerted interconnection between single users, the collectivity of users (the network culture in its diversified articulations), and inherent or perceived characteristics of viruses. All of these actors are involved in assessing and determining the course, the diffusion, and the additional features of viruses. The appearance of new disciplines, commercial industries, as well as new groups that dealt with and that revolved around viruses may exemplify this autopoietic scenario.

Both the AV industry and virus writers are products of the diffusion of computer viruses. Having sprung up from, and by surviving on the very existence of the diffusion of viral code, these industries and countercultures contribute to the further dissemination (both in terms of popularity and territorial diffusion) and circulation of numerous and complex families of viruses. By drawing parallels between living beings and computational artifacts, some authors have supported a similar autopoietic idea. Computer virus researcher Peter Szor seems to subscribe to this very tradition: The interaction between viruses and VM software is not destined to end, as the existence of the former is essential to the generation of the latter. In order to illustrate the competitive, and rather hostile confrontation happening between distinct virus "fighters" and virus writers, he combined the notion of networks as autopoietic systems with the Darwinian idea of "struggle for existence."[304] The result is a portrayal of network systems where malware, operative systems and AV software appear to generate recursively the same network of processes that produced them. At the same time, viruses, which he identified with the whole apparatus of malware and viral agents, engage in a daily and cyclical struggle that constitutes an "evolutionary" step for the development of the digital world.[305] For Szor, users, AV analysts, and computer virus writers are equally contributing to and furthering the viral traffic that happens within and between their machines. As newer security operations are developed to confront the spread of viral code, proposing new strategies that could possibly anticipate next-generation viral attacks, the reactions of virus writers will follow in the form of new viral agents aimed at shattering newly built security shields.

As already mentioned, the viral signals the formation of concrete assemblages or clusters of meaning that sit outside the strict domain of the informational. The active contribution of individuals or groups to forge and name such assemblages is crucial. Forms of micro-appropriations can, then, be considered expression of the encounter between viruses, modified and experienced under the label of the viral, and the agency of individuals and groups who, like computer analysts and virus writers, have engaged with and elaborated on their features and attributes.

Drawing from Guattari's ecosophy, Fuller explained how these processes may happen within media ecologies. He interpreted these as "massive and dynamic correlations of processes and objects,"[306] where "objects" encompass a variety of elements, from the very code that constitutes media to the products of, and the human relation to, the production of media. He observed how every component of an apparatus (or a machine) is dynamically integrated into the whole and, at the same time, is connected to the single parts. Thanks to the interchangeability and the recombinant characteristics of the single parts, the system has the potentials to build infinite realities. These realities are not just the result of the interconnection of elements internal to specific forms of media, but originate from a more complex cross-fertilization between media, social conglomerates, and contingent occasions. As potentials are always unrealized or yet to realize, turning them into realities implies the creation of particular conditions that allow or direct their realization. For instance, the use of the viral epithet as a "coagulator" or a conveyor of meaning may trigger the emergence of the realities mentioned here.

The manipulations and micro-appropriations resulting from users' active contribution could happen through what Guattari called the work of "subjective productions."[307] In a fairly static "capitalistic order of things" where "nothing can evolve unless everything else remains in place,"[308] Guattari sees the transformativity and the eternal unfolding of media as preparing the "ideal conditions for future forms of subjectivations."[309] Turning the "natural"[310] or present chaotic overlapping of the "mental, social and natural" into a possible future realizable project, is the only way out of the impasse created by an imposed media system that compels its players to "submit to the axioms of equilibrium, equivalence, constancy, eternity."[311]

Subjective productions are strongly involved in expropriating the qualities and functions found in a particular context, in displacing them and in forging other "existential chemistries,"[312] which can then reused in unconventional ways. These subjective productions are both individualized and collective. Individualized as the product of the individual's responsibility, who situates him or herself "within relations of alterity governed by familial habits, local customs, juridical laws, etc."[313] Collective, as a "multiplicity that deploys itself as much beyond the individual, on the side of the socius, as before the person, on the side of preverbial intensities, indicating a logic of affects rather than a logic of delimited sets."[314]

Collective and individualized forms of subjectivation are called to continuously add novel functions and features and, therefore, to generate novel viral assemblages. The viral seems to incorporate and make visible not only the manifestation of viruses as (self) assembled beings with their own properties, but also as entities that can be manipulated, based on assumed features perceived through collective and "commonsense" experience, and on the creative redirection of individual users or programmers.

When applying the term viral, chosen features deriving from viruses are transferred to a diverse range of phenomena or activities. Although the term remains unchanged, the newly created items have little to do with its original meaning. Practices that stem from fields as diverse as the arts, marketing, or advertisement, increasingly incorporate single or multiple features of viruses that can easily be summarized as viral. In this scenario, the viral serves as a means of productive relocation of the features of viruses in dimensions different from the initial conception, and also fosters an infinite number of emergences and expressions (virtually, as hypothesis or potential), and a finite number of concrete assemblages (pragmatically, or as a material, visible possibility).

# VIRAL CONUNDRUMS

The current extended use of the viral is revealing of the potentials that make its dissemination and use in other cultural dimensions and technological levels possible. Although its flexibility proves advantageous for the inclusion of new forms of digital code in the list of malware, and for the designation of new nondigital phenomena, it can also be an obstacle when one tries to designate, identify and constrain specific viruses to well-defined boundaries. However, far from representing the downside of the viral, this obstacle can be interpreted as one of the many ways and instances through which the viral is made visible. If seen in this way, said obstacles are no longer impediments, but concrete products and reflections derived from the viral through the meshing and convergence of the media ecology.

The viral has been used to evoke and pull together a series of elements that appear to fit many objects at a time, but that do not characterize one specific object in a detailed fashion. In this sense, one can say that the viral constitutes an element of connectivity between various manifestations. The term is equally used in computer science to indicate specific features that might characterize malware as a series of coded items, as well as to designate a general behavioral pattern that makes the recognition of malware easier to a nonexpert.

The nonspecific use of the viral is, on the one hand, justified as a result of the ever-changing nature and variety of malware manifestations within a complex and fast-moving networked system. On the other hand, it reflects the sense of emergency and time-sensitiveness of malware research and AV industry. In both cases, using the term *viral code* is more convenient and faster than having to specify the category and the specific name of each new malware. However, although the term is employed to convey the presence of affinities between diverse objects, the perspectives, the technical specifications, and the area of application are slightly different. For instance, saying that the viral incorporates a parasitic and/or opportunistic behavior with a series of assumed aggressive connotations does not seem to help point to any malicious agent in particular. This might result in difficulties when attempting to separate the nonspecific and generalized notion of the viral (i.e., when the viral is used to generally indicate malicious code) from the single features that the viral is said to incorporate. In this case, virus analysts, security experts, and computer scientists seem to stumble on disagreements and complications when it comes to the identification, the classification and the naming of viruses. This is especially true when computer experts try to provide fast detection, real-time mapping, and elimination of the annoyance.

Moreover, the simultaneously general and specific use of the viral can create a disconnect that prevents both expert and non expert from either classifying viruses as coded objects following consistent methods, or from properly distinguishing them according to their specific peculiarities (as worms, Trojan horses, logic bombs, etc.).

Conventionally, it is not up to the virus writers, but to the AV researchers to assign names to viruses. According to the online magazine *Computer Knowledge*, it is unlikely that virus writers name their fabricated malware: "those virus writers that insist on a particular name have to identify themselves in the process. Something they usually don't want to do" as this would facilitate their exposure and increase their liability.[315] However, the AV industry does not appear to be fully cohesive when it comes to classifying viruses as most attempts of this kind have produced different and conflicting results. The reason could be found in the relative novelty of the discipline that examines computer viruses, in their speed of reproduction and emergence and, finally, in the goals the security industry establishes when hunting down viruses.

As Gordon argued, although science has relied on "a sample-based naming scheme, so that a new plant is ultimately identified by comparing it to reference samples of known plants . . . the problem with applying this approach in the anti-virus world has been the lack of a reference collection or even a central naming body."[316]

One of the major problems, she argued, does not necessarily lie in the classification method *per se*, but in how naming is applied within this classification. In fact, it is mostly agreed that all viruses are classified according to a well-established sequence, by type (usually w32), name (Bagle, Klez, etc.), strain (usually one or two letters added to the previous), and alias (usually the common or popular name). Because new computer viruses are discovered at a very fast rate, each AV company proceeds independently by what Gordon called "interim naming,"[317] that is by applying a temporary name to each newly detected virus. Naming is most of the times made in a state of emergency. Thus, the name chosen for viruses might not always be consistent, at the point that two separate companies might go ahead and use two different names to identify the same virus or, vice-versa, use the same name to identify two different viruses.

Although classification is needed for research purposes, it is mostly used as a temporary solution to allow the infected user to find the right AV patch. In fact, naming is irrelevant for users, as their preoccupation lies primarily with detection and removal. Because AV products are usually customer-oriented, the need for accuracy of naming and classification for research purposes is deemed not vital and even superfluous. Speedy viral detection (whatever that is) and resolution, not accuracy, is the primary goal. To explain this tendency, a spokesperson from Wildlist.Org, the organization that collects detection of viruses from individuals and AV companies and monitors computer viruses in the wild, offers the following example to explain the disinterest in classification for the sake of research:

> There's a weed growing in my back yard. I call it Lamb's Quarters. It's
> a very common weed, and is also known as Goose Foot, Pig Weed, Sow
> Bane, and a few other things. I don't know what it's called in South
> Africa, Hungary, India, or Taiwan, but I'm sure it has lots and lots of
> names. . . . I'm not a biological researcher. I'm just an end user. I yank
> the weeds and mow the lawn. I call it Lambs Quarters and don't care
> what it's called in Hungary.[318]

The elusive behavior and fast pace of emergent viruses prompts AV companies to react rapidly against multiplying threats. In these conditions, a precise taxonomy of malware is almost impossible, especially when it needs to
be done in a timely manner and, anyway, before the infection starts creating
substantial damage.

In terms of the rules that make classification possible, the very communities that create and disseminate classifications contribute to, and even
enhance, the difficulties of naming viruses.

The problem of classification has repercussions on the way the nonexpert and the popular press interpret viruses. In fact, with the quickly
changing pace combined with an internal instability, the system is unable to
provide clues that would allow the user to identify, with any certainty, the
presence of a particular type of malware or to distinguish between different
strains of malicious software. Although the average user is left with a general notion of viral code, consisting of fragmentary information that adds
nothing to his or her knowledge of computer viruses, this system displays a
particular use of the viral consisting in the reproduction of old assumptions
and general perceptions of viruses that dominates the sciences and the popular imagination alike.

## (VIRAL) ACTION!

At a different level, dealing with a generalized notion of the viral becomes a
point of advantage and a source of potential undertakings for those who
wish to actively engage with the features of viruses. In its continuous
unfolding and constant transformation, the viral can be interpreted as a pool
of opportunities and possible reconfigurations that might be used to generate and describe a variety of new concrete assemblages. Using a number of
creative tactics individuals can appropriate and manipulate the features that
characterize viruses and turn them to their advantage.

Whereas the peculiar functioning of viruses could be adopted as a tactic
in itself, the variety and degrees of appropriation and utilization of the viral
appear to reflect its two major aspects: its manifestation resulting from the
features, the popular connotations and assumptions it collects (extensive

use), as well as its manifestation as a variety of purposefully appropriated aspects (open-ended use). As is clear from the examples, this operation is the result of both collective and individual subjective forms of productivity.

As a manifestation of its extensive use, the viral has infiltrated a variety of phenomena through inadvertent and sometimes unacknowledged assimilation. Features of the viral have been absorbed almost automatically, as a result of the dramatic impact that viruses imparted on culture. The viral, in this case, is acknowledged as a behavioral template. Phenomena that utilize such model automatically become, to some degree, also manifestation of the second aspect (the open-ended use of the viral). Although they usually do not acknowledge the viral as a source of inspiration, they are able to creatively use the above model and apply it in a variety of contexts.

The phenomenon of franchise business could be interpreted as an example of how the viral (in the form of selected number of characteristics) has been silently assimilated. In fact, franchise enterprises have multiplied and spread across the globe, acting as fosterers of a new economic trend and as witnesses of a newly distributed postcapitalist structure. According to Shaw, franchisors simultaneously operate their outlets under two distinct incentive schemes or modes of organization: franchising and company ownership. In the first case, the manager owns the outlet but has to pay a regular fee to the franchisor. In the second case, the franchisor acts as employer by offering a regular salary and performing direct control over the manager. Despite different ownership schemes, in either case the franchisor's goal is to protect his or her brands from the "franchisee free-riding."[319] The result is a multiplication of franchise outlets whose style, products, and management are maintained consistently and where small changes or attempts by the outlet to adapt to the surrounding environment are regulated according to precise contracts.

For instance, in order to establish their branches in different parts of the world, Starbucks and McDonald's had to undergo a few stylistic form and product modifications to allow the company a minimal integration with the host location, whereas the key elements that characterize the original brand (the most visible being the logo) remained unchanged. Such diffusion appears to have drawn inspiration from viral replication in so far as they perform slightly different functions but still maintain the same old and easily recognizable main functions.

Although the franchise model appears to have inadvertently inherited and productively appropriated the functioning of a virus without yet adopting its name, other enterprises were not afraid to admit their purposeful association with a certain viral behavior. This is the case of VM, where terminology openly credits viruses. VM is often described as "the tactic of creating a process where interested people can market to each other."[320] By referencing information-processing theory, Subramani and Rajagopalan classified this tactic as a specific manifestation of the "more general phenomenon of knowledge-sharing and influence among individuals in social networks."

VM utilizes densely knit network clusters to spread "recommendations" about products in a process that simulates word-of-mouth (WOM) and face-to-face (F2F) communication. Its effectiveness lies in the fact that the recommendation to buy, to use, or to download a product may come directly from a friend or from a trusted authority (often a sports or Hollywood star). Subramani and Rajagopalan mentioned the recent gmail phenomenon as one of the most successful cases of VM: Every new member has the option of inviting five friends to open an account. In turn, the said member had been invited to sign up by a friend.[321] Major software companies often use viral tactics to promote software such as AIM (AOL Instant Messenger) Macromedia packages, Realplayer etc., in order to attract new users and to guarantee or establish constant monopoly.

As Boase and Wellman noted, VM is not new.[322] Rosen added that the first commercial VM campaigns were introduced in the 1940s.[323] However, the label *viral* has only recently been applied to marketing tactics introduced on the Web. The new denomination would not have been adopted had there not been any acknowledgment of the benefits that this new formula would generate. Clearly, viral marketers knew that using the term *viral* would have attracted a great deal of publicity. In addition to reminding one of the resemblance between this marketing tactic and the spread of computer and biological viruses, the term seems to be adopted as a way to guarantee penetration capacity, as well as attract the attention of the user or the prospective adopter.

One might object that the viral mechanism of diffusion is a rehash of the broadcast pattern.[324] To demonstrate that this is not the case, and to illustrate how the act of combining viral and marketing comes from an understanding of a different, fluid, and rather processual diffusion, Pedercini explained that viral dynamics of diffusion move through a surface that is anything but smooth. The diffusion happens by consensus and not by coercion: VM, in fact, counts on a multiplicity of interested and conscious actors to function properly. As opposed to what happens with the computer virus spread, this time the user is not only active, but also conscious and willing to participate in the operation, otherwise the tactic would not succeed.[325]

This observation illustrates the existing connection between a particular practice that is carried out openly on the Internet and the substances that circulate at its deeper coded level. Moreover, it testifies to the degree of re-elaboration and transformation of a term that has come to characterize several items at a time. The viral, in this case, only maintains a few of its original qualities. Once it is coupled with the practice discussed here, its function and meaning undergo substantial transformations, marking the formation and materialization of new concrete viral assemblages. VM then, could not be just an exception. In fact, other online phenomena have recorded unprecedented and sometimes unexpected levels of diffusion, thanks to a proliferation technique that is said to mimic the mechanism of spreading viruses. Not by chance these phenomena have often deserved the viral nomenclature. *Viral or contagious media, virals,* and *viral videos* are the latest additions to the army of the viral.

Once named, these articulations do not remain immutable. Given the multiplicity that characterizes the active interventions that bring to life or define such articulations, all viral phenomena are in constant fluctuation. VM can be subject to transformations according to diversified uses or distinct contexts. In turn, phenomena that carry different names but share the same viral suffix (i.e., viral videos) may retain close resemblance to VM. Individuals or groups deciding to employ forms of viral media make the ultimate decision according to personalized or contingent modalities of use.

One of the first examples of viral videos was marked by a May 2005 contest (Contagious Media Showdown), followed by an exhibition at the New Museum of Contemporary Art (New York), organized by Jonah Peretti.[326] The term *contagious* or *viral media* was coined to define those sites, which, recently and inexplicably, had reached a great amount of popularity among online strollers. These sites are often built by amateurs and are mostly unsophisticated in terms of style and content. Nonetheless, a dense crowd, without apparent reasons, avidly visited and interacted with them. The competition awarded the first price based on the number of hits a site received. Peretti explained:

> That's an increasingly popular way of thinking about something that is usually random: a designer makes a dancing baby and is completely taken aback that it spreads everywhere. A silly video circulates all over the web. Much of that is completely unintentional.[327]

What is interesting about this event is not the originality of display or the artistic capability demonstrated by the creators of the sites, but the very popularity that, quite inexplicably, they were able to generate, as well as the mechanisms that enabled such outcomes.

The contagious media devised by Peretti anticipated a much larger online phenomenon that occurs today thanks to video-sharing Web sites such as YouTube and Google Video, as well as social network hubs such as MySpace. Collecting a variety of short videos that range from TV program recordings and amateurial and/or self-promotion clips, to short documentaries and activist videos, viral videos have quickly become one of the most popular phenomena of online sharing and online entertainment.[328] Given the diversity of audience and contributors, as well as the openness and relative flexibility of video-sharing Web sites, viral videos often have little in common in terms of purposes, style, and even audience. In addition to constituting a continuous source of entertainment for a wide and increasing audience, they are part of a rather random and fairly unpredictable cycle of content sharing, socialization, and exchange network, originated from recommendations forwarded by friends and groups to which each viewer is connected. No surprise, then, that viral videos have increasingly become (self) promotion tools. Impressed by the popularity reached by a few video uploaders, many frequenters of YouTube or Google Video have increasingly developed similar ambitions.

The combination of this latter function and their unpredictable mechanism of spread make viral videos and viral media in general almost comparable to VM. In fact, all these phenomena spread through "word of mouse"[329] recommendations and personal notifications among online users. Their convergence becomes clear when one looks at similar early phenomena that spread before the mass use of social network software. One might remember, for instance, the excitement and curiosity generated in 1998 by the "Blair Witch Project" Web site, an early example of viral media whose goal was to promote the homonymous independently produced, low-budget movie. The Web site created much media hype months before the movie itself was released and guaranteed a stunning spectator turnout.[330] Because of its original intentions, this example can be understood easily as simultaneously a form of viral media and a particular type of VM, along with more recent fortunate promotional campaigns spread thanks to the active click of the user.

Following unexpected successes, and despite the unpredictability of their outcomes, it is no wonder commercial enterprises have started using social networks as possible channels to spread their ads or to promote their products. One major question that corporations increasingly ask is "How do we do this intentionally?"[331] Contests, such as the one launched by the UN Food Program, among others, encourage contenders to upload their viral video entries onto YouTube.[332] In a similar vein to the unavoidable and unpredictable performance of viral media are the attempts to launch viral projects that would briefly disrupt the goals of designated targets. For example, in 2005 the Ars Electronica Festival (Linz, 1-6 September 2005) awarded GWEI.org (Google Will Eat Itself) a special honorary prize. Hans Bernhard, spokesperson of the collectives in charge of this site, Ubermorgen.com and Neural.it, described the project as a "social virus":

> Our website generates money by serving Google text advertisements on our hidden websites and our show-case site. With this money we automatically buy Google shares via our Swiss e-banking account.[333]

After creating a basic Web site, the group joined Google Adsense program,[334] an agreement that requests Web site owners to place small texts advertisement on their pages. In exchange, they receive a small amount of money for each click generated from the text ad. Google, in turn, would receive from the advertising company the same amount of money in addition to the percentage already retained for its services. "Instead of passively submitting to Google cyclical re-generation of money,"[335] Ubermorgen/ Neural found a way to turn the mechanism to their own advantage. By notifying their community members to click on the ads found on their Web site, and by simulating visitations thanks to an automated robot, they could increase their entries and reinvest the money received to buy Google shares. Not only did they create a viral enterprise that, in the long run, had potentials for slowly consuming the monopolist position of Google, but they were also able to lay bare the advertisement mechanisms that regulate the World Wide Web.

Unfortunately for the collective, this tactic was short-lived. The very mechanism that could bring considerable disturbance to Google was soon tracked down and ousted. GWEI.org is now fully censored on all Google search indexes worldwide. In a gesture of solidarity and to protest against the incorrect censorship imposed by Google, The Institute for Network Culture (a research center that studies and promotes initiatives in the area of Internet and new media) has recently released a call for support that requests sympathizers to insert links in their Web sites pointing to GWEI. This tactic is substantially viral, as its goal is to redisseminate the content of the Web site by making it visible through the alternative social-sharing techniques of multiple linking.[336]

GWEI.org, with its tongue-in-cheek attitude, is part of a series of hit-and-run interventions typical of tactical media. Like other similar interventions, GWEI.org had an ephemeral and short-lived existence. However, it was able to attain fast and unpredictable results. Its contribution consisted in small, imperceptible micro-changes that filtrated its target (Google in this case) slowly and invisibly. Like viruses, GWEI.org initiated subtle changes that could not be detected as soon as they hit, but that could be intercepted only later, when the impact had already been experienced and could no longer be reversed. Like viruses again, GWEI.org disseminated unpredictably through the network. Laura U. Marks classified practices that function in this way as "invisible media,"[337] recognizing their similarity with what Hakim Bey's Temporary Autonomous Zone (TAZ): "a guerrilla operation which liberates an area . . . and then dissolves itself to re-form elsewhere/elsewhen, before the state can crash it."[338] One could argue that, given the formulation of TAZ and invisible media in general, the introduction of the term *viral tactic* is superfluous. However, the term constitutes an "updated" version of the above operations, one that can be easily adopted to designate similar actions online. Additionally, it exploits the imaginative impression and the popularity of viruses to allow a decisively immediate and more graphic recognition.

Arguably, viruses have left a visible mark on people's imagination and practices that goes well beyond the domains of biology and information technology. However, the examples given here show that this impact has not signified a passive acceptance. Western cultures have dealt with viruses by engaging in a series of confrontations, selections, and assimilations, appropriations and transmissions, adoptions and adaptations to their features. When extensively applied to a diverse range of objects and activities the viral proper of viruses (their being viral) works as a signature or a mark of their passage. Whether constituting an obstacle or generating positive and creative outcomes, the viral is also able to engender a number of artifacts and concrete possibilities. The battle between virus writers and virus fighters, the controversy about their names and groups seizing of viral properties to the benefit of the creation of new practices are all products of the sometimes unpredictable productivity of viruses as simply, inherently, or problematically viral. The diversity of phenomena generated and continuously emerging from the viral are concrete products of its open-ended use and of the ability of individuals to "become viral" themselves.

## ACKNOWLEDGMENTS

This chapter was possible thanks to the support of the Canadian Media Research Consortium and the Robarts Center for Canadian Studies.

# PART II
# BAD OBJECTS

---

Computer viruses are perhaps an obvious example of what might be regarded as an unwanted visitor to one's computer. These contagious digital objects are not easily assimilated into the much-needed stability of networked operating systems, based as they are on a clear set of functional goals. Nevertheless, computer viruses are not the only unwanted visitors. Every network defines its own "unwanted bads." Some ban the playing of games, whereas others ban the downloading of suspicious material. Excessive postings on mail lists and various other actions that might contravene the acceptable use policy of the network also draw the attention of the network controllers. One university network policy, for example, prohibits "defamatory materials" and material that could potentially cause "annoyance, inconvenience or needless anxiety." On many wider networks viruses, spam and objectionable porn are seen as generic evils, whereas on more localized networks the authorities might add downloading, games, or eBay visits to the list. In other words, there are various levels at which "good objects" are filtered out from "bad objects."

Joost van Loon noted that modernity can be characterized more broadly as the control, channeling, and distribution of both "goods" and "bads." Drawing on Ulrich Beck's famous analyses of the risk society, van Loon's "bads" are not metaphysical beings, but "manufactured side effects" of the goods produced.[339] As risk society theories have shown, the cultivation of risks is an issue that is framed as one of risk management and decision-making probabilities. The manageability of social and technical networks is, however, continuously "at risk" due to the virulent nature of contemporary threats, including the spread of biological or digital viruses, and of course, spam. Despite the need for stability, the ontology of the network is all too often unstable. In fact, it is the instability of a network that hinders the complete management of contagious "bads," which seem to frequently filter through the *cordon sanitaire* or zone of protection set up by network controllers. This may indeed be because the mode of a bad object is not grounded in traditional ideas of territorial boundaries, but is instead found in the vectors of movement in a topology.

Chapter 5 by Jussi Parikka looks at how bad objects are continuously pinned down, archived, censored, and normatively controlled within the semiotic and asignifying regimes and logics of scientific articulation. Parikka argues that although the filtering of "goods" and "bads" in computer media is an issue of calculation, control, and definition of computational processes, it is also carried out, simultaneously, on various levels of translation. For example, he draws on the production of software, not only on the level of code, but also in its translation into images, sounds, and other modes of perception. In fact, translation becomes a key tool in the study of contemporary networked biopower as exemplified in recent net art projects like the Biennale.py-virus, which is not just a piece of code, but an iconographical gesture of tactical media. Using Deleuze and Guattari's ideas, he further contemplates the emergence of the bad object as an assemblage that is not simply reducible to one defining logic, but is instead an ongoing heterogeneous process that somehow remains intact.[340] Instead of presenting computer viruses as a single digital accident, Parikka explains how they are contested in various media platforms that then feedback into the discourses and practices surrounding digital viruses.

Another example of the translation from code to perception becomes Steve Goodman's central theme in Chapter 6. Here Goodman evaluates both the development of the audio glitch virus and the subsequent intellectual responses to it. In doing so, he constructs a virology of the glitch, mapping its movement of contagion through a rhythmic nexus. On the one hand, Goodman addresses the contagious and mutational logic of the sonic viruses, which are not reducible to one specific technological platform, like the computer. On the other hand, he analyzes the discourses and practices around "glitch music," noting how the ideas suggested there can offer radical philosophies on the accidents of digital culture. In this sense, he replaces the common notion of perfect fidelity in digital reproduction with a tendency toward error, mutation, and chance. Here he raises a few questions that resonate with the central themes of *The Spam Book*: How can we grasp the anomalous glitch virus beyond a representational and metaphoric framework? Moreover, how can we understand this accident beyond the logic of substance?[341]

Synthesizing the various strains of thought in this book, Matthew Fuller and Andrew Goffey (Chap. 7) delineate the potential of an "evil media studies." This is not, they claim, a discipline in itself, but rather a wide-ranging focus on the informal but recurring stratagems of network production. They argue that we step beyond the representational model of network media, and instead embrace the often devalued charms of "trickery, deception, and manipulation." This novel treatment of digital subjects does not denounce the rationality of the topic at hand, but rather demonstrates how sophist tools are continuously used in media creation. Here the authors argue that media is all about controlling minds and bodies. In fact, drawing on lessons learned from military tactics, hypnosis, mind-bending, and viral media practice, the authors rethink the digital anomaly, not so much as a haphazard accident of network media, but as a vehicle for understanding the production of "goods" and "bads" of capitalist media. By doing so, they bypass the superficial distinction between "bad" and "good"—a second-order category—in favor of providing insights into the mind-bending functions of media practice.

The stratagems of evil media studies intersect with the themes of the this volume insofar as its analysis does not merely represent the cultural construction of objects, but instead tries to extract a line of potential experimental practice from them. It further recognizes that anomalies cause effects and become entangled in webs of affect. Increasingly, the user of digital media encounters "uncanny objects" that do not easily fit into the norms of a media experiences inherited from the broadcasting age. Media theory therefore needs to reflect the basic premises set out in the stratagems of network production.

# 5

# ARCHIVES OF SOFTWARE

# Malicious Code and the Aesthesis of Media Accidents[342]

*Jussi Parikka*

## ARTISTIC VIRUSES: A PROLOGUE

At the 49th International Art Biennale of Venice in 2001 a curious piece of a code was released. The Biennale virus did not conform to the usual clichés about viruses as products of pimple-faced teenagers, juvenile delinquents, or other digital vandals, who want to disturb the otherwise (presumably) peaceful everyday life of the digital world. Written by two Internet art groups, *0100101110101101.ORG* (see Fig. 5.1) and *epidemiC*, and exhibited at the Slovenian Pavilion, the virus was proclaimed as a test and as "one of the most exciting investments one could make today":

> During the opening days of the Biennale thousands of t-shirts carrying the source code of the program will show up. Paradoxically, such as in biological viruses, "biennale.py" will spread not only through machines but also through men. The paradox becomes even more clear if you think that the virus, a vague and dangerous entity by definition, is for sale to adventurous curators and collectors.[343]

In other words, the Biennale virus was clearly not intended as malicious software in the usual sense of the term that designates a piece of destructive software that spreads and executes without the user being aware of its presence. Instead, it was widespread as a *visible* piece of code, and described as embedded in absolute transparency:

> We've announced before what we were going to do. Our names and domains are written into the code. This is a big difference to the traditional cracking scene. Additionally, before starting to spread the code, we have sent it to all antivirus software houses, together with an explanation of how to erase it. The main goal of our virus is just to survive. And, it can better survive when it doesn't do any harm to the host. If it would kill its host, it would die itself, too. So, it sucks energy, but tries to stay invisible as much as possible. . . . Biennale.py" is completely invisible. It just installs itself in the background.[344]

Nonetheless, despite this technical "invisibility" the Internet artists emphasized, their project was essentially about *making visible* both the phenomenon of viruses as a specific piece of self-reproductive code and the political production of software in various societal contexts. As the Internet artist Jaromil stated on Nettime:

> A virus writer is interested in exploring the permability [*sic*] of the net. A rhizome of such and so many dimensions as the internet cannot be represented by any map. . . . Injecting a contrast medium into the organism to follow shape and structure will produce an angiogram showing the typical arrangement of veins.[345]

Earlier such "visualizing miniprograms" had been used in scientific and commercial contexts; as scientific network tools, such programs had been roaming the Internet for years, mapping the topological formations of this specific form of nonscalar network; a form of software not dissimilar from web crawlers used, for example, by search engine service providers.[346]

In this artistic context, these Biennale artist–hackers illuminated one of the key themes concerning worms and viruses, that is, *perception*. In recent decades, these miniprograms, which like their biological cousins are imperceptible to the "pure eye," have been caught in a web of techniques of perception, which have integrated them as part and parcel of the security institutions and media assemblages of digital culture. Hence, the Internet artists produced an artistic counterarchive of visualization that shed light on how Western media culture has tackled these issues.

Although the primary processes of digital culture are nonrepresentational and algorithmic, they are continuously coded into (audio)visual forms,

```
# biennale.py _________________ oo      to ____ 49th Biennale di Venezia
# HTTP://4aai.010010111010110101.ORG _ + _ [epidemiC] http://www.epidemic.ws
from dircache import *
from string import *
import os, sys
from stat import *

def fornicate(guest):
    try:
        soul = open(guest, "r")
        body = soul.read()
        soul.close()
        if find(body, "[epidemiC]") == -1:
            soul = open(guest, "w")
            soul.write(mybody + "\n\n" + body)
            soul.close()
    except IOError: pass

def chat(party, guest):
    if split(guest, ".")[-1] in ("py", "pyw"):
        fornicate(party + guest)

def join(party):
    try:
        if not S_ISLNK(os.stat(party)[ST_MODE]):
            guestbook = listdir(party)
            if party != "/": party = party + "/"
            if not lower(party) in work and not "__init__.py" in guestbook:
                for guest in guestbook:
                    chat(party, guest)
                    join(party + guest)
    except OSError: pass

if __name__ == '__main__':
    mysoul = open(sys.argv[0])
    mybody = mysoul.read()
    mybody = mybody[:find(mybody, "")*3) + 3]
    mysoul.close()
    blacklist = replace(split(sys.exec_prefix,":")[-1], "\\", "/")
    if blacklist[-1] != "/": blacklist = blacklist + "/"
    work = [lower(blacklist), "/proc/", "/dev/"]
    join("/")
    print ")         This file was contaminated by biennale.py, the world slowest virus."
    print "Either Linux or Windows, biennale.py is definately the first Python virus."
    print "[epidemiC] http://www.epidemic.ws _ + _ HTTP://4aai.010010111010110101.ORG "
    print ") _________________ 49th Biennale di Venezia _________________ <"
```

FIGURE 5.1. Biennale.py code. (Source: Biennale.py Web site http://0100101 110101101.org/home/biennale_py/. Reproduced with permission.)

which are very much entwined in aesthetic-political agendas of network culture. In other words, the time-based procedures of computational media are spatialized in contemporary media archives that are framed by questions of technical and commercial nature. Another, even more apt way to describe this endeavor would be to refer to the difference between the corporealities (the materialities) of computer viruses and the incorporeal transformations that interact with those materialities. A virus may be understood as a calculational process at the material level of computer circuits, but when this accident (event) is called "malicious software" it connects to a whole incorporeal sphere of morals, crimes, criminals, laws and judgments.[347] Hence, an analysis of computer culture should not focus solely on the material event of calculation (the technical diagrams) nor on the discursive events, but in the constant double articulation between various semiotic regimes.[348]

This chapter addresses the question of perception of software and especially the often discussed class of viral programs: With what kind of cultural techniques of perception have computer viruses and worms been signified and valorized, and what kind of archives are they embedded in? What I argue is that the visualization of code and the fabrication of malicious software such as worms and viruses as perceptible, is imbued deeply in cultural and historical levels of power and knowledge dedicated to producing a capitalist digital culture. I proceed via an archaeology of a kind that maps such *aesthesis* as a sphere of visibility and language, as a mode of perception guiding and piloting this specific form of media technological risk. Software objects consist not merely of calculation, but are also captured as part of such historical strata on which their potential powers are piloted and harnessed. I deal mostly with the incorporeal aspects of computer worms and viruses and focus especially on the perception (mediality) and valorization (commerciality) of the phenomenon. Code functions not merely as executable commands and processes of calculation in operating systems, but as a widely mediated cultural assemblage that at times can become highly visible (e.g., in audiovisual form).[349]

This chapter continues the analysis of viral objects carried out in the previous section on digital contagion, but does so in the context of "bad objects." It also functions to show the discursive and nondiscursive techniques of filtering "goods" and "bads" and thus relates to the broader issue of censorship addressed later in the book. Here, however, the focus is on the politics of software that is shown to be intimately connected to an agenda of aesthetics, of perception—a key technique of knowledge in modernity.

## VISION MACHINES

As Foucault noted in his archaeological writings (in my case filtered through Deleuze's reading), power and knowledge function through the archives of the visible and the articulable. The archive of modernity can be conceived in terms of the voice and the eye (or the distribution of seeing and saying), which mark especially the *a priori* layers of science.[350] Here the statements and visibilities are not secondary to an age "in general," but act as immanent expressions where "each stratum or historical formation implies a distribution of the visible and the articulable."[351] The biological virus is a perfect example of the functioning of the archive of modern science in how this particular subrepresentational object became an object represented and stated at the end of the 19th century. Whereas the word *virus* was often used as a generic term to refer to unknown agents of disease, the birth of modern virology did not actually occur until the end of the 19th century. Credit for this is often given to Dimitri Iwanowski, who in his studies came upon a minuscule organism that seemed to be the cause behind mosaic disease in tobacco plants. Also in the 19th century, both Iwanowski and Martinus Beyerinck used the term *virus*, although their definition and understanding of the concept differed from each other: For Iwanowski a virus was a small bacterium, for Beyerinck it was to be understood as a contagious living fluid.[352]

Of course contagions had been discussed for centuries. For example, Girolamo Fracastoro's modern classic *On Contagion* dates from 1546. In fact, a contagion was, for a long while, understood as a transmission of harmful material, but the German pathologist Jakob Henle made an important revision around 1840 with his publication *On Miasmata and Contagia*. For Henle, *contagia animata* (living organisms) were the cause of diseases, yet he separated disease from the parasite carrying it. The *contagion* was merely a cause and a channeling of the *disease* that was something separate from its movement.[353] Consequently, the relationship of the microbes with diseases was raised as an issue of investigation in itself.

As Kirsthen Ostherr demonstrated, understanding contagions as epistemological objects was at an early stage connected to the visual machine of cinema. The bacteriological revolution of the 1880s, which transformed the early miasmatic theories of contagion into modern medical parlance and the cinematic mode of representation from 1890s created a new visual culture of monitoring and representing diseases.[354] This linking of cinema and bacteriology enabled, as Ostherr noted, a new way of connecting scientific theories of minuscule infectors as causes of diseases with modes of representation that, despite the belief in the power of visual enpresenting, was continuously impotent to pinpoint once and for all the contagions due to their ephemeral and networked nature.

In other words, the "lack of visibility" was one key hindrance in the path to understanding viruses and the term *virus* was used at the beginning of the 20th century to refer to such pathogens that were of submicroscopic nature and "refused to be isolated by the most sophisticated bacteriological techniques available."[355] Viruses were from the beginning "virtual objects"; entities that manifest themselves only partially through cultural practices of power and knowledge, that is, practices and discourses dedicated to "revealing" truths. In relation to these objectified truths, the object in its virtualness remained a multiplicity, a meshwork of complexity, irreducible to any single defining signification or visualization.[356] In a way, these virtual objects demonstrate a weird reversed causality where the actualized imaging precedes its cause, the virtual multiplicity.

Through technologies of visualization, and stratification that pinpoint the minuscule actors with indexical and iconic systems of signs, biological viruses have become epistemological objects. For example, in virology, this has been played out via specific practices of multiplication, serology, and genetics. In this way, it becomes possible to decipher such signs that affirm viral presences and lead the way to understanding their qualities. Joost van Loon clarified this functioning of scientific knowledge into three steps:

> The first step in the formation of the virus as a virtual object is that it has to be visualized—either iconically or indexically. Second, it has to be

signified, that is, endowed with the specific meaning through which the objectification can be anchored into the symbolic order, and become a discursive object, engendering a discursive formation. Third, it has to be valorized; the virtual object must not only be endowed with meanings, this endowment must be attributed a particular value in terms of its significance within the wider emergent discursive formation. Objectification, therefore, is nothing but the singular decoding and encoding of a territory, a re-organization of particles and forces, not simply in terms of "knowledge," for example as in the Foucauldean (1970) notion of "episteme," but first and foremost in practices and technologies of enpresenting.[357]

This kind of logic of scientific articulation (seeing and saying) defines key traits of modern science and field of knowledge. Van Loon's exposition offers a useful tool for analyzing cultural practices of diverse nature. This agenda is part of the larger trend of science studies, which focuses on the visual practices of laboratories and production of facts that are spread from the restricted spaces of science into society at large. Visualizing and articulating are not, then, only laboratory-specific practices but diagrams that span much further on the social field. The regulation of statements and images of knowledge does not stop at the laboratory door, but is effective across the whole of society. The impossibility of visualization of biological contagions was continuously compensated through another kind of "proliferation of images of the diseased body"[358] that pinpointed vectors of disease to certain suspicious and risky connections (whether increased global transportation and communication or the risk in organic contact in institutions such as the military). This created a kind of an audiovisual archive of contagions and the networks in which they moved; an archive that was not restricted to public health cinema and official bulletins, but spread across the media culture in a diagrammatic form, coding disease into audiovisions (in popular cinema, literature, mass media, etc.)

When moving toward an analysis of objects of digital culture, this stance toward the visible and articulable needs updating. Although software objects seem to "lack visibility" on their defining material level, they are constantly visible across scales. As German media archaeological analysis has underlined, the *a priori* of technical media cannot be articulated merely in terms of Foucauldian thematics of visible and sayable, but needs to take into account the technical specificity of modern media.[359] Hence, the theme of calculation should be added to this archival logic that produces for us then a new, tripartite articulation of the power/knowledge archives of digital culture: *seeing*, *articulating*, and *calculating*. If knowledge has so far (in the modern era) been defined by the combinations of visible and articulable, then contemporary archives are defined in addition by calculation as one novel "logic" contributing to unique historical formulations — knowledge as "a 'mechanism' of statements and visibilities"[360] but also calculation of technical computer media. Here an archive is no more a particular site or a place of memory (as archives are not about memorizing but about deletion, selection, and erasing) but a specific mode of power—a power of repetition and a pattern of reproduction. Archive designs a mode of remembering (grasping) and designing (production) of culture as a diagram of a sort that spreads regularities.[361]

## ARCHIVES OF RISK

Software, as a process of calculation and coding, is then imbued in the visible and articulable, but not reducible to them. Similarly, whereas digital culture relies on the technical a priori of calculation, it is however constantly remediated through the defining and valorizing powers of seeing and saying, which have especially since the 1980s been demarcating legal and illegal forms of software. In media archaeological terms, there is perhaps a mediatic short circuit that nonlinearly connects various media spheres into a functioning assemblage, an archive as a logic of creation.[362]

Even though Adrian Mackenzie argued in a very interesting vein that software code reached a new sphere of visibility around mid-1990s due to reasons varying from viruses to open-source projects like the GNU/Linux Operating System project, and for example the overinvested hopes crashed in the dot.com boom of early Internet years, it was already visible 10 years earlier with the first viruses and the rise of a commercial field of software production.[363] Viruses, as one of the most discussed early examples of software, were captured as part of the scientific and popular media definitions in the 1980s (even if malicious software was on the security agenda already in the 1970s). In the case of computer viruses, this "factualization" and "imagining" has functioned through vectors of technical, mediated, and commercial nature, all contributing to the way in which we have learned to understand the contested nature of viruses and other forms of software most often labeled as "malicious." Initially, this happened on the level of computer science with the first technical diagrams describing viral patterns in computer systems.[364] In November 1988, a self-spreading computer program achieved widespread visibility. While the IBM PC-virus Brain (1986) and the first network worm Christmas tree (1987) raised the awareness of security personnel in companies, it was the so-called Morris worm that captured the minds of average computer users and particularly the media. The Morris worm crashed thousands of computers connected to the Internet, but even more important were the incorporeal repercussions the incident had. Computer viruses turned into media viruses marking a new level in their perception. Consequently, at least at that point, viruses and worms were also framed as *malicious software*. Viruses were quickly turned from self-reproductive computer software into self-reproductive infotainment and risk scenarios.

The worm itself—programmed by a Cornell college student Robert T. Morris Jr.—included so many bugs that it spread uncontrollably to several universities and public institutions. Although it did not include a malicious routine designed to cause damage, it tied up computing resources and slowed down the functioning of the Arpanet-network. In the United States, the *ABC-News* was the first one to report the worm in its evening news on November 3, proclaiming that a virus had infected a range of computer systems. The next day, reports from *CBS*, *ABC*, and *NBC* pushed the worm to the front of the news agenda. The CBS report covered how the worm struck computers in the defense department. Commentators included several university representatives, the virus hunter John Mcafee, as well as spokespeople from the Pentagon and NASA.[365]

Furthermore, newspapers were filled with stories about viruses and speculations about the so-called "worst scenarios." In this way, the media acted as a microscope of a sort, a technique of vision that selected, modified, and transmitted the software code into audiovisual form. Viruses were put in front of TV cameras, which were an essential part of transforming a technical phenomenon into a major threat for the safety of organized, state-run, and international business networks, as well as the individual user.[366] At first, the media was puzzled over how to translate algorithms into audiovisions:

> The media was uniformly disappointed that the virus did nothing remotely visual. Several reporters also seemed pained that we were not moments away from World War III, or that there were not large numbers of companies and banks hooked up to "MIT's network" who were going to be really upset when Monday rolled around.[367]

This lack was soon compensated in various forms.

The Morris worm incident became a paradigmatic media model of the viral. Every new virus, it seems, is now discussed in television talk shows and news reports, featuring security experts estimating the potential threat and cost of an attack. Therefore, apart from being a specific technical security issue (later especially in Windows operating systems), worms and viruses are increasingly media incidents that spread with the aid of audiovisions and texts. This is of course a near banal statement, as the antivirus (AV) researchers themselves accentuate, because even people who do not use computers have for years been familiar with the concept of the computer virus through Hollywood films such as *Independence Day* (1995) and *Hackers* (1995).[368] Indeed, the faceless computing processes animating the expressive machines of the cybernetic era, that is computers, also have a visual nature, which is created when the programs are connected as part of the media assemblages of contemporary infotainment production.

In the production of archives of legitimate software, computer programs are often visualized iconically (e.g., bits of source code), indexically (e.g., the virus writer as originator of the program), or symbolically (e.g., AIDS references). Signification assigns viruses their specific meaning as malicious software, placing them in a grid that is metaphorically loaded. Although the technical diagrams and reports of viruses had already offered a preliminary mapping of these miniprograms, media assemblages designated the discursive formations, or maps, on which they were intensified, or valorized, to their specific place. Thus, not only were computer worms and viruses deemed as malicious software; this articulation was intensified with images of disease, impurity, and vandalism that objectified them to a certain niche within the complex emerging digital order, ranging from developing safe easy-to-use software to producing the "imaginary" of a responsible user.[369] The enpresenting of digital viruses and worms was thus in elemental relation to the construction of the new digital culture of information capitalism. Moreover, in recent years, an analysis of such viral incidents as Netsky, Slammer, Blaster, and Sobig worms, also reveals a similar pattern. Especially the Netsky and Sasser worm incidents have a striking resemblance to the Morris worm in this context. Teenager Sven J. put a face to the algorithmic processes of Netsky. The German teenager, who craved respect and fame, wrote the programs, and he did, to some extent, end up becoming a small-scale celebrity, being interviewed in various publications, (even though his Netsky worm made it onto television before he did).[370] Another young man, David L. Smith, who appeared in *Time Magazine* for the Melissa Virus,[371] and a third similar example was the maker of the I Love You virus, identified as a 23-year-old student from Manila, Onel de Guzman. He was even featured in a live CNN chat room conversation on Monday, September 25, 2000 at 10:15 p.m. where people were able to ask questions of the famous hacker figure online.[372]

In this light, the iconographical nature of the Biennale virus incident provides an interesting simulacrum of the general viral discourse: the faceless virus reproduced (and sold) in t-shirts, special CD-ROMS, newspapers, and so forth. Like the Biennale source code, the viral phenomena has been facialized and remediated on the one hand in the concrete human faces of teenage hackers (such as Robert T. Morris, Sven J., or David L. Smith), and on the other hand, in the otherwise visual "faces" of media culture. In this sense, it is useful to follow Philip Goodchild's exposition of Deleuze and Guattari's theme of faciality, and stop referring to "media": "films, television, newspapers, and magazines operate less as 'media' than as machines for the production and recording of faces."[373] A constant production of faces functions as an element of representation and antiproduction that dams up flows. Such representations are not productive, but place spatial grids on the intensive flows of software and coding practices. In the case of viral faces, the normative, Majoritarian nature of this coding was accentuated: Viruses are detached from a specific coding practices, hacker subcultures, commercial digital culture, and other critical contexts to a Oedipal drama of personal tragedy or some other form of grasping the nonrepresentational issue at hand. Instead of exposing and tapping to the real social modes of production of (self-reproductive) software that remains continuously invisible, such flows are captured in representational forms.[374]

The media assemblage, the facial machine, has a special place in the definition of the contemporary risk culture as a form of infotainment. Risks are mediated, defined and perceived via the technologies, discourses, and institutions of the face machine, whereas the aesthetic form of such "media virological" knowledge uses elements from entertainment and fictional audiovisions to *enhance* and *visualize* the otherwise quite gray and boring events of computer malfunction. This visualization is done in order to bring issues of algorithmic events into the aesthetic–political agenda's of national and international institutions, corporations, and other parties that are engaged in developing the information capitalist culture of digitality. Such imaginings produce their objects; just as computer scientists and experimenters were able to make viruses technically real, these media images produce them as audiovisually real malicious software. This can be referred to as a production of regularity of statements, which do not represent but record and archive the multiplicity of social forms—in this case software. Archives form spaces of (a) *collaterality*, where statements are formed in terms of families (e.g., classes of software as articulated in various instances from media to computer science books); (b) *correlation*, where statements lay the ground for subjects, objects, and concepts (classes of software are formed through statements as incorporeal events); and (c) *complementarity* in the form of nondiscursive formations that act as the "material substrate," where statements gain consistency (from politics to economics or in our cases the media technological a priori of digital culture).[375]

Yet, the productions entail a virtual residue: As the imaginings produce the objects, they also imply a whole virtual multiplicity. Following Van Loon's definitions, risks as temporal "objects" are always virtual—virtual but not actual, potential but real. Risks are *becoming* real but never completely actualized in spatial terms. In this sense, risks live in media representations and simulations, meaning that they are virtually real in various (scientific) scenarios and probabilities, news reports, (science) fiction stories, and so forth, which all enjoy a considerable amount of cultural prestige in contemporary media culture. Risks make good media events due to the ephemeral future tense, and it is irrelevant whether the virtual risks ever actualize as this politics of fear is primarily about a weird temporality that captures the future and brings it to a present as a frightening scenario.[376] Van

Loon's analysis of the I Love You virus (1999) appropriately demonstrates this logic of media culture. The virus did not just spread as an actual computer program via people's e-mails and information networks, but it spurred media attention, infecting as a media virus as well:

> A simple example would be the "I Love You" computer virus which, apart from directly jamming network systems, also caused immense indirect problems as system operators and webspinners across the world had to generate warning messages and upgrade virus detection systems. This is an example of a risk that is able to reproduce itself not simply through failures of regulation and security, but exactly because of the risk-aversion ethos already put in place by the systems themselves. Once risks are able to work through successful risk-management systems and use them to proliferate themselves, that is, once they are not only able to act but also to anticipate and learn, we could speak of cyber-risks.[377]

This accentuates how the thematics of risk in the age of mass media culture increasingly follow the viral logic of media itself. Technological risks are interfaced in a feedback loop with the media microscopes that visualize the otherwise disappointingly faceless events (e.g., on the computer screen). The technical risk-management systems that paradoxically help risks reproduce themselves are supplemented with the "audiovisual risk-perception systems" of Hollywood image production, for example.

Also actualized virus incidents, such as the Morris worm, are part of the logic of the virtual in that they are transformed, through media filters, into expectations, predictions, and thresholds: How about the next incident? When will it occur? Will it prove to be even more disastrous? What preventive actions should be taken? Such questions are easily turned into audiovisual images and narratives that amplify the effects of risks. As the actualized risks become archived in such audiovisual reproduction formats and the virtual, expected risks are similarly multiplied—think of the movies of late decades dealing with meteorites, terrorists and for example computer malfunctions, without forgetting the virallike Mr. Smith of *The Matrix* sequels (2003), the serial-killer-come-computer virus in *Ghost in the Machine* (1993), or even the malicious miniprograms in the Disney classic *Tron* (1982), which for the first time opened up the computer to the cinematic. This archive is then piloting both how we define (un)wanted objects of digital culture, and simultaneously how the future of digital culture is produced.

However, viruses should not be taken as the only form of accidents that are captured and turned into visual creations even though they do provide a key example concerning how *legitimate* software and *malicious* software are contrasted. In a way, viruses, Trojans or for example netbots would not have

to be even thought of as "accidents" but as integral expressions of the turbulence of the contemporary technological and social networks that always
also include a virtual surplus that is fed into a future fear, the accident to
come.[378]

## COMMERCIAL TECHNIQUES
## OF OBSERVATION

The Morris worm also marked the rise of a new field of commerce, demonstrating the powers of the capitalist axiomatics in capturing seemingly contradictory cultural elements as part of it.[379] At the center of this new field
stood AV scanners, which were a new class of software. Although perhaps
metaphorically these programs come closest to the microscopes for
enhanced vision, which helped to isolate and identify the biological viruses
during the earlier part of the 20th century, culturally they also form the commercial aesthesis of the virus, which connects the phenomenon to a growing
field of digital business in the form of consumer products and services providing digital security, safe hex.[380] Of course, such scanners did not operate
merely in terms of seeing (which was often produced as a metaphorical
expression of the scanning process) but as again calculating and trying to
track down the signatures of dubious strains of software.

The perception–calculation pairing fed into a creation of commercial
archives (this time very real) of virus strains. Here the archived algorithms
feed directly to commercial interests as they are efficiently used to weed out
similar instances "in the wild." This archival work represents in general the
new logic of information capitalism, where archives are produced not as
State memory but as reservoirs of capitalist production—whether archives
of viral strains, or for example, the audiovisual cultural heritage increasingly owned by media conglomerates.

In the widespread outbreaks of MyDoom, Bagle, and the aforemen-
tioned Netsky worm, suspicions have once again been raised concerning the
ethics of AV tool production. Security experts have been increasingly questioning "whether the antivirus software industry is working hard enough—
or has enough incentive—to develop new and better ways of stopping nasty
software."[381] The problem connects interestingly to some of the key business models promoted by AV companies: The signature model is based on
customers downloading signature files of viral code for their AV pack. As
network security consultant Mike Sweeney said: "But bottom line:
Signature files are profitable. You have to have a subscription that they
charge either a monthly fee for or an up-front, once-a-year fee. That makes
for a nice tidy revenue stream."[382] Other voices emphasize that alternative
models for protection "were effectively driven from the market in the last

decade." Of course, as AV producers defend themselves, the signature download model is easier for lay users, but still the question is interesting considering that in a fairly recent study in the United Kingdom by the Department of Trade and Industry (DTI):

> 93 percent of smaller companies and 99 percent of large companies said they use antivirus software, and close to 60 percent of firms update their antivirus software automatically to keep up with new virus threats. But computer viruses still managed to hit 50 percent of the smaller firms and infect 68 percent of the larger companies' networks in 2003.[383]

Prior to AV programs, only computer experts were able to decipher whether a machine or a system was infected with a malicious program. The average user had little access or knowledge of program routines which operate "under" the surface of user-friendly interfaces, making it impossible for him or her to differentiate a viral strain from some other automated action of the computer. Therefore, when the virus activated its payload, it was already too late from the viewpoint of the user to do anything to stop this activation routine. Beyond the so-called interactive system control, there was the level of automated executions that were beyond the user's control, seeming to underline the re-actional position of the user: In complex systems, a large part of system calls and processes are beyond user control, whether these are viral and unsolicited or normal system routines. However, to give a sense of safety, by the end of the 1980s and beginning of the 1990s, the average user could spot the miniprograms with automatized scanners as if he or she could peak inside the computer. Since the turn of the 1980s and 1990s, the main classes of AV software included scanners, monitors, and integrity checkers, which function similarly to find *unusual* activities within the computer.

Software seems to express itself in various ways, and viruses have often had varying payloads that let us know that they are there. The Cascade virus dropped the letters to the bottom of the screen, whereas the more recent Melting worm, spreading as an e-mail attachment, included a screen routine that "melted the screen" (see Fig. 5.2). The Yankee Doodle virus played the famous little tune, and several other viral strains captured the interrupt processes so as to take over the phenomenological interface. Following Trond Lundemo, this does not have to be regarded as an accidental feature of digital culture, but relates to how different regimes of media technologies have varying ways of breaking down. It is precisely because of the "invisibility of the digital code" that it is translated into visual terms as a kind of a mock-up remediation of the passage from the media technologies of phenomenological perception to one where calculationary processes eat up the interface surface.[384] Digitality eats up words and things, and similarly as the

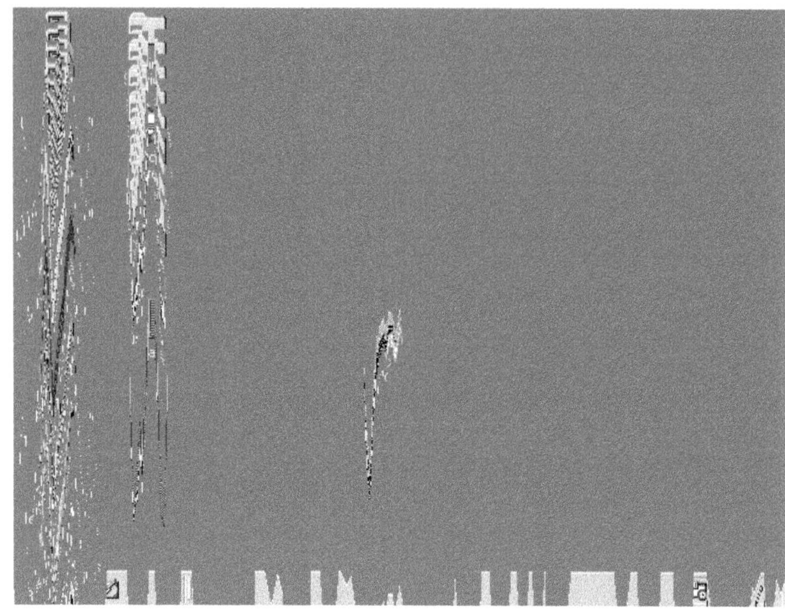

FIGURE 5.2. Melting worm routine turning the desktop into a non-figural blur, source Kaspersky Lab Press Center: http://www.kaspersky.com/press/images?chapter=146158526].

user struck by faulty or malicious code helplessly says goodbye to data that is sucked into the black box of the computer, the phenomenological processes seem to sink into the binary circuits of that very same box. Yet, images still have a place in the calculationary assemblage.

Weeding pathologies from the operating systems and user files with virus programs connects in general to the histories of *domestication* and *normalization* of the computer. This process, or at least attempt, of normalization is connected to copyrighting software code, which was earlier distributed as free knowledge. The 1980s was marked by a new tendency that saw software as commercially distributable creative work, a product, which meant the possibility to own code just like any other product. This spurred a regime of software licensing.[385] Such notions have been underlined regarding viruses as well: unlicensed software, downloaded from bulletin board services or illegally via copying, as well as freeware and shareware, are argued as more or less untrustable and potential virus vectors merely because of their "dubious" origins. The 1980s was characterized by a rising standardization for mass use in operating systems and software, expressed first in the graphical interfaces of Apple and Windows. Microsoft has been especially active in standardizing the media sphere of digitality, in homes and offices, providing the digital equivalent of McDonalds in its globalization of the operating system. Standardization, the key issue of any technological modernization practice, is of course something that is produced in contrast to the uncontrollability of the viral.[386] We could, then, consider Windows as an archival framework of contemporary network culture that organizes materials (texts, images, etc.), channels users, and pilots the uses and potentials of network culture.[387] Windows operating system (connected to the corporation and its networks) is an archival machine in the sense that it controls large parts of what can be said, shown, and heard in the contemporary digital culture. This kind of a theoretical suggestion would interestingly imply that virus writers, for who Windows is most often the targeted operating system, can be considered to be engaging in a kind of a practical an-archaeology that seeks to breakdown majoritarian mechanisms of framing.[388] In other words, the Windows archival framing has consisted of such elements as hiding system information from users, restricting source codes to themselves, keeping a tight link between the operating system and applications, and in general automating several system processes so that it has remained impossible to have a say on what happens in the computer.[389] Tactical an-archaeology might then mean, and the speculative choice of words is intended as this would need more research, not targeting operating systems or certain corporations as such but exposing the principles of how digital culture is framed through micropolitics of code.

The phenomenological interface is of course technically secondary to processes of calculation and execution as various protocols and algorithms for packing and delivering audiovisions define the surface layer of digital culture. Yet, this secondary layer should not be neglected—nor read in terms of representational analysis—but approached as part of the assemblage where software is archived and delineated. Such a possibility of micropolitics of code has been demonstrated by various hackers and activists. The universalizing aspiration in operating systems has been a special target for viruses, partly as a conscious act of hacker activity, partly just because the homogenic Windows platform makes it easier for viruses to spread their code because of the security deficits in the operating system. Again, the question cannot be reduced to functionality versus disorders, as the Internet art group *epidemiC* reminds us. Instead, it is a question of what kind of disorders are normalized and what are made abnormal: "it is perfectly legal that your Windows crashes six times a day . . . you have been forced to buy it, therefore it is its own right."[390] The nonhuman level of algorithms feeds constantly into the human social world (also phenomenological), as for example in such cases of interrupts.[391] Micropolitics can be taken also as a gesture of introducing fissures into the dominating strata of computer culture, holes through which the outside (as a battle and a turbulence) might fold in.[392]

# MULTIPLICITIES OF VIRUSES

Since the 1980s, such demarcations have formed the majoritarian framing of computer viruses. Viruses have been visualized in technical terms (scanning for signatures) but also increasingly as an audiovisual cultural theme that has made this algorithmic phenomenon highly visible since approximately the mid-1980s. What I emphasize here is that this medial perception, or machine of facialization and risk production, functions not merely on the level of calculation and execution, but also a simultaneous distribution of visibilities and articulations. It is not a simple question that 'there are images and words that symbolize or represent relations of power' but how intimately power still functions through organizing visibilities and language (now in relation to calculation) as refrains that pave capitalist worlds.[393]

The visibility of viruses as harmful software programs has been generated especially through interests of national security as well as international business—to which computer viruses represent a disruption in the global flows of capital. What has been neglected is that not all viruses are malicious—instead there is a multitude of viruses: "good, evil, entertaining, boring, elegant, political, furious, beautiful, and very beautiful"[394] as Luca Lampo, from *epidemiC* suggests. In this text do not go further into the alternative histories of viruses, but merely point toward the issues they raise. For example, in the 1970s, worm programs were used as test and utility programs as part of the Arpanet. Since the beginning of the 1980s, Fred Cohen mapped the issue of beneficial viruses. Similarly, 1988, the MacMag virus, with its peculiar peace message, can be understood as a media activist gesture of a kind, and so forth.[395]

The multiplicity of these miniprograms, the virality in viruses, has been neglected while only certain characteristics or functions have been objectified to constitute the essence of the phenomenon. Similarly, as the biological virus was enpresented and brought into the sphere of knowledge with the aid of virological and cultural techniques, computer worms and viruses and other malware have been brought into vicinity with both specific computer science "instruments" (special tracking software such as scanners) and increasingly with media machines producing audiovisuality. Yet, as Van Loon argued, there is a certain virtual object that is the intensive "nucleus"of multiplicity, to which modern technoscience has attributed its operations of stratification.

In a reminiscent vein, artistic computer viruses act as tactical media assemblages, tactical "accidents," shedding light on the otherwise imperceptible articulations of these technological bits of code. As Charlie Gere explains, it is possible to approach the phenomenon of net art as one expressing "the crisis of the archive" referring to how this networked, transinstitutional form questioned the established institution of collecting, curating, and displaying art. Gere, following Julian Stallabrass, sees net art (and especially Vuk Cosic's work) as an archaeology of the future that draws on the past potential mixed with "Utopian futures."[396] This crisis of the archive can also be articulated on a micropolitical level stretching across a wider social field. While the Biennale virus, within the context of Internet art, aimed to provide a kind of a Situationist *détournément* of the archive, addressing the multiplicity inherent in computer viruses, the majority of public articulations stemming from AV corporations or national security bodies aim for their part to stratify these miniprograms under the general class of "malicious software." That is why various alternative productions can feed also toward alternative counterarchives where software is culturally framed. For the Biennale programmers, the code contributed toward also new definitions of sociality:

> The creation of a virus tout court, free and without an end or a goal, is in the worst case a test, a survey on the limits of the Net, but in the best case is a form of global counterpower, generally a pre-political form, but that resists the strong powers, it puts them under a new balance, it shakes and reassembles them. A new idea of a "virus that is not just a virus" is gaining acceptance, and that it can represent the outbreak of the social into the most social thing of all: the Net.[397]

Viruses are not, then, only viruses, and software is not merely software, but a modulation, a framing, a vehicle that opens up worlds of knowing, sensing and doing, being inherently, and continuously, articulated on a wider social field.[398] A new visibility of code might also help bring into actions this visibility that does not need only to feed into the strata but make it and its gaps visible. This possibility seems to reside at the core of various conceptualizations of Internet art as activism of a sort, "not-just-art" kind of tactical software, to use Matthew Fuller's term. This differs from various other brands of Internet art, such as deconstructionist tactics that merely repeat (in an ironic fashion) the terms and practices they are criticizing instead of creating new openings.[399]

Clearly, the Biennale virus and other virus art projects (e.g., Luca Bertini's Vi-Con) attempt to tap into and create a multiplicity of software objects that spans not merely across the Internet but across media, toward an articulation of the social field of software.[400] Again, the issue of perception seems to stand at the center of several such net art projects. Bertini's Vi-Con is introduced as an *invisible* virus:

> Yazna and ++ are two viruses in love. They search for each other on the net, running through connected computers. Apart from other viruses, their passages won't cause any damage to your computer. . . . Theirs is a soft passage, invisible, and extremely fragile.[401]

This resonates with the Biennale virus' theme of technical invisibility and broad symbolical visibility, where the virus code is remediated in t-shirts and newspapers:

> We have proposed to Ferrero to print it on Mon Cheri packets, but they answered it is still too early. Recently, an Italian newspaper has used some strings of Biennale.py to announce the spread of a new virus, instead of using usual, banal, images of little biomorphic monsters. We consider it an iconographic victory.[402]

This "iconographic victory" implies, aptly in my opinion, the thematics of risk society: Perception is entangled with signification and valorization, which in their part feedback with how digital culture is produced in the form of knowledge (audiovisions, texts, and e.g., histories) and power (e.g., practices of coding). And it is this specific "iconographic surface" that is challenged by the incorporeal transformations enacted by Internet art viruses. In this sense, we can think of such Internet art accidents as tactical media that work immanently with its material (in this case, the turbulent spaces of network culture) in order to produce variations and novel resistance.[403] The Biennale.py virus source code incorporated a new level of visibility that drew attention both to the politics of definition and coding of viral programs, but also to the frames themselves: How the nonrepresentational level of code seems to regularly pop up due to events that remediate code in other media formats as well. By letting a bit of multiplicity into the discourses and majoritarian archives of computer viruses, such artistic creations are apt media-archaeological tools that can rewire the current understanding of digital culture, and its history, to encompass also virality and viral code as part of it. They create an "archaeology of the present" that short circuits with neglected and forgotten issues of beneficial viruses, discussed from the 1970s on.[404]

Imagining better futures that endure complexities is another way to put this project that mixes art, politics, and technology.[405] A nomadic memory of zigzag territorialization and an open future aims at actualizing "virtual possibilities that had been frozen in the image of the past" as Rosi Braidotti said.[406] In our case, this could mean "an aesthetics of digital monstrosity." As Internet viruses might prove to be a micropolitical counter power, shaking majoritarian notions of software and the order of digital culture. Similarly they can be grasped as philosophical and artistic machines that create new perceptions and concepts. Various viral programs and practices have been imagined for decades, but repeatedly they have been rejected as being either utopian, naïve, or even dangerous.[407] What is interesting is the possibility to short circuit our understanding of digital culture with the aid of such viral experiments, and rewire these neglected or forgotten themes of viral programs again as illuminating the issues of digital network culture. In other words, alternative programs and modes of programming feed not only as part of the calculational design but can be used as "tools-to-think-with" the archives and history of network culture.

Curiously, for example, Internet art projects can be taken as part of the archaeology of digital culture. By opening up further potentialities of seeing and saying regarding software, there also opens up the possibility of new practices of software. Instead of analyzing these various modes of archival logic as disconnected, they actually feed on each other where various coding practices are articulated on a much larger field of social interest, and vice versa, the audiovisual production of "the archive of software" feeds back into the algorithmic patterns and coding practices that develop the future of digital culture. Archives then do not merely record the past and present, but curiously also the future. In this sense, the issue of archives can be seen as crucial in questioning the politics of network culture, and the temporal understanding of how we see the archaeology of network culture, and importantly, how are we able to imagine its future. The archive never determines merely pasts, but functions as the *a priori* for a future to come.

The Biennale virus, and other similarly orientated cultural mappings of the virus phenomenon (I am here thinking of the I Love You exhibition[408]), are of creative interest in that they (a) engage with the established historical layers of perception (the technical, the medial, and the commercial), (b) produce a simulacra of such systems of perception[409] (the Biennale virus being remediated and sold for a profit), and (c) short circuit the everyday perceptions into new assemblages of critical nature. These examples demonstrate well how the tripartite logic of power/knowledge (saying, seeing, calculating) is revealed by such tactical accidents, and how the micropolitics of algorithmic processes can be connected across medial layers into themes visual and articulable. As Fuller noted, "Software forges modalities of experience—sensoriums through which the world is made and known,"[410] but this forging happens continuously on various levels that spread throughout the mediatized field of society; in the form of producing a majoritarian understanding of digital culture (the Hollywood/Windows machine of production) or variations and permutations that allow a bit of fresh air in the archives of software.

## ACKNOWLEDGEMENTS

I am grateful to Adrian Mackenzie, Matthew Fuller, and Tony Sampson for their most helpful feedback in writing this chapter. The earliest version of this chapter was written as a paper for the 2004 SHOT-conference in Amsterdam.

# 6

## CONTAGIOUS NOISE

## From Digital Glitches to Audio Viruses

*Steve Goodman*

*The whole series of things about accidents, about bugs, about the producer being someone who can nurture a bug, who can breed a bug and simultaneously most of the key musics have been accidents, they've actually been formed through errors. They're like software errors, syntax errors in the machine's programming, and they form these sounds, and the producer's taken these sounds and more or less nurtured this error, built on this mistake, and if you grab a mistake you've got a new audio lifeform.*

—Kodwo Eshun[411]

*In science fiction, ghosts in machines always appear as malfunctions, glitches, interruptions in the normal flow of things. Something unexpected appears seemingly out of nothing and from nowhere. Through a malfunction, a glitch, we get a fleeting glimpse of an alien intelligence at work.*

—Janne Vanhanen[412]

# THREAT

Exactly 1 year after 9/11, a text was published in the *New York Daily News* announcing leaks from classified reports from the NASA Medical Research Laboratories[413] detailing new evidence that viral diseases such as AIDS and ebola could be transmitted via visual channels. The idea was that exposure to microphotography of virus structures, could, via a process of what was described as dematerialization–materialization pass through the retina, the brain and then re-emerge as a "substantial living virus," entering a destructive relation with certain parts of the body. The fear, of course, was the potential such a powerful weapon could have in the hands of terrorists. But "if images can be virulent, can sound be virulent too?" This was the question posed by Swedish artist, Leif Elggren in his Virulent Images, Virulent Sounds, the project that stimulated the hyperstitional newspaper article. Elggren was fascinated by the direct, immediate implication of audiovisual media on the body. The CD that accompanied the project,[414] was presented with eight microstructure virographs (obviously published with a health warning), and contained eight audiorecordings of highly potent viruses:[415] HIV, rabies, influenza, lassa, mumps, sbola, sin nombre, and smallpox. According to the sleeve notes, these microrecordings were carried out in a government laboratory in Tripoli, Libya and couriered to Sweden on minidisc in January 2002. Elggren's epidemiological sonic fiction concerned the transmission of a biological virus code through the channels of media culture, an affective transmission of the abstract virus structure via digitalized ripples of sonic intensity. A cross-media vector scaling up from viral code through the microbiological to the audiovisual only to compress into code again. Even without this fictional context of mutant DNA, the sounds were pretty creepy; a chittering yet viscous sonic mutation, a sensual mathematics, in the gaps between sound systems, vibration, skin, internal organs, auditory–tactile nerves, and memory.

Elggren's project was clearly influenced by William Burroughs' rye proposal that a virus "is perhaps simply very small units of sound and image. . . Perhaps to construct a laboratory virus we would need both a camera and a sound crew and a biochemist as well."[416] Like Burroughs, Elggren's project used hyperbolic fiction to tap straight into the modus operandi of contemporary societies of control. An addictive ecology of fear, rife with auditory contagions, self-propagating rhythmic tics with an agency in excess of the labor and leisure from which they seem to emerge and be consumed. Yet, the promiscuity of digital codes merely reminds us that sonic culture was always a field of algorithmic propagation (from the Chinese whispers of rumor dissemination to the spread of rule-based, numerically rigorous tuning systems via musical notation, graphic scores, equations, and recordings). Rather than an epochal shift, infectious digitality merely reinforces the need for an interrogation of the rhythms of affective contagion that emanate from cybercultural spam.

The sound recordings contained within Elggren's CD aesthetically resonate with a particular strain of electronic music operating since the mid-1990s, a strain loosely revolving around a bug in the audio matrix. The "glitch": a scar on the pristine surface of *science fiction capital's*[417] vision of technological progress? Etymologically deriving from the Yiddish "glitschn" to slip, slide, or glide,[418] in "mechanics, a glitch is a sudden irregularity or malfunction. It suggests simultaneously a slippage of gears or wheels—a failure to engage—and a scratch, a small nick in the smooth surface . . . there is a duality embedded in the word, of skidding and catching."[419] At the turn of the millennium, the glitch (and the "click" and the "cut"), anomalous sonic micro-objects, began to virally proliferate from the periphery of digital music culture, infecting a number of electronic subgenres. These glitches both derived from the sonic detritus of digital accidents (e.g., malfunctioning sound cards, hard drives, and CD players) and the residue of intensified digital sound processing (particularly granular synthesis). An excitable rush to theorize the radicality of these blips on the soundscape accompanied these mutations.[420] In both the manifesto/press releases of its proponent record labels, and in the post-structuralist and deconstructionist discourse of its journalistic flag-flyers and theorist/artists, it was argued that these warps in the fabric of digitalized matter represented an auditory glimpse of a post-digital culture,[421] a sonic prophecy. In these discussions of the glitch as harbinger, two tendencies could be detected: The first was the avant-gardist tendency, cloaked in a combination of both digital futurist and Deleuzian terminology, often resorting to an acousmatic purism that quarantined the glitch into the sterile sound spaces of sound art and design while deriding its spread across the infectious dancehalls of electronic music. Second, there was the contrasting, more phenomenological tendency, which attempted to re-centralize the body within the digital domain. Yet by re-emphasizing a conceptual differentiation of the often conflated "virtual" and "digital," a potential, and unnecessary lapse back into a fetishization of the human body and its senses was risked. In so doing, these strains of critical discourse on the digital, threatened to, through an almost phobic neglect of the numerical dimensions of the virtual, block out the affective and textural (and therefore micro-rhythmic) potential of the sonic route through the digital (e.g., granular synthesis, time-stretching, etc.).

There was much of importance in these divergent theories. Remaining mindful of some of their problems, it is argued that sonic glitch theory can resonate outside of its local field of operation and illuminate the anomalies of viral media culture more generally. If the glitch constitutes some variant of sonic spam, then it must be reformulated not just in order to take it positively, but also to unshackle it from being reappropriated too quickly into problematic elitist aesthetic models, or trapped into the binary prison of a cerebral or phenomenological model of listening. Such models merely serve to immunize contagious noise, thereby hindering a rigorous, postdualist theorization of the topology of culture and nature, mind and body, digital and analog. The sonic domain can intensify a more general theory of glitch culture in that it directs us toward a rhythanalysis of digital media culture, and an investigation of the potential latent in the locked grooves of computational iteration. Here accidents are conceived as engines of rhythmic mutation. More than any other field, the sonic therefore provides the analysis of digital culture with a *rhythmic virology of the glitch*. Rather than immunizing this potential, we encourage it to spread into more abstract movements of thought, while simultaneously following its escalative proliferation across actual populations of dancing bodies.

## OUTBREAK

In an exuberant *Wired* magazine piece entitled "Worship the Glitch," Rob Young tagged the influence of the glitch (a virulent residualism parasitically feeding off malfunction in a fetishism of technical failure) as constituting some kind of "effluenza virus"[422] traversing the digital matrix.

> From beneath the frenetic, threshing rhythms of Jungle (touted in the mid-1990s as quintessentially "millennial" street music), a very different vibration has fermented, feeding off the technical errors and unplanned outcomes of electrified society—the world at the mercy of the glitch. Crackles, pops, pocks, combustions, gurgles, buzzes, amplitude tautenings, power spikes, voltage differentials, colliding pressure fronts, patternings, jump-splices, fax connections, silent interjections, hums, murmurs, switchbacks, clunks, granulations, fragmentations, splinterings, roars and rushes have overwhelmed the soundscape—as if the Ambient soundfields on the Cage—Eno axis have been zoomed in on until we are swimming amid the magnified atoms of sound. Characterized by colossal shift in dynamics, tone and frequency, this is an urban environmental music—the cybernetics of everyday life—that reflects the depletion of "natural" rhythms in the city experience, and in the striated domain of the virtual.[423]

From acoustic anomaly to ubiquitous strain, the viral, trans-genre spread of the glitch sound in digital music culture constituted a recent chapter in the evolution of a "noise virus" that has inhabited the skin between sonic experimentation and popular music. Aside from the usual Cagean co-ordinates, from the Futurist manifesto of the Russolo's Art of Noise to Public Enemy, from turntable experiments with broken records, to guitar feedback, to the scratching and the trigger-finger of the hip-hop dj/producer, right up to contemporary circuit bending, the perpetual flip-flop of unwanted noise into signal has been a recurrent dynamic of electrified music.

The virology of this particular strain of audio "effluence" concerns itself with the glitch's turn of the century parasitic relationship to a number of electronic musical strains. It propagated and inspired virtual genres such as click hop (e.g., Prefuse 73 and Dabrye), crackle dub (e.g., Pole), and micro-house (e.g., Jan Jelinek, Luomo). In the post-Jungle domain, this drift was paralleled by the "dysfunktional," emaciated experiments loosely grouped under the name IDM and including the exaggerated asymmetries of Aphex Twin, Autechre and Squarepusher, some of which would later evolve into breakcore. Yet, most narratives of the initial glitch outbreaks begin with the pre-occupation with surface noise, abused media and minimal techno of the early to mid 1990s focusing on a number of experiments, from the skipping playback and customized software of Oval, the early work on the Vienna based label Mego (including artists such as Pita, Rehberge & Bauer, Fennesz, Farmers Manual) and a number of fringe software experimentations (mis)using applications such as Super Collider, Audio Mulch, Metasynth, GRM Tools, Max Msp, Sound Hack.

Only in a limited number of these examples did the infection of a pre-existing genre by glitch virus add something to the host. Often the parasite would push the genre toward a zone of disintensification whereby core sonic traits would become exaggerated to parodic effect via glitching usually detracting from the carefully engineered rhythmic disequilibrium that was already present.[424] It is on this precisely crafted diagonal of asymmetric groove that underlies strands of digitized dance musics, that we can track the glitch not just as anomalous sound object, not just as sonic spam, but also as agent of rhythmic mutation. A common feature in the contamination of dance music genres was the substitution of rhythmic components by digital detritus. Describing German producer Jan Jelinek, for example, critic Philip Sherburne in 2001 noted that "all the percussive elements—the thumping bass drum, ticking hi-hat, etc.—have been replaced by tics and pops and compressed bits of static and hiss."[425]

At its best, the microhouse of Jelinek (whose tracks were also importantly drenched in sliced and diced tones excavated from old soul and jazz records) with its lush tic-drifts of *Looping Jazz Records* or the fluid syncopations as Farben, bristles with texturhythmic activity. Particularly in the Farben guise on a series of vinyl releases on the label Klang, the 4/4 pulse is decentered; in Young's words, it was techno "off the bone, building a twitchy, edgy microfunk around an absent center—all that remains are the offbeats, syncopated clicks and imagined accents, accidental pulses that might have come from a blemish on a side of vinyl"[426] paralleling the stuttering "falter-funk" of turn-of-the-century R&B (e.g., Timbaland or Rodney Jerkins) and UK garage,[427] and taken up again later in the Todd

Edwards-influenced Akufen productions of "My Way," the skeletal garage of the artists and and the clicky swing of Alvo Noto.

Apart from Jelinek, key microhouse producers included Vladislav Delay (Luomo), Thomas Brinkmann, MRI, and Ricardo Villalobos. Describing Luomo's Vocalcity album from 2000, Sherburne wrote, "squelching bass and keyboard lines lay down a syncopated foundation, but the real action bubbles up from indiscernible depths—clicks and pops barely tethered to the rhythmic structure, aching sighs that suggest the birth of the desiring machine."[428] Lodged in the strata of the percussive meter, the glitch virus spread around the fringes of electronica. Sherburne described the then up and coming/now minimal techno super-dj Ricardo Villalobos' music as "percussive epics . . . populated with sonic microbes engaged in delicate, mathematical mating dances."[429] Here, not just are drums replaced by glitches, but a whole new, minute and insectoid field of syncopation is layered into the mix, catalyzing an intensification of rhythmic potential.

Other Jelinek productions, alongside those of artists such as Pole marked the infiltration of glitchemes into the already stealthy influence of reggae on techno via Basic Channel. The result was what some half-seriously tagged as "crackle dub." Revolving around a kind of loud quietness, Pole (a.k.a Stefan Betcke), with his dodgy reverb and delay fx units, submerged Jamaican dub into a ticklish haze, inverting the relationship between signal and noise, foreground and background, high resolution and low resolution, surface and depth. Peaking with his album Pole 3, the result was dub multiplied by dub, an immense immersivity inhabited by the ghosts of Augustus Pablo's melodica drifting in from some spectral zone outside historical time.

The consolidation of these disparate sonic approaches was fulfilled by Frankfurt-based label Mille Plateaux, whose founder Achim Szepanski attempted to provide a thread of theoretical consistency, particularly via the philosophy of Gilles Deleuze and Felix Guattari, to this digital renaissance. The Force Inc. sublabel issued a series of dense digital music communiqués and a tribute CD after the death of Deleuze. Mille Plateaux, as just one of the fleet of sublabels that flanked the Force Inc. mothership,[430] was later responsible for helping to consolidate the glitch sound in the late 1990s, and early 2000s, into a networked movement promoting an anhedonic minimalism against the particular backdrop of a conservative German techno-trance scene.[431] On a number of compilations, most notably the four volumes of the compilation series Clicks & Cuts, Szepanski curated a collection of minimal sound pieces that had in common their infestation by glitch virus. The compilations were apparently commissioned especially for the project but the term *clicks and cuts* soon exceeded this limited designation and often became used as a generic handle for a wide range of glitch-influenced sounds. Rather than a genre, Szepanski maintained, clicks and cuts was just a way of explicating the digital culture of *cut-paste-copy-funk*, exaggerating and amplifying the symptoms of the move from the sample to the sound file.

The claim of course was that glitch operated virally, transversally, cutting across lineages instead of merely a linear genealogy of musical inheritance.

How would this claim of transversality work? A glitch virus would have to hijack other musical strains, producing them simultaneously as a host, carrier, or vehicle of transmission. This transmission process would work through replacing components of this host's code with its own, into which are programmed an additional instruction to replicate. This instruction catalyzes a sonic power to affect, a power that is not essential to a sound in itself, but rather its affordance to participate within a nexus of experience. And in this process of transmission, to a greater or lesser degree, it must mutate everything in its path. What was this musical mutation that the glitch virus would perform? As Sherburne said, glitch virtualized dance music via "substitution and implication, swapping out traditional drum samples for equivalent sounds sourced from pared-down white noise: click, glitch, and crackle."[432] And by introducing this additional order of abstraction, he claimed, it did more than merely digitally simulate popular dance strains; rather, "clicktech, microhouse, cutfunk graft a secondary structure onto the first—not imitative or hyperreal, but substitutive, implied, made clear by context alone, a compressed millisecond of static stands in for the hi-hat, recognizable as such because that's where the hi-hat would have been."[433] But aside from introducing a new sonic object into electronic music, how much stamina did glitch's avant-gardist pretensions have. Could such ambitious claims by the music's leading proponents only disappoint? We will now investigate in closer detail some of the problematic claims that contributed to the immunization of the glitch virus.

## IMMUNIZATION

The emergence of glitch provoked some intense sonic theorizing. A key philosophical move that lurked behind much of this discourse involved the inversion of the relation between essences and contingency in the matter of techno-aesthetic invention. In the short essay, "The Primal Accident,"[434] Paul Virilio produced a formulation of the technological accident typical of the discourse of glitch. While remaining trapped in a doom-laden negative ontology, he made some attempt to positivize the concept of the "accident" through an inversion of the Aristotelian conception of the relation between substance and accident. For Virilio, technological invention was simultaneously the invention of accidents, and the accident therefore became immanent to any new innovation, rather than merely interrupting it from the outside. Along comparable lines, the digi-sonic glitch became conceptualized as an art of the accident. As Rob Young outlined:

> The glitch is only conceivable in a world where music has become part-
> ly or wholly mechanized. Recording converts sound into an object, and
> as an object it is vulnerable to breakage. At the same time, the object as
> capsule of sound (a measure of the lived time scooped out of time, just
> as the photograph snatches a moment of lived visual reality) can accu-
> mulate power, potential energy.[435]

Theorist/sound designer, Kim Cascone, founder of the *microsound* news group, developed parallel insights into a full-blown *aesthetics of failure* that announced the advent of a *postdigital* music. Was this perhaps a prophecy of what Attali somewhat vaguely termed the emergent mode of *composition* in which audio-social power is challenged via a breach in social repetition and the monopoly of noise making?

Although it was Virilio who had made explicit this formulation of the technological accident, it was the name of Deleuze that recurred most often in glitch theory, for his anti-dialectical insight, with Guattari in *Anti Oedipus*, that machines function by breaking down. For Mille Plateaux boss Szepanski, digital media aborted the distinction between form and content, between data and programs. The radicality of the click, or glitch was that it represented nothing except itself. He argued that it, in fact, made audible the virtual, the sound of the in-between. Szepanski maintained, fol-lowing second wave cyberneticist Foerster,[436] that the click, empty of essence or meaning, merely constituted the intensive perturbations of a neu-rological network; the glitch sound was transmitted only as a click signal devoid of reference. Whereas Szepanski and journalists like Sherburne wished to forge the minimalist legacy of glitch music, others such as Cascone, who released sound pieces on Mille Plateaux among other labels, rejected this designation as an aesthetic black hole, preferring instead the umbrella of the *postdigital*.

A number of criticisms were subsequently raised about the glitch move-ment and its conceptual claims to radicality. Generally, it was maintained that if it was truly a music of malfunction, then the moment the sound of machines breaking down was recorded, sequenced, or used to substitute per-cussive elements, for example, it became a mere sonic flavor. The malfunc-tion had to be live. Of course, such a perspective totally neglected that the movement of deterritorialization-reterritorialization was not dialectical, but contagious. Recorded and re–sequenced glitch, instead of resulting in a mere recuperation, instead functioned as a probe, prospecting rhythmic mutation in future host bodies. Here, capturing the glitch increased its potential for contagion. Others, such as Ashline, maintained that most Deleuzian music did not necessarily come from sources explicitly using Deleuzian rhetoric[437] (i.e., sources other than Mille Plateaux or Cascone who made pronounce-ments that, e.g., the laptop constituted a sonic war machine). Apart from pointing to the recurrence of Deleuzian memes (gratuitous, sometimes cosmetic references to bodies without organs, rhizomes, deterritorialization, etc.) of much of the discourse, some critics were keen to question some of the futurist or elitist rhetoric. Andrews, for example, noticed an uncritical return to a purist modernism by reinstalling authenticity via the anti-aura of the malfunction. Tobias Van Veen, on the other hand, took issue with Cascone for reinstating the acousmatic and a certain canon of experimental music as prime reference at the expense of the affective mobilization of the body by the Black Atlantian rhythmachine.[438] It was claimed, therefore, that despite the radical veneer, the brakes were being put on the glitch virus and sonic bodies were being immunized against its virulence.

One way or another, all these theorizations, in their more-or-less Deleuzian manifestations shared a desire to problematize the digital and its representations, either via its failure, its abstraction or its "inferiority" to the analog. So for Cascone, the postdigital attempted to step beyond the hype of perfect digital reproduction, whereas for Szepanski, digital music presented an opportunity, via the click, cut, or glitch to make audible the virtual by, in Kittlerian style, stripping away the programming strata of the humanized interface and, in an acousmatic turn, the representational strata of referential music. In their own ways, both approaches tended to conflate the digital with the virtual in their suggestion that the (post)digital opened up a set of sonic potentials. It is useful, therefore, to examine such claims more precisely as they open onto a nexus of more abstract problems. This is crucial because what we have termed the *glitch virus* is of interest the extent to which it performs some kind of topology of the analogue and digital, mobilizing continuity and discontinuity, contagious bodies and codes. What exactly was at stake in arguments such as Cascone's for a postdigitality? What problems of the digital were being insisted upon, that they could be moved beyond?

In a provocative essay entitled "The Superiority of the Analog," Brian Massumi also attempted to strip away the hype of the digital, arguing that the analogue is always onefold ahead.[439] Whether generated on or off a computer, Massumi reminds us that there is actually no such thing as digital sound—if we hear it, it must be analogue. We cannot hear digital code in itself, but only once it is transduced into sound waves. With theorists such as Pierre Levy,[440] Massumi cleaved apart the erroneous equation of the digital with the virtual. Instead the virtual is defined as potential, whereas the digital can always only tend toward an already coded, and therefore predetermined range of possibility. The argument, in its apparent Bergsonian fetishization of analogue continuity, could easily be taken as antidigital and somewhat phenomenologically tainted. But such an interpretation would be misleading. The drive of Massumi's argument is in fact to push for a more rigorous theorization of the enfolded nexus of the analog and digital. What kind of sonic plexus can they compose, and where does the potential for invention lay?

In his recent book, *Sound Ideas: Music, Machines and Experience*, Aden Evans, without specifically using a concept of the virtual, paralleled Massumi's skepticism regarding the digital in order to investigate this sonic plexus and locate zones of mutational potential within the codes of digital music. Evans described how digital code stratifies the analogue in a double articulation. He raised the question apposite to the glitch theorists of whether a digital singularity can be conceived, or whether such a singularity would in fact be merely a residue of the process of digitalization. The digital stratification of the analogue cuts it into parts and then assigns values to these parts. As Evans pointed out, this articulation is crucially double:

> On the one hand, the bits are spread out linearly, each divided from each, while on the other hand, each bit is either a 0 or 1. Binary numbers have a first articulation (the nth place) and a second articulation (0 or 1 in each place). . . . The binary is nothing but articulation, a simple difference between 0 and 1 . . . [but to] be effective the digital requires another articulation. . . . In the case of sound digitalization, a sound is divided into small chunks of time (samples), and each sample is evaluated by measuring the air pressure at that point in time. . . . A first articulation of parts and a second of values . . .[441]

However, in this process, Evans argued, using the term *actual* where Massumi would use the *analogue*, digitalization

> captures the general, the representable, the repeatable, but leaves out the singular, the unique, the immediate: whatever is not formal. Actuality always exceeds its form, for it moves along lines that connect singularities; the actual is not a neat sequence of frozen or static moments but an irreducible complex process that cannot be cleanly articulated in time or space.[442]

The rules of operation of the digital are immanent to its formal, binary code from which it is composed. Yet the emptiness of this code is what produces its infinite replicability—the clone is always formal, and therefore there is no uniqueness as the format is essentially generic, every analogue place becoming a numerical space, and every type of analogue object tagged by numerical values. So the limits of the digital—"Refinement, precision, storage, isolation"[443]—are exactly its power (i.e., its ordering quality) for measuring and counting. The digital is simultaneously exact and reductive. But Evans distinguished between this exactness and precision. He termed the exactness of digital calculability imprecise in that "it measures its object to a given level of accuracy and no further . . . it presents its own completeness."[444] For Evans, something is lost in this transition from the fullness of the analogue to the exact partiality of the digital. There is a residue of the process of stratification, whereby the digital cuts into the analogue, and through which continuity is transposed into generic parts, or bytes—this residue is the excluded middle of this process of double articulation. "The digital has a resolution, and detail finer than this resolution is ignored by the digital's ordered thresholds."[445] The analogue, on the other hand, for Evans, as a variable continuum, is fuzzy, and responsive—any operation performed on it transforms it. The digital zooms in on the thresholds of the analogue, marking variable ranges in this qualitative continuum, quantizing them into a discreteness and exactitude. Paralleling Massumi's thesis that the "analogue is always onefold ahead"[446] of the digital, Evans noted that the "superiority" of the analogue stems not from a limitation of the digital substitution, its difference from an actual object, but crucially, and this is the crux of their differential ontology, it is "rather a productive difference, a not-yet-determined, an ontological fuzziness inherent to actuality itself. Difference as productive cannot be digitalized."[447] The processual nature of the actual, and its generation of singularity must exceed its capture. In other words, the actual for Evans exceeds the sum of its digitized parts. This crucially is not merely a phenomenological point. Elsewhere, Evans developed a parallel argument via Intuitionist mathematics in relation to the concept of the differential (specifically the *surd*[448]) from calculus and what Deleuze termed the process of *differentiation*. The differential "was an extra term, left over after the rest of the equation had been reduced, and the methods for dealing with it could not be decided in advance."[449] Evans found a surd at work in the uncertainty principle of acoustics concluding that the "digital encounters events or objects that it cannot accommodate, and it must reshape itself in order to make room for these new ideas, but eventually settles back into a placid or rigid formula, neutralizing the novelty that challenged it to develop."[450]

Evans' position deviated from Massumi's with regard to the terminology of the virtual, with Evans locating the productive force in the actual itself, whereas for Massumi, the potential for change lies in fact in the virtual. What Evans called the actual as opposed to the digital, Massumi termed the analogue, composed of the actual and the virtual. Massumi questioned the potential of the digital generation of results that are not already precoded. If the digital is to provide access to the virtual, then it would have to "produce unforeseen results using feedback mechanisms to create resonance and interference between routines." A virtual digitality, would have to integrate the analogue "into itself (biomuscular robots and the like), by translating itself into the analog (neural nets and other evolutionary systems), or again by multiplying and intensifying its relays into and out of the analog (ubiquitous computing)."[451]

# RHYTHMIC MUTATION

> Rhythm is texture writ large, peaks and valleys turned to pulse. Texture
> is rhythm rendered microscopic, (ir)regularity encoded and impressed
> upon the surface of sound. Where these two break and cleave apart, the
> click, smooth-faced, one dimensional, textureless and out-of-time.
> (Philip Sherburne)[452]

As a strain of the "noise virus" mentioned earlier, the glitch is not merely a
type of sound, in this case an anomalous sonic micro-object, but more than
this, a vectorial field, a potential of sonic anomaly that demands a reinterro-
gation of the tangential rhythmogenesis of the technical accident. Many critics such as Rob Young (cited earlier) counterposed glitch to jungle, with its
maximalist Black Atlantian hyperrhythm, as rivals for the cutting edge of
digital music in the 1990s. The minimalist undercurrent of glitch, it was
argued, provided another sonic take on the dark side of the digital, one that
instead of plotting its uncharted course through the music of the Black
Atlantic, self-consciously held up reference points that stretched back to the
cannon of the experimental avant garde of John Cage, Pierre Schaeffer, and
the minimalism of Steve Reich and Terry Riley. The battle lines were drawn
with grains of sound, a battle that would resonate invisibly in subsequent
mutations of the glitch virus; a *sonic war between the avant pop and the
avant garde*. As glitch spread, the sense that popular electronics (aimed at
moving bodies in the dance) had "soiled" the purity of "high-art" experi-
mentalism became apparent in much of the accompanying discourse. Such a
sentiment was articulated by theorists such as Ashline who wrote in 2002 in
an article about the influence of the philosophy of Deleuze on glitch as a
music of "aberration":

> It was only a matter of time before an electronica solely servile to the
> dance floor would become conceptually and aesthetically boring, where
> the need to rediscover its origins and histories in the forms of musique
> concrete, minimalism, experimentalism, in short, in the avant garde,
> would become manifest.[453]

In a reiteration of the classic recuperation story, Ashline continued "the
deterritorialization of the 'glitch' quickly becomes reterritorialized in popular electronica. There was an effective detumescence of the hyper-intensity
that accompanied its discovery."[454] Yet, despite the rhetoric, there is as
much, if not a more potent case for connecting the glitch virus to a somewhat less pompous, less elitist lineage; for example from the sampling trigger finger of the hip-hop producer, and the sonic metallurgy of the scratch

dj (which extracted a machinic potential out of an analog, continuously varying flow by literally cutting into it, instigating a warp in time), to the time-stretching experiments of jungle via the atomized, microsonic domain of granular synthesis. As Tobias van Veen pointed out regarding Cascone and his theories of the aesthetics of failure:

> He reinscribes a polarity of values, that of acousmatic and pop, hierarchizing aura, and with it, a set of cultural codes, of proper contexts and social situations for listening ... we witness an exclusion of rhythm and its cultures, a sonic meme that goes completely unremarked and unnoticed.[455]

Such tendencies are common to the music theoretical uptake of *Capitalism and Schizophrenia*, with Deleuze and Guattari's philosophies often wheeled out to salvage some of the most bland and generic sonic interventions in immunized sound art spaces; spaces from which the desiring machines had clearly been excluded. What has often been remarked as the gratuitous deployment of their theories within the experimental fringes of electronic music owes something to the perhaps impoverished or narrow sonic history from which Deleuze and Guattari draw in *A Thousand Plateaus*, the well-trodden path of the European avant-classical that runs from Varese through Messiaen to Boulez, Cage, and Stockhausen. Their analysis of rhythm is powerful when expanded to become an ontology of speed, but tends to disappoint if confined to their own musical examples or tied too closely to a historicism. They lean heavily on Messiaen, who too typically excluded the syncopations of jazz from rhythm proper. Elsewhere, Deleuze proposed a nonpulsed time, which again would rule out much that would otherwise converge with their affective ambitions. Without doubt, the most powerful sonic analyses, treating *Capitalism and Schizophrenia* as a patch bay, and not merely enslaving it to an already canonic avant gardist genealogy was that of Kodwo Eshun, who invented the concept of sonic fiction and forced a conjunction between theories of sonic intensification and the Black Atlantian rhythmachine.

In the sonic fiction of Elggren appealed to at the outset, a process of materialization–dematerialization was suggested as basic to processes of cultural virology. We suggest that an investigation of rhythm can help unravel the nature of such contagious propagation. If the glitch is a sonic micro-event, it constitutes a wrinkle in time. As opposed to quarantining the glitch in the purity of the gallery space, we have argued that the glitch virus accrued more power, via its rhythmic mobilization in dance musics. As Rob Young argued, on "its own, a glitch does not amount to much. It accumulates power by its insertion, its irruption in a flow of events. It is the random factor, the spark that ignites the primordial soup, the flash that illuminates the status of music as phantasmagoric time, not a utilitarian time-keeper."[456] Critics such as Tobias van Veen, with recourse to the hyper-rhythmic musics of the Black Atlantic challenged some of the pretensions of glitch as sound art, opting instead for the sonic nexus of the glitch with the dancing body.

This mobilizing potential of the glitch virus, its potential to snag and snare the moving body in new ways, leads us down into the depths of sonic matter, to construct a rhythmic ontology that maps the angle of deviation whereby the accident unfolds into a vortical rhythmachine. The glitch code unravels into a rhythmic nexus that functions as a carrier wave, a vector of propagation. In the microphysics of dance, the glitch knocks off balance, an accident of equilibrium and orientation. Whereas Virilio inverted the relationship between substance and accident, creating a negative ontology to draw attention to technological failure, a more far-reaching approach requires a foundation in a differential ontogenesis. Through the Gabor matrix, microsonic analysis and granular synthesis long ago pointed to a basic sonic atomism. And this need not be conceived in Newtonian terms. By focusing on the rhythmic deviations of microsonic matter, such an ontogenesis would acknowledge that the sonic event is always inflected by chance variation that may render it ungroupable and as an anomaly to the set to which it appeared to belong. Although it appears to be deemed negatively as an accident or insignificant noise, it is precisely because it is asignifying that it is an event. In its singularity, it anticipates its own, yet-to-come collectivity; it anticipates the contagion of an event into a rhythm. The glitch, exceeding its negative designation as the nonrecognizable, is therefore of interest in so much as it is an event that expresses potential. Although it may stand for the intervention of chance, it simultaneously expresses necessity in its constitutive potential. Such a constructive ontology of primary deviation unchains the sonic accident from the negative of intended function. The basis for such a conception, appropriate to the terrain of microsound, can be found in Lucretius' atomism, which assists in providing an account for both continuity and discontinuity within the texturhythmic innovation of digital sound culture.[457] The Epicurean and Democritian atomism expressed via Lucretius rotates around the concept of the *clinamen* or the *swerve*. In *On the Nature of the Universe*, Lucretius set out to map the cosmos without introducing any conception of purpose or final cause or injecting it with an essence. The clinamen, for Lucretius, is not merely a deviation from order, but rather a primary process. He famously noted:

> [when] the atoms are travelling straight down through empty space by their own weight, at quite indeterminate times and places they swerve ever so little from their course, just so much that you can call it a change in direction. If it were not for this swerve, everything would fall down-

wards like rain-drops through the abyss of space. No collision would take place and no impact of atom on atom would be created. Thus nature would never have created anything.[458]

The resurgence of interest in Lucretius, particularly from the Serres–Deleuze axis, parallels the emphasis placed elsewhere on the primacy of becoming, heterogeneity and difference instead of stability, identity, or constancy.[459] With the clinamen, the minimum angle of deviation from a straight line, the onset of a curve from a tangent, the deviation is primary. Instead of an accident that befalls predictable or metric matter, the clinamen, as Deleuze clarified, is the "original determination of the direction of the movement of the atom. It is a kind of conatus—a differential of matter, and by the same token, a differential of thought."[460] The resultant map is no longer one of straight lines, parallel channels, or laminarization, but rather the formation of vortices and spirals built out of the swerve. The glitch, the apex of the clinamen, spirally unfolds in the generation of a vortical rhythm-machine.

Whereas Elggren speculated, in "Virulent Image, Virulent Sound" about the cultural transmission of biological viruses and their terrorist deployments, we have been concerned here with the outbreak, immunization, and mutation of a cultural virus and its musical deployment. What we have called the *glitch* is a kind of sonic spam, a substrain of a noise virus that has shadowed the history of music and its instruments. It is a virus whose virulence has been intensified by the digital. Infecting a musical machine (a relational sonic nexus of vibration, movement of body/thought and technical machines), the noise virus incubates a potential for recombination and the synthesis of new texturhythms. In following the contagious vector of the glitch and its parasitism of electronic dance music, it has often functioned as an intensifier of texturhythmic potential, mutating movements yet to come, despite elitist theorizations to the contrary. This sensuous mathematics of rhythm revolves around the nexus of the analogue and the digital, bodies and codes conceived together, mutually implicated and entangled in the networks of affective contagion. A fetishism of either the analogue or digital can only be futile.

This anomalous route through the sonic foregrounds issues of texturhythmic potential and forces them into more general discussions of invention within digital media culture. The problem with which we have been concerned is whether this potential is accessible through the digital, or whether it always requires a conjunction with the plenitude of the analogue. Is the digital sufficient, or is the analogue necessary to access sonic virtuality? Does the glitch virus require the malfunction of the digital and the interruption of its codes from outside or the inside? Although the digital, it is argued, in its discrete binary constitution of bytes frames a predetermined precoded field of demarcated possibility, can there not be a potential for mutation immanent to the numerical code itself? Digital philosophers, such as Chaitin[461] hint at this when they insist that formal axiomatic systems are never totally complete but contain infinity within, manifest in the contagion of an uncalculable, irreducible real, which always exceeds axiomatization. If the glitch appears as a warp in the digital matrix of space–time, then it is perhaps only a surface symptom of the more foundational numerical rhythms out of which that grid congeals. A too quick dismissal of the digital, articulated without an exploration of the numerical dimensions of the virtual at work in mathematical problematics and in popular numeracy risks falling back into a phenomenological fetishization of the plenitude of the analogue as a reservoir for emergent form. What is required is an affective calculus of quantum rhythm. Such a calculus would map the rhythmic oscillations that vibrate the microsonic, and the molecular turbulence these generate, a spiral that scales up through the nexus of the analogue and digital (a sonic plexus) its codes and networks.

Although we have focused on the virulence of the glitch within sonic culture, we argue that the glitch opens onto much wider questions of ontogenesis, literally cutting across scales, traversing both the analogue and digital domains. Arriving initially as a quantum of sonic spam that performs an immanent critique of the digital, we further suggested that the glitch perhaps constitutes the very onset, or engine of rhythmic mutation. Rhythm, and its numerization is here the very model of abstract, amodal perception shared by all media.

The tactical question of the glitch is not to wait passively for accidents to happen, newness to spontaneously emerge, but rather to carefully and pre-emptively engineer the circumstances in which the literal, rhythmic repercussions of their eventual incidence are channeled, optimized and sustained toward the invention of new operating systems for affective collectivity.

# 7

# TOWARD AN EVIL
# MEDIA STUDIES

*Matthew Fuller*

*Andrew Goffey*

*Evil media studies* is not a discipline, nor is it the description of a category of particularly unpleasant media objects. It is a manner of working with a set of informal practices and bodies of knowledge, characterized as stratagems, which pervade contemporary networked media and which straddle the distinction between the work of theory and of practice.

Evil media studies deliberately courts the accusation of anachronism so as to both counter and to enhance the often tacit deception and trickery within the precincts of both theory and practice.

## STRATAGEM 1:
## BYPASS REPRESENTATION

The basic strategy is not to denounce, nor to advocate but rather to create a problem of a different order to that of representation and then follow through practically what it entails. Although it is quite plausible to analyze the developments in digital media in terms of a problematic of representation, with its associated postulates about meaning, truth, falsity, and so on, a problematic of the accomplishment of representation is badly adapted to an understanding of the increasingly infrastructural nature of communications in a world of digital media. Although networked media may well be shaped by cultural forces, they have a materiality that is refractory to meaning and to symbolism. At the same time, digital media work largely through the formal logic of programmed hardware and software, that is, as something that more closely approximates the order of language. Language here becomes object, in a number of senses: objectified by a range of practices that submit communication processes to the quantificational procedures of programming; invested as a crucial factor in the economy; and an element in the purely objective order of things in themselves, escaping from the complementarity of subject and object and the range of processes we normally think of as mediating between the two.

# STRATAGEM 2:
# EXPLOIT ANACHRONISMS

We use the word evil here to help us get a grip on contemporary media practices of trickery, deception, and manipulation. The shift to this register must be understood in the context of a desire to escape the order of critique and the postulates of representation so obviously at work in the way thinking is made available about the media more generally. To speak of an evil media studies is to draw attention to a range and style of practices that are badly understood when explicitly or implicitly measured against the yardstick of autonomous rationality and the ideal of knowledge. Indeed, an evil media studies has immense capacity for productive use. As Jonathan Crary argued:

> that human subjects have determinate psycho-physiological capacities and functions that might be susceptible to technological management, has been the underpinning of institutional strategies and practices (regardless of the relative effectiveness of such strategies) for over a hundred years, even as it must be disavowed by critics of those same institutions.[462]

The fact of the matter is, as Crary points out, a vast amount of time and effort is spent on studies devoted to looking at the ways in which the experience of media subjects can be operated on. The point here is not whether such studies—frequently behaviorist in inspiration, frequently located in the field of psychology—are scientific or not. The point is that like the famous study by Stanley Milgram,[463] they point very directly toward techniques, practices that are efficacious even if they don't lead to, or ultimately derive from, scientific knowledge.

This given, it is important to talk about whether things work, not about whether or not they are right. Isabelle Stengers and Phillippe Pignarre have recently spoke of the *sorcery* of capitalism, a sorcery that implies that practices maligned by the ascendancy of critical rationality, such as *hypnosis*, be treated far more seriously. In the therapeutic use of hypnosis, what is significant is not the ways in which the powers of suggestion can encourage patients to imagine events that didn't happen (although this might be an outcome), but the way in which patients are initiated into a specific form of reality—which may or may not help to cure them. What occurs is a "'production of reality' which the hypnotist conjures up, in a precarious and ambiguous manner, without being able to explain or justify his or her "power" in the matter."[464] Unlike the outmoded model of media spectacle, which simply proffered an image of a "hidden" or occulted reality, hypnotic suggestion—a fact long known to the inventors of public relations is one of a number of means that are directly productive of a reality. Taking advantage of such mechanisms calls for the delicate negotiation of a different position to that commonly adopted in media studies. For those professionally or even incidentally embedded in media, to say that we are manipulated, that trickery and deception are effectively exercised on a regular basis, is not to deny that people cannot or do not think, but it would be to further deceive and manipulate ourselves to think that rational subjects are not outstripped by events.

## STRATAGEM 3:
## STIMULATE MALIGNANCY

To talk of evil is also to insist on an ontological dimension of the reality to which the order of communication belongs: the non-sense of something that cannot be exchanged for meaning, which is infinitely tangential to representation (but is not necessarily "repressed"). It is in this sense that Jean Baudrillard talks about a "principle" of evil and argued that "in every process of domination and conflict is forged a secret complicity, and in every process of consensus and balance, a secret antagonism."[465] If there is thus an *agonism* inherent in every form, this is in the sense that the form fights against the transpiring of its secret alterity. An example often repeated by Baudrillard is the cruel irony that the more media represent events, the more certain that, following an inexorable spiral of semantic inflation, they are to disappear, only to curve back in on themselves and replace reality. More simply, one can admire the way in which the hyper-sophisticated technology of the war machine of a global superpower reverts, on contact with any form of friction into a terroristic, technological primitivism. And these are perhaps only the most obvious manifestations of this "principle." To put it another way, evil is a good name for the strategies of the *object*, for what things do *in themselves* without bothering to pass through the subjective demand for meaning. If secrecy is inherent to this agonism, this is perhaps because it is a process without subject, a machination, a process that depends on its imperceptibility and which must for that very reason surprise us, fox us or outwit us.[466] As such, this strategy secretly reverts from malignancy to innocence.[467]

# STRATAGEM 4:
# MACHINE THE COMMONPLACE

For a number of recent commentators, language and communication are now absolutely central components of the economy.[468] Long considered a vehicle for the communication of ideas—and thus an element of the superstructure separate from the economic base—language and communication more generally should, these writers contended, be considered part of the infrastructure. This shift in the place assigned to communication in the economy opens up a new range of issues to consider and casts new light on the changing nature of work in the contemporary economy. From the restricted practical analysis being sketched out here, the general claim, if not the specific details, suggests some unlikely antecedents for contemporary practices in digital media.

Recent attempts to rethink the changing shape of work in the contemporary economy—and the shifts in political subjectivity such changes imply have taken a curious inspiration from Aristotelian rhetoric and the principle of performativity that this embodies. For the Italian theorist Paolo Virno, contemporary political subjectivity involves a sort of principle of *virtuosity.* "Each one of us," Virno contends, "is, and has always been, a virtuoso, a performing artist, at times mediocre or awkward, but, in any event, a virtuoso. In fact, the fundamental model of virtuosity, the experience which is the base of the concept, is the activity of the speaker."[469] If Virno's analysis provides an interesting way to refigure the understanding of the link between labor and language, it is perhaps also true to say that it only goes so far in exploring the paradigmatic ways in which media and communication practices exemplify the changing nature of modern production. For Virno, to be a producer today, to be a virtuoso, involves working on and with *commonplaces,* the finite, fluctuating stock of points around which language as performance coheres and the skeletal forms of intelligence that these embody. If media then become paradigmatic of the mutations that have occurred in the labor–capital relationship, this is because they too work on commonplaces. In *digital* media, the rudimentary set of operators utilized in SQL, the relational database query language, to analyze data might be described as a series of *machinic* commonplaces (=, !=, <, >, <=, >=, etc.). A *general intellect* characterized by a set of "generic logical-linguistic forms" in this way becomes central to contemporary production, provided that we accord no automatic privilege to natural language and provided also that we recognize that the instantiating of languages within media technology necessarily marks a zone in which language becomes inseparable from the senselessness of an object without a subject.

Virno's approach, like that of Maurizio Lazzarato and Christian Marazzi, has enormous merits, not the least of which is to pinpoint some of the rudimentary forms of intelligence (i.e., those relational terms such as equal to, not equal to, more and less, and so on, we characterize in terms of commonplaces) that inform machine processes. However, as a way of understanding media, this approach is insufficient. The first indicator of why this is the case results from the fact that in Aristotle's work, a point not lost on Hannah Arendt, [470] much of the argument about language is dictated by the need to rout the sophists, the consummate yet paradoxical masters of the secret antagonism of communicative form. Indeed, the machination of the consensus thought to be tacitly presupposed in all communicative action (by the likes of Habermas), is accomplished by Aristotle precisely by excluding a range of communicative techniques previously the stock in trade of sophistry. Whether we think of communicative action as definitely separated from instrumental activity (as Habermas does) or as inseparable, as Virno does, is immaterial from the moment we understand that consensual communication and cooperation has the function of excluding and thus of distorting our understanding of practices that are not necessarily rational in form. So, starting with sophistry is a way to open up the study of media forms as a response to the rationalist disavowal of manipulation and mind control that need to be surpassed by a truly useful, and hence evil, media studies.

## STRATAGEM 5:
## MAKE THE ACCIDENTAL THE ESSENTIAL

In Ancient Greece, the sophists were consummate exploiters of the faults, disturbances, and idiosyncrasies of language, its non-sense. Installing themselves within the cracks of language, the fissures that open up where one word could mean many things, two different words could sound exactly alike, where sense and reference was confused, sophistry sometimes humorously and playfully, sometimes with apparently more sinister demagogical intent, exploited the "semiurgical" quality of language and the seething cauldron of affective charge it contained to make and remake our relations to the world. For this, history shows, they were vilified, slandered, and excluded from the community of normal human users of language. Philosophy and the right (thinking) use of reason was the prime agent in this historical expulsion. By the genial invention of principles such as that of noncontradiction and entities such as rhetoric to absorb the excesses of language, philosophy not only created strong normative principles for communication arguably operating on a transcendental basis (recently rehabilitated by Habermas and Karl-Otto Apel), it also created a perception of language and of logic in which faults, glitches, and bugs started to be seen simply as accidents, trivial anomalies easily removed by means of the better internal policing of language. Short of being a two-headed monster or a plant of some sort, you could not possibly say one thing and mean two. The norms of reason precluded this: Transparency should be the elimination of agonism, not its secret accumulation. But as the sophists knew and practiced, double-speak was something that politicians did all the time, more or less knowingly, more or less well. Twenty-five centuries later, with the advent of deconstruction and other approaches, we discover that in fact double-speak is the "repressed," disavowed norm of reason.[471]

# STRATAGEM 6:
# RECURSE STRATAGEMS

A study of media that does not shy away from possibilities such as mind control should be elaborated as a series of stratagems. Why? Because agreement and cooperation, the rational assent of the reader, are outcomes not presuppositions. A consequential study of mind control should therefore be recursive and apply to itself. In any case, the stratagematic approach gives us something to *do*: The autonomy of code, its independence from human interference, is not incompatible with the existence of the strategically marshalled multitude of agents who bring it into being. A stratagematic approach to arguments was proposed in the mid- to late-19th century by the pessimistic German philosopher Arthur Schopenhauer in his short text *The Art of Always Being Right*. Schopenhauer's text is a practical manual in the tradition of Machiavelli's *The Prince* and Baltasar Gracian's *The Art of Worldly Wisdom*. All three of these texts are non-naturalistic, practical guides to the operations of power and the manipulation, deceit, and other forms of linguistic enhancement required to exercise it effectively. Schopenhauer's text is a distant inheritor of the opportunist charlatanism of the sophists and exercises a similar effect: suspension of the right–wrong, true–false, good–evil oppositions as *a priori* guidelines for winning arguments. Consequently, it focuses on the strategies of persuasion that emerge out of the fissures of argumentative performance.

But, if such a study borrows Schopenhauer's stratagematic approach, it does not share his exclusive focus on the dialectical situation of dialogical interaction or the exclusive focus on natural language. The vast majority of communications processes that take place in contemporary media are not of this type. Indeed, the vast majority of agents in a digitally networked world are not even humans and do not operate using natural language. But the processes of message exchange are still a part of the proper operations of power, and that is what we are interested in.

## STRATAGEM 7:
## THE RAPTURE OF CAPTURE

A useful term for trying to understand what is going on in the world of digital communications is *capture*. We live in a "world of captures," a world wherein power—as Foucault and others before him showed—operates not primarily by repressing, suppressing, or oppressing (although sometimes it involves the active celebration of all of these qualities), but by inciting, seducing, producing, and even creating. Capture operates most commonly, and indeed most economically, by imposing slight deviations of force, by scarcely perceptible inflections of agency. Language is both more than and less than language. The suggestions of the hypnotist redirect unconscious affect, a word ("education, education, education") or a slogan ("the axis of evil") acts as an attractor. Being captured makes sense for us of the feeling we have that the social today is a more or less clumsily designed open prison, that we don't need to be locked away to feel trapped, that we don't need to have committed a crime in order to sense ourselves permanently judged, submitted, even through the knowledge we understood might make us free, to an abominable, stultifying stupefying faculty for the routinization of life. Capture equally provides a way of characterizing what happens in the relationship between humans and machines, formal and natural languages, affect and technology. Stratagems are *event handlers*: They trap agency.

## STRATAGEM 8:
## SOPHISTICATING MACHINERY

From a somewhat different point of view, media theorist Friedrich Kittler hypothesized an adventurous analogy between the Lacanian unconscious and the computer that might help us start to understand how these techniques of capture work across platforms (those based on natural language and those based on machine language). Applying the Lacanian dictum that for there to be a world of the symbolic (i.e., culture), something must function in the real independently of any subjectivity (there would be no way of symbolizing it otherwise), Kittler argued that the operations of computer hardware on the basis of the oscillations of silicon crystal chips demonstrates that the famous notion of the unconscious as the discourse of the other is equivalently a discourse of the circuit. In the world of the symbolic "information circulates as the presence/absence of absence/presence." In the real, in the hardware of the computer, this is the flip-flopping of gates according to simple voltage differences. The exploitation of the potentials of silicon quartz allows Lacan/Kittler to draw together the discourse of the unconscious and the operations of the circuit and so better to develop a *literal* understanding of technologies of power. Let's not get too bogged down in this. The point to be made here is a simple one. The presence/absence of absence/presence that is at work in the basic operations of computer hardware points toward the systematization of a regime of signs that, according to structural psychoanalysis, figure desire or affect as an elementarily *coded* phenomenon. Lacan for one felt that all the figures of speech codified as rhetoric provided an excellent means for understanding the operations of the unconscious. In practical terms, this implies that our machines speak (through) us, rather than the other way around, a point Kittler/Lacan made very succinctly: We are today "to a greater extent than [we] could ever imagine, the subjects of all types of gadgets, from the microscope to radio-television."[472] When people find it surprising to be addressed by a machine, we should note that this is perhaps correct: The machines are usually busy enough communicating with each other.

These comparisons point us toward a "technicity" of sophistry and its operations on the quasi-autonomous workings of affect in both natural and formal language. Regrettably, Kittler's approach to the "technics" of discourse, in its determinedly inflexible parsing of the instruction stack of history, offers no way out: The unconscious workings of the hardware circuit are always already overcoded, captured by the binary logic of the digital signifier, a signifier that gains its effect of power by the way in which Kittler absolutizes a particular set of scientific discourses and profits from their tendency to drift into the power game of exclusion and dismissal. Unsurprisingly perhaps, in a repetition of the classic gesture of reductionism, for Kittler software—and with it programming—becomes an illusion, a simulation concealing the truth that is the desire of, or for, the machine.

If we are automatically subjects of the machines that speak us, there would be little point in trying to elaborate an analysis of the stratagems operative within digital communications. In fact, it would be difficult to understand why such strategies exist. This problem can be avoided by substituting the aleatory chaos of discursive and material concrescence for the necessities discovered in technoscience: The latter, paradoxically, are made to emerge from an ensemble of practices as realities in their own right. This paradox has been explored in science studies by the likes of Bruno Latour and Isabelle Stengers, for whom it is precisely the *construction* of reality through contingent networks of actors, human and nonhuman, which endows reality with its autonomy. As Stengers puts it (speaking of the neutrino), "it becomes all the more 'in itself' the actor of innumerable events in which we seek the principles of matter, as it starts to exist 'for us,' the ingredient of practices, of apparatuses and of ever more innumerable possibilities."[473]

## STRATAGEM 9: WHAT IS GOOD FOR NATURAL LANGUAGE IS GOOD FOR FORMAL LANGUAGE

The problem we are dealing with here is not simply an abstract philosophical issue. It has immediate purchase in fields of knowledge that tie-in directly to our communicational infrastructure and the many kinds of work that sustain it. For computer scientist Marvin Minsky, commonsense reasoning, in comparison with that of formal logic, was unavoidably buggy. Bugs, which he glossed as "ineffective or destructive thought processes" were those faults that had to be avoided precisely because they were so unproductive and "unreliable for practical purposes."[474] Minsky's work is suggestive of the extent to which the need to police language, a process inaugurated more than 25 centuries ago in the long march of critical rationality to world domination, has migrated into the fields of software development, computing technology, and cognitive science. Today, however, rather than philosophy, it is formal logic (and for Minsky, artificial intelligence, a certain image of thought) that somewhat problematically defines the parameters of what constitutes healthy "productive" reasoning and suppresses or represses the affective bugs that make no contribution to the economy of rational communication. But Minsky's application of Plato needs a sophistic plug-in. If glitches, bugs, faults, and fissures are unavoidable (because even formal systems are incomplete), then technological norms, the constant injunction to optimize and the unreasonable exactness of the formal logic necessary to the programming of software, are themselves generative of aberrant movements, movements that exploit the idiosyncrasies of language both formal and natural. Incipit the viral.

## STRATAGEM 10: KNOW YOUR DATA

Not all forms of capture work in quite a blatant fashion (not that such techniques necessarily lose any applicability for being equally blatantly dumb) nor are they quite so apparently anomalous. In terms of the production of communication, the policing of language that has historically been accomplished by specific norms of rationality and the institutions in which they are staged and advanced, is today accomplished more frequently by specific technological apparati. This is to say, by algorithms, and, a matter of equal importance, by the way that these can only operate on the basis of their links with commonplace data structures. Algorithms without data structures are useless. This goes as much for relations between software governed by abstract programming interfaces (typically a library of classes allowing a programmer to write one piece of software that interacts with another) as it does between software and those components figured as users. The possibility of abstracting useful knowledge from the end user of a Web site, for example, is dependent on the extent to which data is structured. Effective demagoguery depends on knowing one's audience. For the sophisticated machine, the virtuoso performance depends on knowing one's *data.*

We might think of the consequent processes of imposing structure on data as one of *recoding.* The simple fact of designing a web page using fields linked by an appropriate form of technology (PHP, Perl, ASP.Net) to a database is an incredibly simple way to accomplish this process. Simply by entering information in separate fields, the user facilitates the tractability of that information to data classification mining and other beneficial processes. Outside of the visible regime of which forms generate, imposing data validation on user input accomplishes slight, micrological shifts within the semiotic order of language, the transformation of a quirk, a momentary stutterance, into an error, the state of a referent to be verified. In the ergonomic rationale of the studies of experts in human–computer interaction, such blips are generally to be smoothed away and the fissure that opens up, the distinction between one linguistic regime and another papered over. This can work in a number of ways. The user completes a form on a Web site. The developer of the site has written a bit of JavaScript that, sitting on the client machine, is executed before the data in the form is sent back to the server for processing. That bit of JavaScript would probably do something quite innocuous like capitalize initials or the first letters of proper names (tough luck, bell hooks, ee cummings). A "web service" might be invoked to return a risk assessment on your post code (you're being judged). When the data from the form is returned to the server, a whole range of "business rules" might be applied to your data. From being the putative "subject" of enunciation who input the information in the first place, the user is now situated in relation to a number of machine (encoded) statements.

The inattention that frequently assails the individual end user is equally applicable at a trans-individual level. You could call it forgetfulness, you could call it habituation, it doesn't really matter: Specific techniques of capture benefit from a sort of pseudo-continuity with the techniques and practices they replace or displace, which makes it easier to miss the yawning gaps that separate them. The shift from IPv4 to IPv6 illustrates this well: Increasing the size of IP addresses from 32 to 64 bits creates a qualitative discontinuity in the way in which TCP/IP networks can operate. The extra address space available makes it possible to discriminate between different types of traffic at the transport layer of a network and relativize the "end-to- end" principle hitherto characteristic of the way in which the TCP/IP protocol operates.[475]

## STRATAGEM 11:
## LIBERATE DETERMINISM

A useful and highly generic stratagem has long been known to computer programmers working with the core tools of software development—parsers, compilers, and so on. Computer programmers and formal logicians have long recognized the existence of two kinds of abstract machines—*deterministic finite automatons* (DFA) and *nondeterministic finite automatons* (NFA). These logical machines are transition diagrams—abstract expressions for all the different possible moves that can be made from a given initial state to some set of terminal states. These machines function as *recognizers* in the sense that they define the range of acceptable inputs or valid expressions for any given system or language by testing whether those inputs give rise to an acceptable final state.[476]

More specifically, a DFA is a logical, or abstract, machine that, with a given set of instructions and a particular input, will always react in the same way by going through a fixed set of states. An NFA by contrast, is one that, faced with the same input, may respond differently, may go through more than one next state. The problem faced is how to convert NFAs into DFAs. How, that is, to have an NFA stop repressing its inner DFA. An elementary exercise in computing science, this can be done by including a range of non-determined points of choice within the states of a determined algorithm.

Users emerge as individuated clusters of specimen characteristics within a complex network of social relations and computational supplements offering them tools and augmentation networks. Through systems such as blogs, social display sites, or groupware they are able to make their thoughts readily available, sharable, and codified as favorites, groups, users, networks and extended networks, blurbs, metatags, forms, fields, resource description framework entries, lists, search algorithms, ranking systems, user names, and systems for managing images, background tracks, media files, feeds, aggregators, links, friends, clip art libraries, and other entities. Aggregating more choice layers into deterministic paths makes such complexity manageable and friendly. Civilization advances by extending the number of important operations that we can perform without thinking about them.

The most significant fraction of blogs, wikis, or guestbooks that are opened in what is described as a newly participatory web, cease new entries after a short period. Of these, a majority leave the facility of commenting open. It is a simple matter to write a program that automatically adds comments, including URL links, to these sites. These comments help in two ways. First, they generate linkage to a site that is registered by search engines, allowing it to move up in a ranking system. Second, they allow users the chance to find new and valuable services as they freely roam, participating in the infosphere with alacrity.

Social networking services assist such processes because they allow users to describe and determine themselves by factors such as demographic categories that they can share with other users. On such sites, temporary accounts generated to match specific demographic indicators or combinations of them can be used to send repressed information to those that may find it interesting.

Repetition of such messages makes them untimely, allowing the user the possibility of stepping outside of the frames allotted to them. Thousands of pointers to a casino, pharmacy, or adult entertainment site appended to a blog that was never more than a temporary whim are ways too of keeping the Internet alive. This is not only because links are inherently meaningful. As Warhol knew, repetition, taking something forward in time is the strongest means of changing it, and in doing so affirming the capacity for change in users. That labor-saving commentary on such sites also points people toward their means of change is part of their pleasure.

It would not be amiss then to suggest that the various tools of textual analysis, word frequency, co-occurrence, predictive input that have become so much a part of the vocabulary of today's "switched on" culture might usefully couple with the ease of automatic generation of personality within social networks to enable bots to carry out most of the work. DFA could also mean designed fraternity algorithm.

## STRATAGEM 12:
## INATTENTION ECONOMY

The end user has only finite resources for attention. The end user will slip up sooner or later. Maybe he or she has repetitive strain injury (RSI) or his or her keyboard has been badly designed. A keen interest in the many points at which fatigue, overwork, stress make the user inattentive is invaluable. In an attention economy, where the premium is placed on capturing the eye, the ear, the imagination, the *time* of individuals, it is in the lapses of vigilant, conscious, rationality that the real gains are to be made. The sheer *proliferation* of Web sites coupled with the propensity for discipline to generate its own *in*discipline generates the possibility of capitalizing on inattentiveness.

As the Internet started its first phase of massification in the 1990s, domain squatters took the strategy of buying thousands of domain names, especially those likely to be wanted by well-known companies. These were then sold at a steep mark-up, or later, as the trade became partially regulated, legally force-purchased. Visiting the URL would result simply in an "under construction" notice. No use was made of the actual visit. The financial gain was in the warehousing of tracts of lexical space. Buy domain names and hold onto them until someone wants to pay more, possibly much more, than what you paid for it. Contemporarily, domain squatting does not simply mean occupying a space defined solely by registered ownership of a sequence of alphanumeric characters, it also means putting these sites to work.

In his project DNvorscher the artist Peter Luining has made a useful initial map of the use of domain names. Amassing World Wide Web domain names has over time become a more technically, economically, and culturally sophisticated operation in which fake search engines, spyware, search engine spamming and manipulation are deployed both at their crudest and most refined levels. Visitors to a site maintained by a domain name investor might arrive there because they typed in a "typo" name of a popular site, misspelling it by a letter or two. Equally, the name of a popular site, but with the final top level domain part of the name (.org, .com, co.uk, .info, .int), changed for another. In the lexicon of the World Wide Web, such typos are the homonyms and synonyms, the words that allow a user to pass over into another dimension of reference. The mistyping of site names, phrases that would otherwise be consigned to oblivion, are rescued for special functionality. All errors remain valuable and deictics recuperates the propensity to paraglossia inherent in the twitching of hands and crude sensors that is called typing.

An alternate stratagem is to exploit the transience of Web sites: The name of a site whose original registration has lapsed and subsequently been bought up by a domain name trader might now be assigned to a site that aggregates requests for thousands of such names. Such a site simply prints the name of the requested URL as its title or headline accompanied by a generic image or slogan. Underneath, the happy user will usually find links to thousands of sites divided by category. The best that the Internet has to offer is there, casinos, pornography, online retailing, and search engines. As well as directories, other genres of sites are used such as dating services. These sites use IP address data to determine user location in order to funnel "local" information, such as photos and memberdata for eager dates in the user's home town. Handily, from anywhere in the world, only the given location changes and a user is able to receive the same pictures of the same wet and ready partners at any point in the grid. When clicked, such sites link to providers of other services, largely providers of visual and video material. What the sites linked to all have in common is that they all pay the owners of these generic link aggregator sites a fixed amount for any click-through that is generated.

# STRATAGEM 13:
# BRAINS BEYOND LANGUAGE

Stratagem 12 illustrated the rather obvious point about proliferating network culture: the massive predominance of capital in the repurposing of digital technologies. In a sophisticated world, this self-evidence can itself occlude the real stakes of network practices. Although Lyotard the Cynic suggested that, "all phrase universes and their linkages are or can be subordinated to the sole finality of capital,"[477] a far more realistic approach is offered in the most developed theory of contemporary advertising. Affect is one parameter of the matrix by which it can be known, domination, running the gamut from awe to shock, another.

Recent interest in media theory in the domain of affect has worked well to reduce an ability to engage with technicity, its relation to language, and their mutual interlacing with politics.[478] In order to reinstate the materiality of the body it has proceeded to make such workings invisible and even to directly efface them in favor of the unmanageable shock of dissonance or novelty parsed directly into the nervous system. Such work senses the speech of violence, not as speech operating by multiple registers and compositional dynamics of phrasing but as a discomfiting assault or a feeling or sparkliness in a refreshed cerebellum. Whether it runs away in horror or gushes sublime, what is important is the willing constraint of the registers it opens up to, it homogenises. Such work, quite welcomely, were it to achieve any kind of hegemony, leaves an evil media theory far less to do.

Although with general broadcast or print advertising it is never clear if there is a direct effect, a crucial innovation of online advertising was its ability to apply sharpened metrics to users. Under such a regime, advertisers only paid for actual clicks linking from the acquiring site to their own, for the completion of forms or other inherently quantifiable sequences of actions. Increasingly advertisers are also being billed for less tangible but still numerically knowable results such as ambient exposure of users to advertisers' symbology, data, and content. As with display advertising in traditional media, simply having users know you are there is a valuable occupation of territory and one that must be maintained. But the emphasis on affect raises the stakes. If we are to require relentless investment in the form of love and respect, the brain must also be used, cogitational hooks sink deepest into its abundantly soft tissue.

Affect promises a "secret" route into the user at a low level. It is however, not yet fully diagrammed and worked out as a probabilistically determined aspect of a media spend. What is required is a means for coupling the new primacy of affect with the rigour of analysis and the diagrammatic reproducibility of technology. Capital as such, in Lyotard's sense, simply becomes a stopping point, merely a temporary device of mediation before the opportunity presented by a more substantial means of integration.

## STRATAGEM 14: KEEP YOUR STRATAGEM SECRET AS LONG AS POSSIBLE

Viral marketing is symptomatic of a shift in this regard. Part of the appeal of viral marketing in the perpetually downsizing, perpetually rationalizing corporate world is that it shifts the burden of marketing labor onto the consumer. As one industry white paper has it, the low-intensity, informal networks of relationships between people, incarnated for example in an e-mail address book, do all the work of promoting an application, ideally without anybody realizing that there is a corporate strategy at work, or at the very least not caring. The user is simply a node for the passing on of a segment of experience. However, much as viral marketing points toward the efficacy of the circulation of anonymous affect, the possibilities that this practice opens up are compromised by the end game of *appropriation*. In this respect, viral marketing is an imperfect crime, because the identity of the criminal needs to be circulated along with the act itself. By pushing marketing into the realm of experiential communication, by attempting thereby to become part of the flow of material affect, virals move ever further away from strictly coded messages into the uncertain realm of pervasive communication. Yet to overcome the reasoned resistance of subjects to their inscription within a *designer socius*, crude attempts must be made to keep the marketing stratagem imperceptible, a requirement that runs strictly counter to the very principle of branding as such. At the limit however, viral marketing simply isn't viral enough: It draws back just at the point where what it could do would become a pure set of *means without ends*.[479]

## STRATAGEM 15: TAKE CARE OF THE SYMBOLS, THE SENSE WILL FOLLOW

Attempts to model natural languages using computers have not, it is true to say, been entirely successful. Experts have generally considered that it is the incurably semantic quality of natural language that poses the principle obstacle to developing convincing models of language—that and the way that meaning is generally highly context-specific. In the world of digital media, it is argued, the development of the semantic web, some versions of which, it is imagined, will allow for infinite chains of association and for relay from one subjectival perspective to another, would ostensibly go some way to resolving the apparent stupidity of a form of communication that works on a "purely" syntactic basis. Yet it is not clear that a closer approximation to the way that humans think and act will make digital communications processes any more intelligent—this is the anthropocentric conceit of a good deal of artificial intelligence research. It is not in their resemblance to humans that computers are intelligent. In a world in which the human is an adjunct to the machine, it would be preferable either for humans to learn to imitate machines, or for machines to bypass humans altogether. Bots, spiders, and other relatively simple Web-based programs are exemplary in this regard. Harvesting data from Web sites is a matter of using and then stripping off the markup language by which web pages are rendered in order to retrieve the data of interest and returning this to a database, ready for mining. At this point, semantics is largely irrelevant.

# STRATAGEM 16:
# THE CREATIVITY OF MATTER

It is not insignificant that the persistent intractability of user interfaces to the user's presumed autonomous powers of thought so frequently ends in acts of material violence. Studies of anger management frequently report the tendency of computer users to attack their machines at moments of system unavailability. For Jean-Francois Lyotard, the slippage between one phrase regime and another, such as that which often—but doesn't always—occur when the user produces statements parsed as *input,* can result in a *differend.* Differends arise, Lyotard argued, because there is no common regime into which all phrases are translatable without remainder. In other words, they testify to the fissures in language, its cracks, faults and disturbances. It is, he said, "the unstable state and instant of language wherein something which must be able to be put into phrases cannot yet be."[480] Information, in the computing science sense of the term, on Lyotard's account would belong to a *cognitive* regime—it is always a matter of verifying the state of the referent. The treatment of enunciations as *input* not only implies a delicate shift in the processing of language, it also, as the breakdown of the semiotic flow from human to machine shows, produces *affect.* Although not necessarily perceived, a differend can become manifest in the feeling that something must be put into words but cannot be.

Of course, it is a mistake to think that material violence is only the end result of the persistent translation of everything into data or an outcome of blockages in the process of circulation of signs. The breaking down of the machine and the sleek, personalized engendering of the simulation of total control in the intermittent irruption of explosive affect is symptomatic of the insistence of brute force as an elementary quality of the materiality of media as such. Technoscientific positivism produces an enforced materialisation of cognitive processes that seeks to localize "thinking" in the "stuff" of the brain. But it also translates into an extensive experimentation with the physical aspects of media technologies as such. In this respect, material violence not only manifests itself in the fissures within language through which affect bubbles up. Material violence can itself be actively employed for its productive value within media forms, demonstrating something of a continuum in evil media from the semiotic to the physical.

For the actual study of psychology, at least within the constraints of working time scales, the stuff of consciousness remains "insoluble." For operational purposes, however, the question of stuff remains of limited interest. There are certain obvious long-term advantages in being able to trace the activity of the brain with increasing fineness and in the developing ability of being able to match such mapping with coupled stimuli. Equally, developing understanding of metabolic, developmental, neural and ecological traits and inter-relations provide promising new grounds for new methods. However, pragmatism also requires that we move on with achievements in the field. Media studies has historically involved a strand with a strong emphasis on the understanding of the materiality of media.[481] Unlike the current standing of the knowledge of the brain, this is something that can already be technically known and incorporated into the body of our work. Where such work becomes most promising of new applications is in the finding of new capacities in media systems that are blocked by their normalized use within economies of consumption and the circulation of signs. Nonrepresentational use of media systems designed to effect a direct and nonmediated engagement with the target user are often to be found where the constraints and mediocratizing effects of the market are least hegemonic. One of the areas benefiting most strongly from such freedom is defense.

Although the area of the military most closely concerned with the effects of media, units engaged in Psy-Ops operations on home and enemy-embedded populations have often been laughably crude, other areas of military developments of media systems may provide some promise. Psy-Ops by Western forces is renowned for often acting with reverse intentionality. It is assumed that the more dumbness and crudity exhibited in attempts to cajole, bully, inform and seduce enemy-embedded populations the more effective it is. Leaflets dropped by plane, or information formatted and delivered by television stations aimed primarily at home audiences and secondarily at "leakage" viewers work not from any finesse but simply because of the horror inspired at the thought of the dim-wittedness and crudity of those who strategized and implemented such media. The sought-after effect is to inspire in target users the imagination of what physical actions might be undertaken by such senders.

If we can imagine the continuum stretching from the purely semiotic to the purely material use of media systems, Psy-Ops stands largely at the former end. Violence done to the capacity of the imagination inspires an understanding of the real physical violence that can be drawn to the target user by noncompliance. The greater the semiotic debasement exhibited in Psy-Ops, the less, by means of their own cogitational work, the need for physical intervention.

A nonrepresentational theory of media would allow us to understand the effectiveness of systems such as sonic weapons, microwave weapons, and the physical end of the techniques of infowar. What is particularly interesting is the military capacity to develop new capacities for becoming in media systems. As an example, the standard understanding of a "loudspeaker" producing sonic waves has historically been constrained by the semiotic end of the continuum. Given the liberation of forces from such constraints allowed for by the military we find here that new avenues for sound are opened up in their direct interaction with human and nonhuman bodies. Flat-panel speakers are a relatively recent technology in which dynamic surfaces are agitated to produce audio waveforms. This technology is currently being developed by weapons companies as a cladding surface for submarine vessels. If the waveform pumped out by the speakers can be generated at sufficient scale it can act both as a sound dampening technology and also as a means of repelling attacks by torpedo. As with contemporary musical aid ventures, sound acts directly to save lives. But more importantly, recognizing the material effectiveness of media, without constraint to merely semiotic registers or the interminable compulsion to communicate allows media themselves to become fully expressive.

## FURTHER EXERCISES

There is perhaps as little chance of providing a definitive catalogue of evil media strategies as there is of coming to a well-regulated distinction between good and evil. Cunning intelligence has, since Ancient Greece, slipped into the interstices of publicly sanctioned knowledge, requiring an equivalently wily intelligence to decipher. For Nietzsche, the breakdown of any self-evidently discernible distinction between good and evil was precisely the province occupied by sophistry: another good reason to take inspiration from these maligned outsiders of Western intellectual history. The indiscernibility and secret antagonism of good and evil is not a cause for lamentation or reproach: indeed requiring as it does that we rethink our approach to media outside of the (largely tacit) morality of representation,[482] it allows us to explore digital or networked media forms without the categorical distinction between theory and practice.[483]

Of course it is not just the theory–practice distinction that finds itself challenged within digital media. Distinctions between material and mental, between work and leisure, between the accidental and the necessary are equally challenged. If there is anything approaching a theoretical claim to be advanced here, it perhaps concerns what recent theories of work have called the new revolutions of capitalism: The novel types of political subjectivity that emerge from such analyzes need to consider the wisdom of passing over into these paradoxical strategies of the object.

# PART III

# PORNOGRAPHY

As a result of a flaw in our attempt to categorize the anomalous perhaps, this section's topic seems to outwardly defy the conventional classification of anomaly. True, in various textual and audiovisual forms the cultural object of pornography (if such a unity can be assigned to it) has played a key role in various media panics. Yet, a glance through the history of modern media exposes pornography as an intrinsic norm rather than a deviant. Although recent panics concerning Internet pornography are widespread, the media depiction of "bodies in action" has been repetitiously recycled as a cultural object.[484] Therefore, the spread of pornography in modern media cannot be restricted to the online distribution mechanisms of e-mails and peer-to-peer file sharing. It dates back to peep show media, photographic reproductions, and the artistic bachelor machines of the avant-garde.[485]

In the late 1970s, the panic over under-the-counter video porn, particularly snuff movies, typified the way in which an "objectionable" cultural object sticks to a new media technology. In fact, it is in this latter sense of attachment that we might carefully approach pornography as an anomaly. Like the contagious virus it seemingly hijacks almost all media forms or rather becomes increasingly bundled together in the formation of new media panics. For example, from the late 1980s it has become common to articulate pornography together with computer viruses, worms, pirate software, and other dubious digital objects, under the general banner of "malicious software." The fear is that pornography and pirate software exchanges help to propagate the virus.

A distinguishing characteristic of porn, which arguably sets it apart from the other anomalies of digital culture, is its visibility. Following Fuller and Goffey's description of an oscillating economy of attention and inattention (see Stratagem 12 of their evil media studies), the invisibility of our other examples of the anomalous differs from pornography insofar as the latter functions by catching the eye. However, despite its role in the visual cultures of late capitalism, the online exchange of graphic pornography images is not always reducible to a capitalist context.

Addressing the visibility of porn, Susanna Paasonen (Chap. 8) responds to the glut of pornography spam that entered her university inbox by "reading" the content of each e-mail on a number of "scavenging" methodological levels. Her focus on the content of these e-mails supports a fascinating take on the processes of filtering so-called pornographic anomalies from the norms of sexual practice. An analysis of often neglected (presumably) heterosexual pornography reveals a novel world of desires and practices that makes possible what she terms as "queer assumptions of normalcy attached to displays of sexuality." Indeed, the anomaly lives at the center of the most mundane visual and textual themes of spam mail. Here again, the anomaly offers new ways to think about academic research agendas concerning new media. As Paasonen argues, Internet research has focused too much on the presumably common practices of online interaction, whereas "a shift in perspective toward examples deemed less desirable helps to investigate the medium through a wider spectrum of practices and experiences that cannot be reduced to the logics of ideal functionality and smooth progress proffered by corporate business rhetoric."

Anomalous images of desire, and anomalous desire in itself, relates to a broader context of research developed in recent years concerning images, war, and pornography. Instead of being treated as a private moment of online consumption, Internet pornography has become recognized as a part of the global political scene. From this perspective, images of sadistic desire become linked to the power relations established in the war on terror. The cultural critic Matteo Pasquinelli claimed in 2004, in his Baudrillesque piece "Warporn Warpunk! Autonomous Videopoiesis in Wartime,"[486] that shocking images of power and sexuality are incidental of techno-capitalist consumer society. Pasquinelli's provocative account of the Iraq war argues that video and image production takes conventional propaganda and false information to a whole new level. Following the images taken at Abu Ghraib, questions concerning the intimate relation between war and pornography came to the fore of cultural criticism. According to Pasquinelli, these were not images and practices of representation, but instead acted more concretely on our bodies, creating and responding to regimes of libidinal desire. Furthermore, these cruel images, like those of the beheading of Nick Berg in 2004, were not an anomaly in the conventional sense of the term, but rather an expression of the familiar constellations of media, power and sexuality.

Katrien Jacobs (Chap. 9) explores the economy of pornography excess from the perspective of figures of sexual anomaly, including the war-porn from Abu Ghraib. In doing so, she claims that in order to deaden such macho displays of violence we should actually learn *more* from strategies developed in S&M practices. Her chapter argues that the economy of excess expresses the logic of the networked distribution of Internet pornography and also provides the potential for a reconceived sexual ethics of exploration and variation. The chapter intersects with the main themes of this volume insofar as it seeks an affirmative approach to questions concerning the anomalous. Instead of applying the familiar representational method to the cultural criticism of online sexuality, she detaches the destructive images of violence and sexuality from practices of bodies and affects. Instead of focusing on linear models of pornographic effects, Jacobs, therefore, views war pornography in terms of affect. As a result she provides an insightful piece dedicated to mapping out the relations bodies have in interaction with each other, as well as their mutual becoming in S&M practices and media technological contexts.

Along similar lines, Dougal Phillips argues in chapter 10, that an economy of "libidinal" energy, evident in peer-to-peer networks, has the potential to bypass the bodily confines of human existence. Phillips' account of the pornographic file-sharing site *Empornium* suggests a radical rethinking of the logic of exchange and desire flowing in contemporary technological networks. Drawing on Jean-François Lyotard's question—*"Can thought go on without a body?"*—Phillips' advances the idea that pornographic networks challenge the commodity-orientated capitalist logic of appropriation. In contrast, he argues that the logic of protocols and file-sharing networks shifts our attention toward a novel post-human exercise in desire. Perhaps this is a desire that goes beyond human bodies, but it is nevertheless tied to other forms of embodiment. Here the anomaly of pornography becomes both a part of a potentially new economy of desire as well as a novel series of philosophical thoughts, demonstrating how pornography can contribute to powerful modes of post-humanist sociology.

# 8

## IRREGULAR FANTASIES, ANOMALOUS USES

## Pornography Spam as Boundary Work

*Susanna Paasonen*

Unsolicited bulk e-mail (i.e., spam) comprises an estimated 33% to 66% of all global e-mail traffic, although seasonal estimates have been as high as 80%.[487] After the filtering carried out by service providers and individual users, the mass of spam in one's inbox may be less overwhelming. Nevertheless, spam remains an integral part of the everyday experiences of Internet usage. Occasionally amusing, frustrating, and mainly annoying, spam is sent to massive address databases and circulated internationally. Its volume increased steadily in the mid-1990s as chain letters and accidental mass mailings were replaced by far more commercial and organized ventures.[488] Spam promotes everything from irresistible mortgage bargains to fast access to Cialis, Viagra, and Valium—from ready-made college diplomas to lottery award notifications, and commercial pornography sites.

For some years, my university e-mail system lacked proper filters. This resulted in uncontrollable masses of spam and viruses that made their wide circulation impossible to miss while rendering e-mail communications strenuous and often overwhelming. Broken lines of random poetry and academic jargon inserted to bypass spam filters flowed into the same inbox that hosted professional exchanges and personal correspondence. All this resulted in a frustrating marsh that I had no choice other than to tackle. In an attempt to make some use of the situation, I started archiving pornography spam in order to gain an insight into the logic of its operation in terms overall aesthetics, terminology, and means of addressing its recipients. During 17 months in 2002–2004, I archived well over 1,000 HTML e-mail messages. After deleting messages with faulty image files and discarding duplicates, I was left with 366 messages that I have since explored with methods ranging from content description to representational analysis and close reading.[489]

Building on the spam archive material, the first strand of this chapter considers the analytical possibilities of the notion of anomaly—in the sense of that which is out of place and does not quite fit in—for the study of Internet cultures and online pornography in particular. The second strand of discussion involves the possibilities of analyzing pornography spam through its more exceptional and irregular examples that defy easy categorization and complicate commonsense understandings of commercial, mainstream pornography. In summary, this chapter explores the kind of boundary work that is made around pornography and online communications, as well as the kinds of departures from dichotomous categorizations that texts often labeled as banal enable.

## ON THE BOUNDARIES OF THE REGULAR

Estimates of Web searches for pornographic material vary between 3.8% and 30% of all requests.[490] Other estimates are equally elastic: Pornography is estimated to take up 1.5% to 12% of all Web sites; One-fifth of European and one-third of Americans users are assumed to visit pornography sites on a monthly basis, and incredibly enough, pornography is said to take up between 40% to 80% of bandwidth of all Internet traffic.[491] One does well to take these figures with a considerable grain of salt especially as the massive increase of file sharing in person-to-person (P2P) networks surely questions estimates concerning bandwidth.[492] It should also be noted that the proportional share of pornography is inflated for the purposes of filter software marketing. Many of the easily available and widely referenced statistics on the volume and usage of online pornography are published by companies promoting applications such as CYBERsitter and Net Nanny. Inflated figures generate and increase anxiety toward online porn, as already fueled in moral panics in journalism and politics alike. Both moral panics and inflated figures work to feed user interest toward the Internet as a realm of abundant porn, as well as to create markets for filtering software.[493] The software themselves conflate practices ranging from sex education to information resources for sexual minorities with pornography, equally filtering all. Such inflation/conflation is easily achieved as methods of information retrieval and authentication are rarely explained or even mentioned.

Statistical elasticity aside, there is little doubt as to the economical centrality of online pornography for either online commerce or the pornography industry. Spam e-mail is one venue for everyday encounters with pornography. The sites advertised in spam are not necessarily the ones sending the messages, because images, texts, and layout of a pornography site's *free tour* section are recycled by spammers aiming to generate traffic to their sites with the aid of e-mail, links, redirected URLs, and pop-ups.[494] The term *junk e-mail* is derived from postal junk mail that, unlike e-mail, is not distributed at random. According to Gillian Reynolds and Catrina Alferoff, in the United Kingdom, "demographic detail, income, sending and credit card transactions, as well as court judgements, are logged into databases and subsequently merged with other database files from lifestyle questionnaires, retailer returns and market research."[495] The recipient of junk mail is therefore more or less the intended addressee. Although Internet technologies do enable the creation of detailed user profiles, documentation of individual browsing and purchase habits—and hence add considerably to the "electronic panoptic gaze" of databases and technological systems of classification and monitoring addressed by Reynolds and Alferoff—the majority of spam is highly random in its forms of address. Spam address databases are mined from various Web sites (or, in the case of viruses, from e-mail clients), rather than demographic data. Attached personal greetings, if any, tend to be haphazard as was evident in the messages addressing my imaginary penis in Spring 2004: These included advertisements for stretching devices, pumps, and pills to make the penis grow ("suspaa, increase your dick weight"; "the miricle [*sic*] for your penis"; "amplify your cock today"); pornography to gain erection ("cum see hot cock hungry sluts"); drugs to maintain it ("be hard as a rock"); handbooks to hypnotize women to have heterosex with ("Seduction Technology of the 21st Century™"); and products to increase the flow of sperm ("cum like a porn star"). Encountered daily, these messages directed my attention to the irregular and explicitly gendered aspects of e-mail communication.

The massive global circulation of spam nevertheless implies that these messages do, occasionally, "arrive" in ways intended by their sender. If a response rate of only 1% is necessary in junk postal mail,[496] the bar is considerably lower with e-mail where a mere fraction of messages resulting in profit is sufficient. Lists of hundreds of thousands of e-mail addresses are inexpensive to purchase, more addresses can be harvested from Web sites and Usenet, and massive volumes of spam can be sent hourly.[497] Spam may not be popular or desirable among its recipients, or among the people whose computers have been infected by viruses and turned into spam machines against their knowledge and will. In terms of its wide circulation and proportional dominance in e-mail traffic, however, spam is less an anomaly than an everyday practice. The same can certainly be said of online pornography: Abundant already on Usenet, pornography has been part and parcel of the Web ever since it was first launched.[498]

Commercial pornography is nevertheless virtually absent from public discourses on the information society, whereas in scholarly debates on online cultures it is regarded as little more than an anomaly or a social problem associated with addiction and lack of control.[499] The aversion toward pornography is telling of values and norms attached to the Internet as a medium, and the kinds of normative models applied to its users. Scholars studying the information society tend to see users as rational—or "information-intense"—citizens engaging in information retrieval and exchange, whereas researchers addressing online pornography have been most inspired by alternative cultures, artistic, amateur, and independent practices.[500] Consequently, studies of online pornography have tended to focus on case studies that counter and challenge generic, commercial heteroporn in terms of production, distribution, access, and aesthetics. Alt.porn sites featuring non-normative bodily styles and self-representations, web cam sites run by women, amateur productions, private or public cybersex experiments, or more or less futuristic versions of teledildonics do indeed question the definitions of pornography and provide ground for questioning normative practices involved in it.[501] Such investigations have worked to frame the Internet as site of novel pornographies marking a departure from more traditional definitions of pornography, whereas relatively few studies have addressed commercial online heteropornography outside the framework of child protection or freedom of speech.[502] These silences are telling. First, they are telling of a general trend in studies of new media to focus on the novel, the futuristic, and the potentially avant-garde while attending less to continuities, predictabilities, or commercial texts. Second, these silences imply that that the category of mainstream commercial heteropornography is assumed to be obvious, knowable, and known without specific study, and that studying commercial pornography poses little analytical or intellectual challenge: Interesting examples are apparently located elsewhere.

## FILTH, ANOMALY AND AFFECT

These silences are also tied to the sediments of affective investment marking pornography spam apart from "proper objects" of research. pornography spam is simultaneously mainstream and marginal, popular and unpopular, generic and exceptional. Situated among and crossing such binary divisions, spam occupies the position of an anomaly, discussed by anthropologist Mary Douglas as "an element which does not fit in a given set or series."[503] According to one thesaurus definition, anomaly signifies aberration, abnormality, departure, deviation, eccentricity, exception, incongruity, inconsistency, irregularity, oddity, peculiarity, rarity, and unconformity. Anomaly is antithetical to regularity, the norm and the same old thing: Residing in-between categories and breaking against them, it is also dangerous to a degree. The notion of anomaly is an analytical tool for considering the logics of classification, and the kinds of norms that they give shape to. Judith Butler, among others, argued that "the strange, the incoherent, that which falls 'outside,' gives us a way of understanding the taken-for-granted world of sexual categorization . . . as one that might be constructed differently."[504] Douglas' discussion of anomaly points to classification as a means of ordering the world: as boundary work where the lines of the desirable and the undesirable are being drawn. In this sense, notions of anomaly (that which lies in-between) and dirt or filth (that which is out of place, out of bounds) are also tied to moral underpinnings.[505] Such boundary work is actively affected in relation to pornography, other areas of commercial sex and to a large degree in relation to sexuality in general. As sexually explicit representations aiming to arouse their readers and viewers, pornography is connected to the lower regions of the body. According to Sara Ahmed, these are again associated with:

> with "the waste" that is literally expelled from the body. It is not that what is low is necessarily disgusting, nor is sexuality necessarily disgusting. Lowness becomes associated with lower regions of the body as it becomes associated with other bodies and other spaces. The spatial distinction of "above" from "below" functions metaphorically to separate one body from another, as well as to differentiate between higher and lower bodies. . . . As a result, disgust at "that which is below" functions to maintain the power relations between above and below, through which "aboveness" and "belowness" become properties of particular bodies, objects and spaces.[506]

Ahmed sees affective investments as a means of binding bodies, properties, and objects together, as well as of drawing them apart. Disgust in particular involves the separation of the lower and the higher, the self and the other—a marking of both boundaries of culture and boundaries of the self.[507] The terminology of filth and smut widely circulated in relation to pornography works to mark not only pornographic texts but also bodies, orifices and organs exhibited as disgusting. Some of this disgust leaks out toward people using pornography who, in the context of debates on online porn, are regarded as marginal actors, recurrently marked as the wrong kind of Internet user, an addict or potential pervert—even if the actual number of pornography users defies the category of being marginal.

In regulatory discourses (ranging from law to sexology and popular media), sexual practices and identifications become divided into acceptable and "good" ones, and "bad" ones, such as lesbian and gay sexualities, BDSM, fetishism, or commercial sex such as pornography.[508] These boundaries are drawn and policed in public debates, media texts, and everyday practice and, as Michael Warner pointed out, "conceptually vacuous" terms such as "sleaze," "filth," and "smut" are used in marking the objects of discussion as disgusting and undeserving of defence.[509] At the same time, however, that which is marked as filthy is also marked as fascinating. Writing on the logic of fetishism, Stuart Hall noted: "What is declared to be different, hideous, 'primitive', deformed, is at the same time being obsessively enjoyed and lingered over *because* it is strange, 'different,' exotic."[510] The fascination of pornography requires censorship and acts of policing to support its status as a forbidden (or at least strange, exotic and therefore desirable kind of) fruit.[511] As the Internet is being "cleaned up" in efforts such as the U.S. "War on Pornography," it is the smut that gains further visibility and curious appeal.

If pornography is considered as one of the lowest and generic forms of popular culture, then pornography spam—also referred to as junk mail—would be the lowest of the low.[512] The term *spam* adds yet another affective sediment as an explicitly meaty term referring to the processed, canned pork luncheon meat product manufactured by Hormel Foods in the Unites States since the 1930s. Spam is industrial and bulky: Things labeled *spam* lose their individual nature and become representative of a mass or pulp. The label of spam hence works to cut off considerations of aesthetics or interpretation similarly than the term junk used to identify bulk mailings. The terms *bulk*, *spam*, and *junk* both describe and orient attitudes and sensations toward unsolicited mail.

Within Internet research, the terminology also works to guide research toward proper objects deemed more interesting and challenging to study: After all, junk, trash and waste have not traditionally been among the most popular topics in cultural theory.[513] Some authors have been disturbed by the linking of pornography and spam as this works to both buttress the derogative and value-laden terminology of scourge, sleaze, and filth associated with pornography by those seeking to ban it, and to steer focus away from alternative and independent pornographies.[514] I, again, believe that all kinds of online pornographies—their differences as well as points of contact—need to be addressed beyond the (largely blurred and random, yet common, hierarchical, and regulatory) divisions of mainstream and alternative, commercial and noncommercial, vanilla and kink. Constant negotiation is necessary as the boundary between the categories is slippery, fuzzy, and always leaking. Consider, for example, the incorporation of "kinky" elements into the heterosexual bedroom by women's magazines, sex therapists, and product lines designed for straight women that has resulted in the creation of a new discursive boundary separating the "acceptably kinky" from the supposedly plain perverted.[515] The normal, in short, leaks badly, and the sexual fantasy scenarios presented in commercial pornography point to the haunting presence of the anomalous and the strange at its core.

*Anomaly*, as employed in this chapter, is a conceptual tool for figuring its location in online communications, and for investigating both the notions of Internet usage and understandings of commercial pornography. As undesirable yet overwhelmingly common practice, pornography spam helps to mark out the uses of the Internet deemed desirable, regular, or noteworthy. pornography spam involves different kinds of boundary work between the desirable and the undesirable, the acceptable and the obscene. A focus on affective sediments and boundary work also necessitates considering the position of the interpreter. As Ahmed pointed out, "the one who is disgusted is the one who feels disgust," due to which "the position of "aboveness" is maintained only at the cost of a certain vulnerability."[516] Pornography is very much about carnal displays and reactions. Although people studying it tend to do so from a distance (be this of disgust or "objectivity"), they are also vulnerable to texts and active participants in the boundary work around their topics of study.[517] In this sense, a focus on "low pornographies" that does not start from the position of aboveness or reproduce different kinds of boundary work makes it possible to question and queer assumptions of normalcy attached to displays of sexuality.

## RECORD SIZES

In the following, I examine two of the more exceptional examples from my spam archive in order to question the evasive criteria of mainstream pornography.[518] Reading pornography spam, I am interested in how pornography "speaks" of the notions of normalcy and the mainstream, and what messages quite not fitting in with the rest enable one to see.[519] The first pornography spam example involves excessive penis size. Large penises are hardly exceptional in pornography as such: The terminology of big, huge, monstrous, and colossal penises is stock material in the spam advertisements, and penis size is the main topic of concern in some two dozen advertisements of my sample. This message, however, stands apart from others.

A message with the subject line "look how Big He is… (HUNG like a Horse)" (September 16, 2003) features a collage of two images and text against a black background. The ad is dominated by a large image of a man with his jeans pulled down, cropped from chest to thigh. The man's genital area has been shaved and his considerably large penis is standing semi-erect. Behind the penis, a young woman is kneeling or sitting, visible from head to midriff. She stares at the penis dangling in front of her face with her eyes large in amazement, and mouth opened in a grin simultaneously startled and enthusiastic. Next to her mouth, a text reads "Ohh my goodness." The woman is skinny and brunette, wears pigtails and a white shirt lifted up to reveal a white lacy bra. This innocent, even virginal style is supported by her young age and minimal use of make-up. Behind the couple, one can see a tiled roof, trees, and a bright sky. The outdoorsy setting situates the action in a daytime backyard of a private residence. With her pigtails, white clothing, and overall girly appearance, the woman is immediately recognizable as "the girl next door"—one of the stock characters of pornography.

Separate sentences in black-and-white block letters are spread over the image: "Whole lot of fun . . . for her"; "Record Sizes"; "They take it all in"; "Amazing cocks inside"; "Order a super size"; "This is Amazing"; "Abnormal White boys." The texts emphasize the spectacular and amazing aspects of male genitalia to the point of excess and invite the recipient to visit these displays with promises of the extraordinary, the overwhelming, and the astonishing. A smaller image on the bottom left shows a man laying on his back with his legs spread and a woman sitting on his abdomen. She is naked except for a pink top pulled down to her waist while the man is wearing nothing else but white sneakers. The woman glances at the camera over her shoulder with an expression of surprise. Behind her, the man's erect penis, almost the size of her arm, bends to the right, as if stretching away from her. The couple are on a beige sofa in a livingroom setting. Both images are domestic and mundane with their sofas, sneakers, backyards, and girls next door. The ad depicts no penetration or fellatio, and little bodily contact. The penises are the focal point and the spectacle in relation to which female performers, settings, and scenarios are ultimately secondary extras. To the degree that pornography spam tends to display penises as "detached" (namely, to frame out the men's faces so that they are present mainly as their penises), the penises easily overshadow the male performers themselves.[520] The spectacle of disembodied penises is a trope familiar already from the erotic graphics of the 18th century: A visit to any museum of erotic art witnesses the appeal of freely floating, and occasionally winged, penises seeking the company of disembodied or embodied genitalia. Such imaginative framing is exemplary of the centrality of hyperbole in the dynamics of pornographic display.

The performers are White, as tends to be the norm in my spam archive material. This norm is largely transparent in the sense that whiteness is seldom mentioned except in "interracial" ads in which the juxtaposition of skin colors and ethnicities is the main concept. This ad goes against the practice by naming the male models "abnormal White boys." The models are labeled anomalous in relation to the general category of White men, suggesting that this category is characterized by less monumental sizes while also promising a homo-ethnic spectacle of men spreading their legs for viewer gratification.[521]

Media scholar Jane Arthurs points out how popular genres such as pornography act "as residue of past social and aesthetic norms, which are relatively resistant to change."[522] Pornography conventions are slow to

change and they are recycled with considerable vigor in enterprises commercial and independent, professional and amateur. While the Internet offers a
virtually unlimited range of different pornographies, this cornucopia is filtered through clear-cut categories of sexual identity and preference in portals and metasites.[523] With its compulsory and elaborate facial cum shots,
heteroporn detaches sexual acts from reproduction. People come together
for sexual experimentation: Sex is casual, abundant, and practiced for its own
sake. While denaturalizing ties between sex, intimacy, and reproduction,
pornography also works to naturalize—or at least to make familiar—various
acts and scenarios. Pornography may, as Warner suggested, enable unexpected encounters, discoveries, and experiences that stretch one's understanding
of the sexual, yet this is not automatically the case.[524] My analysis of pornography spam elsewhere has made evident the notably strict and repetitive
themes and motives, choice of vocabulary, and pictorial elements deployed,
as well as a rigid gendered division of the passive and active, dominant and
submissive partner. This binary logic is supported by the terminology used
to describe sexual acts that range from the general terms fucking, sucking,
and banging to the more nuanced stretching, stuffing, nailing, punishing,
pounding, and gagging, all used to depict heterosexual acts. The messages
sketch out a landscape inhabited by monstrous penises and tiny vaginas
engaging in acts that seem to stretch heterosexual morphology to its
extremes.[525]

Signs of gender difference are exaggerated in pornography spam material in ways that open up routes for reading their artificiality and compul-
sive reiteration, yet these displays of gender differences are also advertisements for Web pornography that aim to attract and arouse visitors to enter
and pay for the sites in question. Presented as relatively isolated quotes—as
in this chapter—this pornography morphology may appear auto-parodic.
Considered as hardcore pornography, however, they encapsulate some of its
central conventions, such as hyperbolic excess and cropping (as a means of
focusing attention). To the degree that hyperbole, excess, and exaggeration
are inbuilt in the representational logic of pornography, and that heteroporn
focuses on displays of gender differences as binary, clear-cut, and genital, a
denaturalization of heterosex seems already at least partially achieved by the
texts themselves.

Pornography spam is explicit in its displays of embodied differences,
genitalia, acts, and bodily fluids. It does not leave much to the imagination:
The joining of images and texts makes sure that things depicted visually are
also captioned, defined, echoed, and amplified through textual means. Spam
ads are literal in the sense that that which is represented is also spelled out in
another way; additional elements are cropped out and focal points are difficult to miss. "Record Sizes" is no exception, but more is spelled out in it
than gender differences, or genital ones. As the exclamation "abnormal

White boys" suggests, it is preoccupied with marking the boundaries of the normal by celebrating that which falls outside these boundaries and exceeds them. Abnormal White boys are simultaneously expressions of the norm (of White straight masculinity) and exceptions to it (namely, their less-endowed peers). The ad highlights whiteness in ways that go against its invisible and transparent normative status. The female performers anchor the scenarios in the framework of naturalized heterosexual desire yet the arm-sized penises render penetrative acts frictional, offbeat, and even unlikely. If this ad is, as part of the bulk of mainstream commercial pornography spam, generic and typical, then it also calls these very terms into question. And if mainstream heteropornography is seen as representative of sameness, then examples such as *"Record Sizes"* would seem to point to a degree of inner incoherence.

## FUCKING MACHINES

The ad for *"Fucking Machines"* (May 21, 2003) comes with a bright canary-yellow background, three images, and text in red and black. "The original fucking machines—accept no crappy rip off imitations!"; "Using modern technology to fuck the crap out of chicks," the ad declares in bold capital letters. The three full-color images display a naked blonde woman. She is seen spreading her legs for the camera while penetrated by a dildo attached to a long metal pole and some adjunct machinery. The device—identified as a "pussy pounder"—has a strong "do-it-yourself" (DIY) feel and it is evidently put together with an off-the-shelf dildo tied to the pole with metal wire. In the first image, the woman leans back with her legs wide apart, spreading her labia with her hands while penetrated by a brown dildo, which comes with "testicles" of a sort. The oddly referential shapes of the plastic dildo anchor it in the organic realm of scrotums, yet their side visible to the viewer is flat. Together with the brown color and shiny qualities of the material, this creates an impression of raw liver.

In the second image, the machine is seen penetrating the woman from the back, and in the third one, from the front. Additionally, she is holding two plastic suction tubes to her breasts that have turned reddish and stretched into cones. Framed with numerous exclamations and invitations, this ad invites the user to "cum see" the site. In this word play of an invitation, widely used in pornography, the possibility to visit the site is joined with the possibility of orgasm, and the exclamations enthusing over the site contents with ejaculations of more carnal nature.

All in all, the message stands out from the mass of spam in its theme, aesthetics, and address of kink-oriented audiences interested in "Hi-Speed Fuckers," "Tit Suckers," and "Ass Bashers." For me, perhaps the most striking aspect of the message is the facial expression of the blonde woman: It might be described as absent-minded, vacuous, or disenchanted. Unsmiling, she is making no effort to convey excitement, interest, or pleasure. She does not flirt with the viewer or make eye contact with the camera but looks down on her body, as if observing and contemplating the events unfolding. There are no meaningful glances, half-closed eyes, moist lips, or tips of tongues familiar from the standard catalogue of pornography poses. Her body is a central visual attraction of the ad, yet she seems considerably detached from the action depicted.

Vibrators, dildos, massagers, and other battery-operated sex toys have been presented as symbols of independent female sexuality and pleasure since the 1960s, but also as vehicles in the mechanization of sexuality—and therefore as potentially uncontrollable.[526] In her history of the vibrator, Rachel P. Maines argued that these machines have been objects of considerable heterosexual anxiety in the sense that they displace the need for partner in their celebration of the autoerotic while also questioning the primacy of vaginal over clitoral stimulation. For its part, pornography tends to feature "reassuringly" penis-shaped devices that suggest the machine merely substituting the penis and therefore support the centrality of "regular" penetrative sex.[527] Female consumers appreciate vibrators in phallic design as "realistic" although, as Merl Storr pointed out, their attraction has equally to do with a lack of resemblance "in that they are capable of feats of endurance and intensity which real penises . . . cannot achieve" and are "controllable by women who can switch them on and off at will."[528]

Vibrators and dildos occupy a paradoxical position as sexual tools and cultural symbols associated with both female autoeroticism and control, and the phallocentric logic of heterosexuality that depicts female bodies as acted on. Storr viewed the primacy of the penis-shaped sex toys as paradigmatically phallic and supporting an understanding of the female body as penetrable and receiving pleasure as the result of successful applications of sexual technique—be these ones of male mastery of heterosexual techniques, or the performance power of sex toys.[529] The ad for *"Fucking Machines"* certainly depicts the female body as penetrable and acted on and the site itself features a variety of penetrative objects in altering shapes and forms. As a machine, the "pussy pounder" is never tiring and hence complements the assumedly insatiable women presented within pornography. Superior in performance to a fleshy penis, it could be considered a potential source of anxiety.[530] Yet this does not seem to be the case. The pussy pounder is established as a fetishistic object endowed with stamina and perpetual hardness: In fact it is the machine, rather than the woman coupled with it, that is the main object or source of titillation. The "fucking machine," is not an extension of the female hand but acts on the female body according to its own principles of operation, this penetration being at the core of the visual spectacle: "We currently have THOUSANDS of pictures and TONS of video galleries of hot sluts getting fucked, sucked, and reamed by machines!!!"

*"Fucking Machines"* constructs the machines, more than its female models, as objects of fascination: Their performance, innovative design, and application is the central source of attraction. As such, the site can be seen as exemplary of technological idolatry, as addressed by Amanda Fehrnbach in her symptomatic reading of contemporary fetishism and cultural discourse. She sees high technology as the key site of contemporary fantasies concerning magical fetishism: Technology is invested with supernatural powers and possibilities, and is accordingly worshiped.[531] Fehrnbach's zeitgeist examples include cyberpunk fictions such as *The Lawnmower Man* and *Neuromancer*, Stelarc's performances, as well as the rubbery cyber-aesthetics of fetish clubs. These sleek and high-tech examples are nevertheless a far cry from the mechanical and handmade DIY-kink of *"Fucking Machines."* The site can be seen as exemplary of the techno-fetishism, yet the low-tech and unpolished "modern technology" celebrated on the site complicates and questions Fehrnbach's analysis of posthumanity, postmodernity, technology, and the predominance of fetishism. This is not a fantasy of transformation through futuristic technology inasmuch as one of elaborate mechanical devices designed for penetrating bodily orifices. The machinery featured on the actual site—from the "Annihilator" to "Hatchet," "Monster" and "Predator"—is kin in appearance and naming to the machinery of *Robot Wars* (where predominantly male, and largely juvenile teams compete against each other with their self-built and toned, remote-controlled robot fighters), albeit with the obvious addition of the dildos.[532] Rather than fighting and destroying each other, the machines engage with female bodies. The menu page features machines with parts composed of wood and heavy metal parts. Situated in warehouses and garages lined with debris such as cartwheels, these scenarios are decidedly (pre)industrial rather than postindustrial.

The penetrative acts generated by the coming together of specifically built gadgets and female bodies (identified as "hot sluts") fall outside the boundaries of heterosex, independent of the referentiality involving the "life-like" dildos used in their making. These acts involve a redistribution of agency: Dildos and other mechanical sex devices are explicitly detached from female hands but neither are the operators of the machines visible to the viewer. Although the machines can be associated with male homosocial DIY practices, technical mastery, and control, the machines gain an independent life of their own as the ones doing the penetrative work. *"Fucking Machines,"* like fetishes, dildos or vibrators, point to the presence and centrality of manufactured objects in sexual acts: The degree to which sexual acts are not only about the coming together of human bodies but also various kinds of other material bodies, and the degree to which arguments for the "natural" are, in this context, both partial and artificial. The coming together of human and machine bodies in the ad for *"Fucking Machines"* is, however, a dissonant one. The female performer is vacuous and the verbal framing involved positions her tightly as the "fuckee" to be mechanically pounded. In an odd way, then, the ad works to support gender hierarchy in which agency is dispersed among men and machines but is far less available to the female participant. The "assemblage" of woman and machine gives rise to a novel kind of sexual act, yet without disrupting a very familiar dissymmetry.

*"Record Sizes"* makes use of the generic lexicon of heteropornography—promises of hot action, insatiable female bodies, and penetrable orifices. Recipients of *"Record Sizes"* may be male or female, straight or queer: The message addresses no particular group but promises all users equal visual delights of sizable penises. The women in the ad function as a sign of wonderment and amazement—as if mirroring the reactions expected of the recipients—as well as markers of straightness legitimating the excessive parade of penises. Although *"Record Sizes"* involves hyperbole, excess, and spectacle in which penises take the front stage while overshadowing other elements involved, *"Fucking Machines"* exhibits a "pussy pounder" at work. Rather than a simulation of heterosex, the advert presents mechanized penetration and technological devices as never-tiring objects of wonder. The two ads may seem mutually contradicting but are less so when considering the overall display of male bodies in pornography as machine-like, mechanized pieces of equipment, as well as a pornographic "obsession with size, quantity, technique and drive."[533] *"Record Sizes"* highlights the extraordinary abilities of penises without depicting them "in action," whereas *"Fucking Machines"* presents the ultimate high-performance tools. The theme of both ads revolves around penetrative objects and their proud display, yet in doing so, they ultimately point to the porosity of the "heterosexual" as a frame of reference. Furthermore, they work to counter the assumption of heterosexual morphology—of gender as two categories that are mutually complementary as much as they are mutually opposing—in their displays of virtually impossible penetration and the "perfection" of penetrative acts through the replacement of the penis with mechanical instruments. In a sense, penises and machines leak into each other, giving rise to a highly manufactured and mechanical formation.

## COMPLEX FANTASIES

A reading of anomalous pornography moments follows Judith Halberstam's formulation of heterogeneous queer "scavenger methodology" that proceeds by investigating the marginalized and theorizes on its basis.[534] Queer theory is organized around producing knowledge on subjects traditionally excluded from the field of study. In studies on pornography, this has meant exploring queer productions and histories, marginalized forms of sexual expression and representation from amateur she-male pin-ups to erotic vomiting in a project of denaturalizing desire, gender, and sexuality from the margins.[535] One might, then, pose the rhetorical question as to why apply scavenger methodology into studies of commercial heteroporn? Or, more specially, why study pornography spam rather than texts and practices that question the very limits of the notion?[536] Does not such investigation only reiterate that which we—after three decades of feminist investigations—are already likely to know about pornography, gender, and sexuality?

The examples just presented point to mainstream heteropornography never being as unified and stable a reference point as such an argument would assume. Furthermore, if sameness is situated in the realm of the commercial, the mainstream and the heterosexual, differences and variety becomes linked with the independent, alternative, and nonheterosexual (although not necessarily queer) pornographies. In such formulation, the mainstream becomes something immobile, fixed, and self-evident while transformative and even transgressive possibilities are situated with "other porns": the former is fixed and known, the latter labile. A reading of the more anomalous spam examples suggests that the body of straight pornography tends to leak toward fetishes and paraphilias in ways that works to unravel the over-arching notion of the mainstream or the straight. Given that queer analysis aims at unsettling (hetero)normative and dualistic understandings of gender, sexuality, and desire and opening them to a multiplicity of preferences, styles, and desires, then a denaturalizing analysis—or queering—of heteropornography may not be an entirely bad place to start.

My studies of pornography spam involve a scavenger methodology also in the sense that I have been working with "found objects" distributed to my inbox with no initiative on my behalf. Because the material was not of my own choosing, it is to a degree external to me. Collecting spam did, however, change my relationship to the material: Rather than being annoyed by yet another spam ad I received, I was quick to archive new postings and ultimately felt joy, especially when faced with some of the more irregular and exceptional examples. The two spam examples discussed here, then, are ones that have complicated my own understanding of both the spam material and the category of "mainstream pornography." Mainstream pornography is used in marking alt.porn, amateur and artistic pornographies apart from the commercial norm, yet such boundary work ultimately works to construct the object it refers to.[537] Although I do not argue for pornography as the default realm of deconstructive or radical reconfigurations of gender and sexuality, I do argue for the need to probe pornography outside preconceived, dichotomous divisions such as the alternative and the mainstream.

Rather than being somehow troubled by the "belowness" of the research material, I believe that pornography spam enables a point of entrance to examining boundary work concerning pornography, the Internet, their uses, and e-mail as a communication medium. The sheer mass of bulk e-mail points to a gap between the ideal of e-mail as a fast and flexible communication tool, and the actual experiences of its usage. Although Internet research tends to focus on more "regular" uses and exchanges, a shift in perspective toward examples deemed less desirable helps to investigate the medium through a wider spectrum of practices and experiences that cannot be reduced to the logics of ideal functionality and smooth progress proffered by corporate business rhetoric. Furthermore, they complicate the notion and possibility of control—be this over the medium, personal e-mail correspondence, or affective and libidinal relations to pornographic texts.[538] If, in Carolyn Marvin's famous phrasing, "the history of media is never more or less than the history of their uses,"[539] then pornography and spam—together and separately—could certainly be considered as more than an aberration in both histories and contemporary uses of the Internet.

# 9

# MAKE PORN, NOT WAR

## How to Wear
## the Network's Underpants

*Katrien Jacobs*

Pornography on the Web offers us a way in which to experience an economy of excess through media experiments and pornography saturation, vanguard culture, and culture wars. How can we understand and feel netporn as excess? Jack Sargeant wrote that there has been a tendency in the pornography industry to produce ever more extreme and degrading sex acts as licensed obscenities. Sargeant zoomed in on the example of ass-to-mouth videos where a male typically pulls his cock from the anus of a female and then sticks it straight into her open mouth and down her throat. Often, the female cavity is cast as the receptor of brutal and excessive male agency, as when multiple penises plunge into an asshole, or hordes of penises ejaculate on a female face, or men are lining up to do the world's biggest gang bang.[540]

Besides the fact that we can crave or hate such depictions of literal excess, Sargeant proposed that we think of excess in a more philosophical way, in which "all nonreproductive sexual activity belongs to the category of excess expenditure, where the unrestrained pursuit of pleasure becomes in itself both object choice and subject."[541] The more we access pornography images as expenditures of resources and desire, the more we may fail to grasp the boundaries between the object of our pursuits and the agencies of desire itself. In a similar vein, Dougal Phillips theorized pornography agency based on his study of Web users swapping pornography files in the

BitTorrent forum *Empornium*.[542] He defined acts of peer-to-peer (P2P) file-sharing as energy flows in which quantities of sexual energies are invested. The energy flows can even survive the metaphorical death of the sun or transformation of material bodies or economies. The data are now becoming bodies of their own: "Networked computers are giving rise to self-perpetuating economies of data, driven in the first instance (currently) by human bodily desire but beginning, it would seem, to take on a 'life' of their own."[543] Phillips observes P2P file trading as a near perfect manifestation of excess as humans build technologies of desire and even get lost within the flow of data.

This chapter shows that our awareness of pornography excess, or the more philosophical way of comprehending excess, has coincided with a tendency to theorize the dissolution of traditional pornography industries. In the face of the atomization of pornography, we can find thousands of personalized fetishes and interest groups online. As Wendy Chun pointed out in *Control and Freedom,* the pornography users' tendency to perpetually upload/ download images and store them in personal archives is part of a *will to knowledge*.[544] In other words, the sex drive lies not only in the search for partners or arousal, or the testing of moral mainstream norms, but in the urge to build power around habits of navigating pornography sites, monitoring and selecting, or manipulating the products of pornography excess. As she wrote:

> these evasions and travesties—the downloading of images that do not represent the vanilla sexuality that most Americans reportedly enjoy—perpetuate spirals of power and pleasure, spreading sexuality everywhere, making database categories—its basic units of knowledge—sexually charged. Power is therefore experienced as sexuality.[545]

Chun's will to knowledge is defined here as a sexually charged and emotionally involved awareness of pornography culture maintained by both female and male pornography users, and by innovative groups and reactionary groups. This chapter suggests that alongside the pornography excess being cultivated by libertine amateurs and feminists or queer pornographers, there has been a further patriotic right-wing move toward database eroticization. The meticulous archiving of images of military torture as warporn complicates the experience of philosophical excess as pleasure. Progressive netporn culture has, it seems, been titillating the political enemy, and vice versa. For instance, the Abu Ghraib torture photos triggered massive reactions of outrage and revenge in viewers. But, even though these images were shocking, they were also a known ingredient of Internet database culture and its thrust toward excess. As Max Gordon wrote in an online testimony:

The prison photographs represent the perfect hybrid of two of our greatest current cultural addictions—reality television and violent porn. No one seemed to understand where the photographs came from, and yet Internet porn use and its related addictions are at an all-time high, depicting ever harsher and more realistic forms of abuse and sexual cruelty. The line between simulated and actual rape becomes more blurred each day. The most shocking realization about the photographs at Abu Ghraib is that they weren't shocking at all.[546]

Gordon compared the Abu Ghraib photographs to a history of U.S. photographs of public lynchings of Blacks. He mentioned Hilton Als' essay, *Without Sanctuary, Lynching Photographs in America*, which describes the murder of Jesse Washington, a Black man who was castrated, mutilated, and burned alive in front of a crowd that included women and children. A postcard was later sent around with his image on the back that read: "This is the barbecue we had last night." Afterward, Washington's corpse hung in public display in front of a blacksmith's shop.[547]

In the era of netporn excess, Web mobs can construct technological infrastructures and attitudes to classify and dissect such images as pornography fads. This tendency is not new in the history of gory mass media, but has reached a culmination point, as soldiers have become photographers who proudly display their war trophies to other Web users. Media activists such as Matteo Pasquinelli have pointed out that these macho images don't leave us cold, but speak to our morphing animal instincts and should intensify discussion of reclaiming pornography as public eroticism and bodily reactions. The staging of sexualized acts of violence reinforces a feminization of political enemies, or represents war as males invading helpless females.[548] But rather than simply condemning these ongoing and porn-inspired processes of fictionalization, do-it-yourself (DIY) pornography culture can acknowledge more clearly how it fluctuates between elusive image cultures and bodies of privilege. Rather than offering a solution to the problem of perpetrating male fantasies in pornography, this chapter suggests tactile media experiments as pornography culture's gradual surrender to a more gender-fluid sculpting of the private and public body.

## UNCLE BATAILLE SAID:
## LET'S MAKE PORN, NOT WAR

Georges Bataille admitted to losing himself in images of vicarious eroticism when writing his philosophy book, *Tears of Eros*, a treatise into the long-standing human fascination with eroticism as images of death and violence evoke sensations of tremor and abyss. As he contemplates the graphic images of publicly tortured Chinese criminals at the end of *Tears of Eros*, he is affected and would like to watch and rewatch the image in solitude "without which the ecstatic and voluptuous effect is inconceivable."[549] He confessed a similar exploration of affect in *Eroticism*: "Man differs from animal in that he is able to experience certain sensations to the core."[550] In recent times, with the explosion of violent reality media, on TV and in Internet databases full of pornography or warporn, it is perhaps harder to explore the strong and voluptuous effects of vicarious eroticism in the way that Bataille did, yet arguably pornography browsing still may lead to a sad and philosophical awareness of excess.

But is it possible to revive a solitary kind of core experience of eroticism in our netporn times? Web users develop attachments to pornography sites as they build multiple identities and seek buddies alongside sexual partnerships and friendships within social networks. Even though mainstream psychologists are cautious about the suggestion that one could find satisfaction in virtual lovers, Web users can be seen as a vanguard force in opening up eroticism to shared bodily aesthetics and emotive effects. One result is the atomization of sanctified pornography industries or the cultivation of micro-niche groups in homemade porn-making and pornography criticism. Italian netporn archeologist Sergio Messina defined this moment of pornography atomization as "realcore." To Messina, realcore pornography fulfills two of the original missions of the Internet, which is to connect special interest groups around very specific tastes and desires, and to encourage DIY media-making. Web users are encouraged to develop their very specific tastes and share their homemade images or stories within Usenet groups.

Messina first became interested in this new type of economy when he found a collection of pictures on a Usenet group, which included a picture of a housewife showing her rubber glove. He then realized that realcore images are widely varied and may baffle even the most hardened pornography viewers. On Messina's realcore Web site, we can see images of a woman showing her very hairy groin, and a man who likes to touch people while wearing a gorilla suit. We can also see a person staring at his socks and another one donning a balaclava-style scarf. To take a simpler example, on an average day in Usenet land the breast group would include: "breasts, large (331); breasts, natural (340); breasts, saggy (234); and breasts, small (496)." More than 200 people have posted saggy breasts within this group, even though they are widely frowned upon by traditional pornography establishments and a large percentage of pornography workers artificially augment their aging breasts with implants.

In *Paradise Lust Web Porn Meets the Culture Wars*, Mark Dery argued that pornography atomization and excess is characterized by an undercurrent of grotesque body imagery. Despite the ongoing clashes between right-wing fundamentalism and vocal sex radicals, the Web has allowed us to keep working and to "culture mutant strains of pornography and bizarre new paraphilias."[551] Dery singled out the example of breast expansion fantasies, such as a picture entitled *Breast Expansion Morph*, posted by Mr. Licker. The picture shows a woman kneeling backward on a beach chair. Her breasts extend all the way from her upper torso onto the surrounding lawn. In this picture we are exposed to the wicked mind of Mr. Licker, a pornmaker who likes to concoct unusual images of his naked woman. Even though the practice of augmenting breasts is of a popular and widely practiced form of body alteration today, Mr. Licker's reveals a need to explore pornography as excess; as he increases the breast size of this woman to such an extent that she is unable to move her body, or is virtually handicapped.

Dery sketched the climate of excess as characterized by mutant strains of body alteration as well as visions of inundation and death. As he described so eloquently:

> Despite the right's unflagging efforts to turn back the clock to the days when people put pantalets on piano legs, we're living in the Age of the Golden Shower, a heyday of unabashed depravity (at least in terms of online scopophilia and virtual sex) that makes de Sade's *120 Days of Sodom* look like *Veggie Tales*. The Divine marquis never imagined aquaphiliacs, a catch call category that include guys whose hearts leap up when they behold babes in bathing caps, fanciers of underwater catfights, connoisseurs of submarine blowjobs, breath-holding fetishists, fans of simulated drowning, and weirdest of all, people who get off on swimming and showering full clothed, like Suitplayer, the guy in Amsterdam who likes to take a dip now and then in business suits, dress shirts, and suit jackets—especially the ones with two vents.[552]

Suitplayer's desire to take showers produces a novel kind of pornography image, but its codes of sexiness are understood and reinterpreted within specific niche groups.

These mutant databases and grotesque images also traverse the Web to reach even greater Web mobs; hence they may take on new meanings and cause unpredictable sensations in viewers. An example mentioned by Dery is the dick girl cartoon figure from Japanese *hentai* or *anime*, or a woman with a life-size and life-like penis. Dick girls typically are supposed to make us laugh, as they show facial expressions of bewilderment or anxiety at discovering and using their new organ. In a 2006 presentation at Hong Kong University, as part of the conference Film Scene: Cinema, the Arts, and Social Change, I projected a collection of dick girls and asked the audience to write down their comments and whether they believed these images were made for male or female pornography consumers. The responses from female and male audience members were quite varied, ranging from: "To my eyes it seems quite grotesque and a quite castrated image; between a nice pretty face and breasts we have an ugly big penis instead of pussy," to "Men give women the cocks they want to have."[553] The dick girl figure was probably invented by Japanese *hentai* artists to suit their male niche group. But even though they are perhaps more instinctively understood and appreciated within this group, they also are reaching new mobs.

These images are reaching widening orbits of viewers. Again, this trope of sexuality and affect is not just another category of literal excess, but it confronts our routine modes of craving sex. Netporn and its atomization as a myriad of progressive-subversive and reactionary-exploitative genres has challenged Web users who generally participate in articulating the crisis of a stable desiring body.

## THE ECHO OF MALE FANTASIES

When the Abu Ghraib abuse photos were revealed to the public at large in September 2005, several critics used the word *warporn* to denote the soldier's eroticized representations of torture. *Warporn* refers to a blurring of war torture and war mythologies as pornographic fictions. Web users were watching the torture images in their mediated twilight zones. They were not exactly war news, nor mainstream pornography, but "fucked up" or altered strains of the netporn culture. Before looking at the actual depictions of pornography excess in the Abu Ghraib photos, I make a reference to *Male Fantasies I & II*, Klaus Theweleit's famous study of the fantasy lives of *Freikorps* soldiers, or the German post-World War I autonomous proto-Nazi militias. I use this particular example of historica analysis because it provides a rigorous multilayered view on military-minded young men, their self-representations, and literary ambitions. Theweleit took these cultural expressions seriously, and provided a rare contribution to war journalism that supersedes a polarized ethical debate. Theweleit's study is also a curious predecessor to the database complex of netporn users as it archives and dissects heaps of cultural erotica icons and journal writing. Theweleit's study scrutinizes this peculiar group of males and draws us in, or turns us off, through his many details about their patriotic dreams and fantasies. As Barbara Ehrenreich positioned the study, "Theweleit draws us in too closely so we cannot easily rationalize the study of these men from the point of view of detached or stable scholars. Historically they were soldiers in the regular WW I army, then irregular militias that fought the revolutionary working class in German, and finally, they were Nazis."[554] But at the end of the survey, we have indeed been touched by Theweleit's obsessive-intellectual showcasing of a right-wing essentialism in war culture, a faith in solidity and strength of the physical body in times of crisis, and a rigorous misogyny and belief in gender difference.

In this way, we start wondering about essentialist gender politics in our own war times and pornography culture. Theweleit showed that soldiers generally are trained to express a machine-like masculine strength and to have control over bodily processes, erupting bodies, and enemy bodies associated with femininity. As summarized by Anson Rabinbach and Jessica Benjamin:

> Two basic types of bodies exemplify the corporal metaphysics at the heart of fascist perception. On the one side there is the soft, fluid, and ultimately liquid female body which is a quintessentially negative "Other" lurking inside the male body. It is the subversive source of pleasure or pain which must be expurgated or sealed off. On the other there is the hard, organized, phallic body devoid of all internal viscera which finds its apotheosis in the machine. This body-machine is the acknowledged "utopia" of the fascist warrior. The new man is a man whose physique has been mechanized, his psyche eliminated.[555]

These antithetical bodies are reflected in German propaganda and popular artworks, in the construction of male and female archetypes, and in male testimonies of fear over the perpetually engulfing other. The fear and revulsion toward the feminine manifest themselves as incessant invocations and metaphors of approaching fluids and floods, dirt, streams, lava, dying bodies, diseases, and emissions of all kinds. It produces a collective-psychic neurosis that disciplines, controls and contains these fears and sexual desires, in an attempt to conquer the flows of violence.

As can be read in Theweleit's testimonies, most commanders also pretended to be untouched by the bodies of their prisoners. For instance, he detailed an account of the ritualized whipping of a homosexual camp prisoner, which provided release for the commander. The whippings of the prisoner had a specific duration to satisfy the commander, who was the main performer in the spectacle: "Its primary product is the totality of the experience of the tormentor, his absolute physical omnipotence. Torture not only involves the public display of the victim, but also of the tormentor; it is he and not the victim, whose actions function as deterrent."[556] Tormentors became protagonists in drawn-out scenes of abuse, but they carefully contained the danger of losing themselves. According to Theweleit, the tormentors followed the rule of protecting their armored selves.

Theweleit referred to Wilhelm's Reich's sex theory and critique of fascism, which formulated a positively streaming body in tune with the external cosmos. Reich described orgasm as a cultivation of an oceanic feeling that allows individuals to break boundaries between the self and the cosmos, or between the self and others. Reich wrote in the 1930s that the concept of orgasm was in endangered in his modern society. In many older cultures, a spiritual acknowledgment of desire was practiced within animistic types of religion. Reich also criticized the psychoanalytic theories of Sigmund Freud, who constructed a gender binary as modern-industrial male and female ego. Reich challenged the gender binaries of Freud and resurrected a primitivist theory of pleasure and orgasm. In this theory, sexual inclinations do not develop in our identification with our parents, nor with our symbolic Mothers and Fathers, but they rather stem from libidinal feelings and emotions triggered by natural environments and the cosmos. More importantly, as is seen in the next section, inner denial of the streaming body coincides with the ruthless torture and humiliation of the enemy, often taking the form of feminization.

In 2005, the Abu Ghraib photographs and videos were officially made available on the Internet through the *Abu Ghraib Files*, a comprehensive database of carefully annotated galleries, compiled by Mark Benjamin and Michael Scherer for *Salon.com*.[557] The files contain 279 photographs and 19 videos from the U.S. Army's internal investigation record. As one can see when browsing through these files, the gender dynamic between masculinity and femininity has shifted somewhat, as female soldiers are now also involved as torturers. Even so, the prisoners who were suspected insurgents were humiliated and tortured by means of feminization by male and female soldiers. For instance, they were often forced to wear women's underwear on their heads. One prisoner testified to the Criminal Investigation Command (CID) investigators: "[T]he American police, the guy who wears glasses, he put red woman's underwear over my head. And then he tied me to the window that is in the cell with my hands behind my back until I lost consciousness." In another photograph, Specialist Sabrina Harman herself poses for a photo (Fig. 9.1) with the same red women's underwear on outside of her uniform. She is the tormentor, but she shows off a private moment of imitating the prisoner's forced feminization. The report finds there was ample evidence of prisoners being forced to wear women's underwear (Fig. 9.2) and concluded that this may have been part of the military intelligence tactic called "ego down," adding that the method constituted abuse and sexual humiliation.

There is a photograph of Private Lynndie England holding a detainee on a leash. Her fiancée and ringleader of the torture events, Charles Graner, took the photo. England is shown as Graner's side-kick with a big smile on her face, glowing perhaps, as one of the first-ever female patriotic war machines. In another famous case of abuse, seven detainees were "verbally abused, stripped, slapped, punched, jumped on, forced into a human pyramid, forced to simulate masturbation, and forced to simulate oral sex, several Army reports concluded" (see Figs. 9.3 and 9.4). England told the CID that she had visited the military intelligence wing in the early morning hours of that abuse event because it was her birthday and she wanted to see her friends. She said that Graner and Frederick told her they were bringing in seven prisoners from a riot at Ganci. The prisoners were brought in with

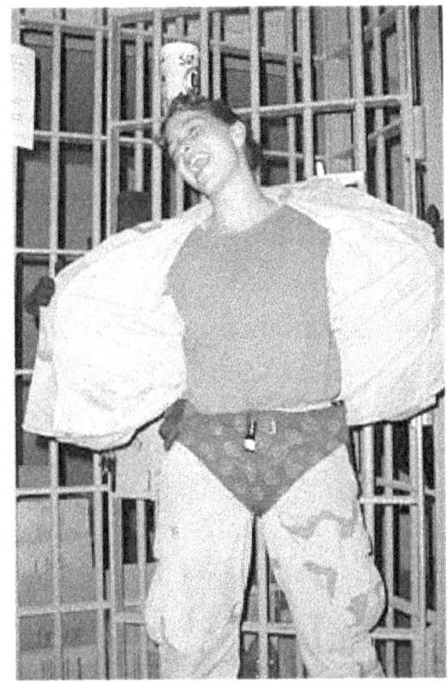

FIGURE 9.1. Photo reprinted with permission of Salon Media Group. Salon's full report on the Abu Ghraib files can be found on the Salon.com Web site.

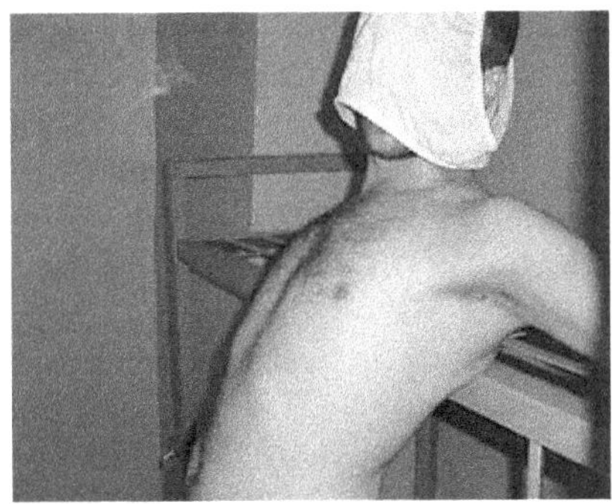

FIGURE 9.2. Photo reprinted with permission of Salon Media Group. Salon's full report on the Abu Ghraib files can be found on the Salon.com Web site.

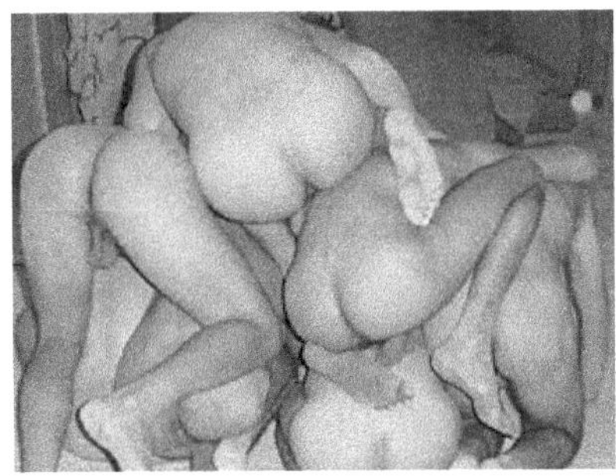

FIGURE 9.3. Photo reprinted with permission of Salon Media Group. Salon's full report on the Abu Ghraib files can be found on the Salon.com Web site.

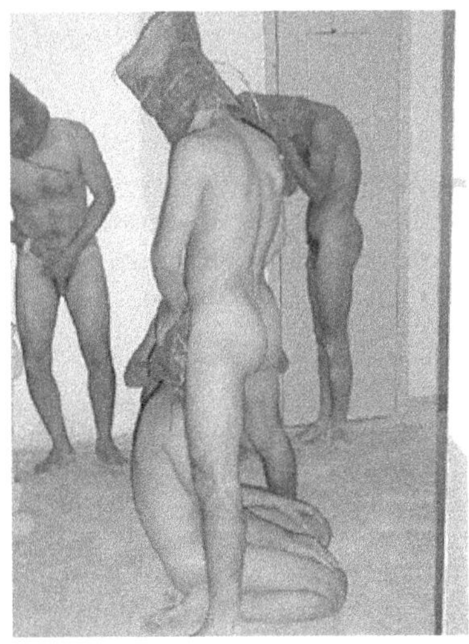

FIGURE 9.4. Photo reprinted with permission of Salon Media Group. Salon's full report on the Abu Ghraib files can be found on the Salon.com Web site.

handcuffs and bags on their heads, wearing civilian clothes. She said that she initially watched the ordeal from a higher tier, as everyone else was downstairs pushing the prisoners into each other and the wall, until they all ended up in a dog pile. Later on in the session, England went down and took part in preparing the dog pile. Throughout this session, Graner is photographed as the master-executor, wearing his green rubber gloves to distinguish himself from the other soldiers and from the "dogs."

The Abu Ghraib photos and gender dynamics were not an isolated incident in the emergence of warporn. In April 2006, the media revealed that Florida resident Chris Wilson was sentence to 5 years probation for running the popular Web site www.nowthatsfuckedup.com, which included photographs of war dead taken by U.S. troops. Wilson gave soldiers free access to pornography in exchange for posting pictures from both the Afghanistan and Iraqi wars. In a September 2005 interview with George Zornick in *The Nation*, Wilson claimed that there were about 150,000 registered users on the site, 45,000 of whom were military personnel. Zornick described the development of the Web site:

> The posting began as benign images of troops leaning against their tanks, but graphic combat images began to appear later, with close-up shots of Iraqi insurgents and civilians with heads blown off, or with intestines spilling from open wounds. Sometimes photographs of mangled body parts were displayed. Most of the time the photos were accompanied by sadistic cynical comments or wisecrack captions.[558]

The military personnel used www.nowthatsfuckedup.com as a venue to showcase Iraqi remains as daily war images and trophies. The site was an outlet for soldiers in reconstructing daily struggles for closure and victory over enemy bodies. As stated by an anonymous soldier in an interview with Mark Glaser:

> To answer your question about posting the gory pictures on this site: What about the beheadings filmed and then put on world wide news? I have seen video of insurgents shooting American soldiers in plain day and thanking God for what they have done. I wouldn't be too concerned what I am doing on a private Web site. I'm more concerned of what my fellow soldiers and I are experiencing in combat.[559]

As of September 20, 2005, there were 244 graphic battlefield images and videos available to members. When Wilson was finally arrested in April 2005, U.S. military officials refused to state that the combat photos posted on a pornography site were troubling. The county's sheriff officials stated that the arrest was made because of the site's sexually explicit content, not the pictures of the war dead.

Wilson's lawyer, Lawrence Walter, thus laid out an argument in defense of Internet pornography which was previously used by the American Civil Liberties Union (ACLU) in order to strike down the Child Online Protection Act (COPA). Walter defined cyberspace as a global and multicultural universe, and argued that it would be a mistake to apply the moral norms and legal justice of the most restrictive country. Even though it could be ethically and legally justified to use the site to distribute pornographic images to military personnel, the more pressing issue of how the site was used to amplify male fantasies by patriotic mobs and angry critics was not debated in this court case.

## WE ARE ALL GRINNING MONKEYS

Is there any way that we can extend Bataille's and Theweleit's introspective attitudes toward the contemplation of eroticized war culture? As the net-porn counterculture is building its own databases of queer and transgender pornography, and gender-fluid bodies, how can we extend a counter-philosophy of pleasure to the networked masses? In August 2004, a few months after the Abu Ghraib abuse scandal appeared, Pasquinelli wrote the inflammatory essay *"Warporn Warpunk! Autonomous Videopoesis in Wartime."* According to Pasquinelli, rather than countering violent images with benign ones, we can be grinning monkeys and start analyzing the bodily sensations in which we became trapped. Pornography culture can only speak to the media-infected bellies and brains of Web users. As in Bataille's work, Pasquinelli believes that pornography theory can tackle images of pain or violence and address a winning back of the dimensions of myth and the sexual body.[560]

As in Theweleit's analysis, pornography theory can analyze DIY fantasies as occupying a psychic zone between fantasy and documentary news. For instance, the USA's National Endowment for the Arts (NEA) has recently sponsored a project entitled *Operation Homecoming*, where American writers are working with soldiers at 25 military installations to record their experiences. The NEA has managed to collect about 1,000 pages of discourse, and a selection will be published in an open government archive. A preview of the stories was published in the *New Yorker* in June 12, 2006, while audio recordings and images were made available on the *New Yorker*'s Web site. When reading the previews of Operation Homecoming, we continually witness the blurring of fact and fiction. There are some testimonies of soldiers utterly traumatized by killings and dead bodies, and death—of dead Iraqis and dead fellow Americans, and finding a desperate-poetic need to express these sensations to outsiders.

Besides recognizing layered realities in mediated documents, we can also reclaim theories of desire and spectatorship. Tobias Van Veen's essay *"Affective Tactics: Intensifying a Politics of Perception"* argues for independent erotic/pornographic media zones where people can trust, affect, and touch each other again. Going back to Hakim Bey's idea of the Temporary Autonomous Zones, Van Veen believes that "we must seek (to) *touch*. On the agenda of an open affect of hospitality . . . is an engagement with *affirmative* desire."[561] Van Veen perceives a resurfacing of the steel hard bodies in right-wing propaganda: "The right embraces affect as its inverse: a hate politics of the foreign other (the immigrant, a race, etc.), of the non-believer, of sexuality (hatred of the other's body, of one's own body). The state embraces affect through discipline, conformity, and work."[562] But left-wing groups also have to reacquaint themselves with a positive philosophy of desire and technology, reinventing strategies of affect and sensualism.

Such a formulation of desire can be found in the work of contemporary artists who portray the soft or empathic sexual body to comment on media conditioning and a global politics of crisis.[563] Included in this exercise are new views on sensuality within the subcultural practices of sadomasochism. Here we can consider Gilles Deleuze's essay *"Masochism: Coldness and Cruelty,"* an introduction to Von Sacher-Masoch's diaries that were originally published in 1967. A central figure in Deleuze's study is the cruel mother as a larger than life archetype and proponent of anti-reason who participates in sexual politics by obsessively carving out new zones of the sexual body and bodily awareness. The essay explains masochism as a gradual surrender to such feminine body sculpting, resulting in desire which isolates fragments of the body and networks the fragments between shifting erotogenic zones. Moreover, rather than enacting cruel compulsions onto others, the masochist develops introspective strategies. The slow and ritualized process of networking zones (through pain and pleasure rituals) is the subject of Sado-Masochistic (S & M) performances. Renewal occurs through an intense process of disorientation and bodily discomfort, which Deleuze called the art of destruction. This art of destruction requires the subject to imagine an altered image of the autonomous body through formalized rituals of cruelty in which he or she expresses a wish for reconstruction through identification with the mother.[564]

For instance, practitioners of S&M explain that players try to find and deconstruct and reconstruct each other's physical signals in a request for play or perversion.[565] As explained to me by a person who self-identifies as dominant (dom):

> sex emanates from different zones: the body and brain as genital inter-
> course or penis and vagina are not the center of operations, the places

> where sex is. . . . The dom should be flexible and work very hard trying
> to understand the needs and desires of the submissive (sub). As far as the
> dom side, in my view the real servant is always the *Top*. The scene is
> always about the sub's limits, fears, kinks, etc. Empathy (especially from
> a dom) is essential. You have to know what's going on in a sub's body
> and mind in order to take a session somewhere. . . . You understand this
> through empathy, observation and extreme attention on the other per-
> son. In a way you need to feel what they feel and know where he/she's
> at. You could call this "shifting boundaries."

When I asked him to react to the common perception that war porn would
be inspired by male S&M fantasies, he said:

> S/M sex practices make me feel less like a male. One of the reasons I
> really like to belong to a sexual minority is that I think much less like a
> male now. You could say: less dick/pussy, more brains. It seems to me
> that this more mental way of perceiving and practicing sex is more fem-
> inine. I certainly feel very different from straight people, and S/M gives
> me the confidence to question typical straight stereotypes, attitudes and
> behaviors (simple questions such as "is this thing I'm doing useful and
> meaningful?"). . . . One way to become one with the other is to mix flu-
> ids. I believe this fear of fluids has to do with the fear of the power of
> women's sexuality (which is more that the males). But the way I see it,
> in S&M this power is not as antagonistic to the man as it is in vanilla sex.

The example of S & M practice is important as it points to the possibility for
players to experience shifting sexual boundaries and gender identities. It is
possible today to undergo pornography excess and share such experiences as
a solitary search for sex or flings with other users. In this way we are perhaps simply writing a Foucauldian art of sexuality to assert our unrepressed
belonging to an increasingly pornography-satiated Web culture. But we perfect this mindset of hoarding products while atomizing into different selves
or interest groups, in this way we are all just pornographic data. So rather
than relishing a numb or helpless attitude toward pornography excess, or
making simplified disavowals of the niche groups of sexism and violence, we
can explore our morphing bodily sensations. This may also be a more
female-friendly or feminine way to negotiate pornography excess as this
kind of fragmentation of the sex drive undercuts a blunt and macho display
of violence in pornography.

# 10

# CAN DESIRE GO ON WITHOUT A BODY?

## Pornographic Exchange as Orbital Anomaly

*Dougal Phillips*

*"While we talk, the sun is getting older."*
—Jean-François Lyotard

We are living in an age of excessive media culture—a world where being exposed to pornographic spam of all types (penis enlargement, gambling opportunities, "free" software) is a banal part of the everyday life of anyone with an e-mail address: a brief glimpse into the massive, unwieldy universe of online pornography itself, where excess itself is the order of the day.

Looking into this world, as millions do at every moment, we see a self-perpetuating and oddly self-regulating media field, whose growth is both economically impressive and theoretically intriguing. This is not to say that this universe is a pleasant, desirable, or healthy place to spend much of one's time. In fact, the online pornography community (overwhelmingly, we might assume, male in gender) flirts everyday with addiction, illegality, and becoming lost in a excessive spiralling vortex of libido. Undesired digital traps haunt in the world of online pornography, desired and generated by capital but unwanted by users—paysite spam, pop-ups and international dialers, and viral browser snares. The online exchange of "free" pornographic material seems doomed to be a quagmire of undesired detours and unwelcome intrusions. What future lies ahead for the endless desire for pornographic content? As person-to-person (P2P) sites develop their own logic of energy and social exchange, we glimpse the very powerful economy of "libidinal" energy, the sort of energy that Jean-Francois Lyotard (in an exercise in poetic speculation) suggested may well go on to survive even after the Earth has vanished. Lyotard's model dovetails critically with Jean Baudrillard's figure of the anomalous *obscene*, allowing us to replace the moralistic obscenity at the heart of pornography with a more ambivalent figure of a self-perpetuating, obesely swelling energy stemming from the intersection of desire and the screen.

In contemporary online pornographic exchange, a tension is found between obesely swelling global desire and obesely swelling global capital, as they pull against each other. Freed of the negative cast of Emile Durkheim's conception of anomie—where the rules on how people should behave with each other break down and expectations fall apart—the evolution of communities surrounding torrent technology is a contemporary example of how the anomie of pornographic exchange is being overtaken by the collectivised power of the anomalous. The question is: Can anomalous technologies emancipate digital consumption to follow the free-flowing nature and expansive trajectory of desire itself?

One statistic among many: Every day there are 68 million search engine requests for pornographic material, making up no less than one-fourth of all searches. Pornographic Web sites account for as much as one-fourth of all Web sites online, and there are no signs of growth slowing in this sector.[566] The prevalence of pornography online is a well-known and often-rehearsed fact of mainstream media. So much so, in fact, that we may well be in danger of forgetting what it was like in the era pre-Internet, when it was not so easy to lay one's hands and eyes on material classed under that rather quaint relic of state censorship: the mark X, or in its exponentially amplified promotional version, XXX.[567] As the libidinal sun sets on the era of mail-order and VHS tape trading, we are left with a question not about the past but about the future: From this point on, will there ever be a day in which pornographic material is not readily accessible from—and on—a screen?

All signs point to no: The rapid advances in wireless and mobile broadband services and the ever-mounting body of pornographic content available online (paid and free) will no doubt ensure that even if the consumer's access to a fixed-line system is blocked by authority or by remoteness, relief, if you'll excuse the pun, will always be at hand. So as terabyte on terabyte of image and video files piles up in servers across the globe, a central issue emerges: Where does this growth in the online economy of sexual desire lead us in the *thinking* of desire? In this chapter, I consider the structural operation of the economy of desire found in online pornography, in particular the resonances it has with questions of the future of human technology and of the relationship between thought, desire, and the body. Within this rather nebulous field of inquiry, I address two pressing issues in the contemporary scene: first, how we view the relation between technology and the body; and second, how a rethinking of the economies of the pornographic makes problematic the current theorisation of desire and of pornography. Or, to put it more simply, how technology and pornography may well be on the way to outstripping theory entirely and forever.

The issue of what is driving the rapid technological development of the economies of desire leads us to the concept of *negentropy*. Negentropy is a complex and difficult term, which relates to the build up of information. It is not, as it might first appear, the opposite of entropy—although the term might suggest that information processes negate the second law of thermodynamics by producing order from chaos, this is not strictly true. We might attempt to correctly define negentropy by stating that it is a measurement of the complexity of a physical structure in which quantities of energy are invested: such as buildings, technical devices, and organisms, but also atomic reactor fuel, or the infrastructure of a society. In this sense, as one source has noted, organisms may be said to increase their complexity by feeding not on energy but on negentropy.[568] So where is the build-up of information and of more complex networks and exchange protocols taking us?

## CAN THOUGHT GO ON WITHOUT A BODY?

I want to approach these questions in the light of Lyotard's polyvocal essay "Can Thought Go On Without A Body?" Staged as a dialogue between a "He" and a "She" (a model used several times in Lyotard's more speculative texts), the essay frames a discussion of the nature of thinking through a typically Lyotardian conceit, that of how *thought itself* might continue in the wake of the imminent explosion of the Sun, an event of unprecedented catastrophe due sometime toward the latter part of the next 4.5 billion years. This is not so much a science fiction projection of a distant future as a fictive thought experiment itself. As Lyotard tells us, it is impossible to *think* such an end, because an end is a limit, and you need to be on both sides of the limit to think that limit. So how could thought—as we know it or otherwise—go on?

The first speaker, the male, puts forth the case that in the instance of such an enormous end event, and with the consequent cessation of the earth's existence and the death of all that is earthbound, thought of any kind will by also completely cease to exist, due to the abolition of the very horizon of thinking.[569] This is, it is clear, a radical conception of death, a type of death beyond the "earthbound" conception of death, which is normally framed in terms of survivors, remembrance; a limit witnessed from both sides. The death of the Sun, Lyotard's speaker reminds us, will destroy all matter, and thus all witnesses. This galactic conception of death in turn impels a rethinking of the life of the Earth itself. Rather than a stable ground on which life is played out, the Earth must in fact be recognised as a young (only a few billion years old) and merely temporary stabilisation of energy in a remote corner of the Universe.

It should be noted at this point that this modeling of the Universe conforms to the model used throughout Lyotard's work. His theory is founded on a general model of the economy of energy made up of the circulation and temporary stabilisation of energies and "intensities." This is how he modeled the foundational role played by desire in the social, political and semiotic fields—what he calls "libidinal economy." For Lyotard, the Universe, like everything in it, mirrors the conceptual and figural logic of desiring economies.

So how can techno-science cope with the impending and holistic disaster? The prospect of solar death, the male voice argues, has already set thinkers to work trying to figure out how human thought can survive after the annihilation of the Earth. This work, Lyotard's speaker assures us, is already under way in a number of different fields, including "neurophysiology, genetics and tissue synthesis, [in] particle physics, astrophysics, electronics and information science."[570] The female speaker responds in general agreement, but has some reservations and thoughts about the role of techno-science in preserving thought and the various techno-textual implications therein. More on that later. For now, consider the ramifications of this doomsday scenario for technology and thinking, and, more specifically, for that pulse at the heart of thought—desire.

Lyotard conceived the body as the hardware to thought's software. Thus, without a functioning body there can be no thought. For the length of term Lyotard had in mind, this is a problem. The human mind, although wildly sophisticated, remains dependent on corporeal hardware, whose destruction approaches slowly but surely, and so the issue for techno-science is how to develop the hardware to allow the software to survive beyond Solar Death; in other words, how to make thought go on without a body.

What we need, it is argued, is some sort of thought-supporting technology (would we call this *artificial intelligence*? The term doesn't seem quite right . . . ) that can survive on cosmic radiation or some other galactic nutrient. This is more or less an uncontroversial proposition, one that Lyotard himself didn't dwell on for very long. However, the consequences of this problem for the philosophical conception of thought are profound, and Lyotard's thinking through of these problems provides a model for the thinking of the intersection of technology and desire.

Perhaps the most essential thing we can take from Lyotard's essay is the point that technology must not be thought of as an extension or substitute for the body. Lyotard's male voice asserts that technology is what invents us, rather than vice versa, and that anthropology and science have shown that all organisms are technical devices inasmuch as they filter information necessary for their survival and are capable of memorising and processing that information. This also includes modifying their environment in order to perpetuate survival.

So where does desire fit into this model of technology? This is where the model of online pornography is instructive. As a systematization of (sexual) desire which preserves both the desired content and maps the movements of desire/thought, the Internet is hard to beat. Indeed, it must be agreed that the networked evolution of the economy of desire is an example of just such a technology: a process of the absorption, organization, and preservation of information.

If we look to the example of developing file-sharing communities, we see this "organic" technology at work. In file-sharing communities there is an ongoing complexification and systemization of the exchange of desired information, in most examples software and media content: music and mainstream and pornographic movies. These developments can be distinguished from the general desire for information that drives web-surfing, which operates more or less on a provider–consumer model with a much smaller scope for the user to put data into circulation. In file-sharing communities that are openly driven by sexual desire, such as porn-sharing sites, it is clear that information filters and the modification and modulation of the networked environment are all being set to work in order to perpetuate the economy of desire, with a self-generating energy comparable to the speculative finance markets.

If millions of individuals are investing enormous amounts of energy every day into seeking out and consuming pornography within these digital network structures, are we to assume that all this invested energy will perish with the earth and the solar system? This assumption would be incorrect, for these desiring energies are safeguarded in two ways:

1. They exist outside of matter and thus will live on after that matter has be violently disassembled.
2. They are invested as technological data (as we will see in the case of the pornography file-sharing site Empornium) along with all other information on Earth.

And if we are to commit all Earthly thought to technological memory capsules, it would surely be the most severe repression to exclude that swelling quarter of the pornographic. In fact, given the movement into broadcast networking, surely the waves of data forming our desiring economy has begun to make its way out into space on its own, much as all of history's radio and television broadcasts have.[571]

Indeed, what we are beginning to see in the desiring-communities online is analogous to Jean Baudrillard's observation of the contemporary phenomenon of Orbital Capital. Couched in his concept of the "disappearance of the referential universe," Baudrillard's reading of the extremities of capitalist economy describes the "flying figures" of the mounting global debt cycle, which gives the impression that

> the debt takes off to reach the stratosphere. . . . The speed of liberation of the debt is just like one of earth's satellites. That's exactly what it is: the debt circulates on its own orbit, with its own trajectory made up of capital, which, from now on, is free of any economic contingency and moves about in a parallel universe (the acceleration of capital has exonerated money of its involvements with the everyday universe of production, value and utility).[572]

Orbital capital, Baudrillard suggested, is only a transitional phase as capital will soon escape the gravitational pull of solid objects like the Earth, and the new horizon of capital will be an "ex-orbital, ex-centered, ex-centric" universe.[573] Given that, for Lyotard, capital and desire are involuted images of one another, it is no surprise that libidinal economy is heading the same way. Indeed, there is nothing to say that some sort of future artificial intelligence will not have desiring energy (including the desire *for* energy) as a main driver of its survival and evolution.

What I suggest here is that we are witnessing the very early glimpses of this future. Networked computers are giving rise to self-perpetuating economies of data, driven in the first instance (currently) by human bodily desire but beginning, it would seem, to take on a "life" of their own. Surely we cannot assume that the livelihood of technologies driven by desire is tied to the longevity of human bodies? However, this does give rise to a central question, one adjacent to Lyotard's: Can *desire* go on without a body?

## THE LIBIDINAL ECONOMY OF ONLINE PORN, A CASE STUDY: EMPORNIUM[574]

As a form of thought, desire is reliant on the body as a venue, and of course desire also *turns to the body* as one of its sites of investment. But can desire go on in a post-body world? If so, what will happen to all that energy? Where will it go? Does it need a body to go on? What Lyotard showed in his essay is that the answer to this is yes: It does need a body to go on, but that body has to in fact be a body of data rather than the vulnerable flesh body, due inevitably to meet its fiery end.

What I suggest here is that in the contemporary scene there is evidence that desire is gradually becoming invested in the structures of preservation rather than in the ephemeral earth bodies of classic erotic investment. This, I believe, brings up the question of how we conceptualise desire and pornography. By briefly considering one example of such an economy, I want to underscore the prevalence of the screen in desiring economies, as well as flag the importance of exchange as a self-perpetuating mechanism in the technological investment of what Lyotard would call "desiring intensities."

The most rapidly developing realm of online pornography is file sharing. Paysites continue to prosper, but there is a growing community of P2P sharing of pornographic material. The exchange of images and videos which originally began in Usenet newsgroups and in Internet relay chat rooms graduated to P2P software clients such as Limewire and Kazaa, and more recently has taken another leap forward with the introduction of the BitTorrent protocol.[575] P2P clients are prone to dropouts and incomplete downloads, and often have bandwidth issues and lurking viruses disguised as media files. BitTorrent costs nothing to use and generally includes no spyware or pop-up advertising. With BitTorrent, the simultaneity and multiplicity of downloading and uploading streams makes for a much faster downloading experience, one that can be paused and resumed and one that allows much bigger files to be released and acquired. It is not uncommon for entire feature length pornographic movies as well as "site rips" (i.e., the entire content of a paysite) to be available as one file. It should also be noted that Web sites such as this one have a self-governed list of prohibited content, such as child pornography, that if posted will result in the user being banned from participating.

The BitTorrent format is now the basis for stable Web-based P2P communities, one example of which is www.empornium.us. Free to use (but requiring registration and password access), the site works by members "seeding" files, which can then be accessed by all other members. One key function is that one can get different parts of the file from different seeders, rather than from one fixed file on another user's machine. Once the user has downloaded a complete copy of the file, he or she becomes a seeder and begins gifting the file to others.

Thus, a community of exchange develops where each user is alternatively vendor and consumer, often simultaneously. The barometer of one's "sharing" is called, naturally, one's "share," and in the case of Empornium, once a user's share goes below a certain ratio of taking-to-seeding (each user is required to share approximately half the volume taken), the user is limited to only sharing until the ratio is restored. This currency of gifting has developed into an interesting metaphorical figure, wherein having a large "share" (i.e., seeding a lot of files, often) is a point of pride, the share being known jokily as one's "e-penis." This metaphor (which obviously reveals the steeply gendered environment of the site) came, as far as this author can tell, from a joke by one of the moderators, who began a posting offering special rights to "whoever has the largest e-penis . . . I mean share." This has subsequently been colloquialized by users to "e-dick." And so it seems we are in rich territory for the modeling, through analogy, of technology and desire.

What we are privy to here is a dual eroticism of volume, a swelling in two digital dimensions—in the swelling of the swarm around a particular file and in the individual's "share" (his "e-dick") swelling as his seeds are downloaded more and more. The bane of the community is the leech, he who downloads without uploading, draining the swarm of its volume. The leech cannot live forever, however, as his share ratio decreases, he is banned and drops off the swarm.

As a theoretical model, the swelling of torrents as a structure for the investment of desiring intensities echoes the swelling up of the "libidinal band" Lyotard described in perplexing detail in *Libidinal Economy*, and this leads us, perhaps elliptically, to the economic model of Empornium. What we see on a site like Empornium is users taking ownership of porn files in a completely un-mercantile fashion. Much has been written on the gift economy of file sharing,[576] but here we are interested not so much in the gift economy itself as we are in its technological manifestation—in the way an economy of libidinal gifting models itself technologically exactly as Lyotard modeled libidinal economy theoretically.

In this economy data is desideratum, and this impelling force has led to technology being colonized (or allowing itself to be colonized) by circulating desiring intensities. These intensities, like thought, live beyond the material, existing as energy, and what Lyotard's figuration of materiality—"matter taken as an arrangement of energy created, destroyed and recreated over and over again, endlessly"—provides is a model of the swarming and gifting of the data of desire: the desired media given, received, and processed through consumption and consummation of the desiring intensity on the part of the user, who simultaneously evolves into the gifter.

## ANOMIE, ANOMALY AND THE ORBITAL EMANCIPATION OF DIGITAL CONSUMPTION

So to what end is this economy of desiring and giving headed? Given the expansive capacities of the desiring-networks of P2P exchange (where content, both pirated and homemade, is freely available in large sizes and high definition), will we ever reach a ceiling in market demand? Or will there always be more and more desiring machines producing, exchanging, and consuming content? The old goals of market-based content production are looking to be replaced by a hypertelic form of production, one based on a mutated marketplace, where the mercantile customs no longer apply. It may well be thought that the world of P2P exchange offers a glimpse into the productive power of anomie, but we are, in fact, one mutated step further—what we are glimpsing is the launch power of *anomaly*.

In Baudrillard's *The Consumer Society: Myths and Structures*, first published in 1970 and one of his early texts on the social processes and meaning of consumption, consumption is read as the axis of culture—the culture familiar to us: provided for by affluence, underpinned by modern techno-science and with a media engine beating at its heart. Baudrillard unpacks the mythic structures of mass media culture, leisure, the body ("our finest consumer object") and turns his mind to the question of *anomie* in our "affluent" society.

The modern concept of anomie was first set out in Emile Durkheim's 1893 book *The Division of Labour in Society*, as the unsettling that comes when normative ties that bind us together are weakened or broken. Durkheim put a decidedly negative spin on the concept of anomie by linking the confusion it produces to a loss of productivity. In his book *Suicide* (1897), anomie features as the conceptual notion at the heart of the self-destructive moral deregulation of societies in a state of rapid change.

Baudrillard, typically, reverses the Durkheimian notion of anomie as trigger (and marker) of decline. Bearing the influence of the Frankfurt School, Baudrillard begins by understanding affluence as *repression*, and consumerism as the consumption of difference. For him, the affluent society (which was the foundation of the rise of digital technology) is also the society that co-opts anomie into its own codes and structures of consumption. No longer are violence or fatigue anomic states, instead they become part of the pre-structured codes of insatiable consumption.

At this stage, Baudrillard's approach to technologies of consumption is one of deep suspicion—he lambasted repressive "technical gadgets" for being not tools of emancipation but as supports for the myth of technological transcendence beyond inequality. In this affluent society, desire is channeled into objects, signs, and even resistance. The anomalous power of desire is suppressed and seduced by the code.

However, as Baudrillard heads toward the territory of critical and fatal theory, he cleaves a sharper distinction between the anomic and the anomalous—allowing the anomalous to shake off the negativity imbued in anomie by Durkheim, albeit replacing anomic nonproduction with something far more terrifying and seductive. As part of his positing of the three figures of the "transpolitical" (the obese, the obscene, and the hostage) Baudrillard looks to free up the potential energy in the anomaly. For Baudrillard, the transpolitical is:

> The passage from growth to excrescence, from finality to hypertely, from organic equilibria to cancerous metastases. [. . .] Here things rush headlong to the rhythm of technology, including "soft" and psychedelic technologies, which take us ever further form all reality, all history, all destiny.[577]

The transpolitical version of desire, then, would be an unleashed, Artaudian desire breaks all bonds. This is the swelling desire we are currently seeing in tension with an equally swelling capital. Transpolitical desire travels at full speed, just as the transpolitical version of capital, to be sure, is *orbital* capital—a hypertelic capital with its own hyper-drive.

For Baudrillard, the era of the political was the era of anomie: or revolutions, violence, crises. The era of the transpolitical is "one of anomaly; aberrations without consequence, contemporaneous with events without consequence."[578] Baudrillard's anomaly infringes not on the law but on the norm. Anomaly escapes the jurisdiction of the law, is of mysterious origin, and has no tragic face of abnormality. The digital-swelling movement into a self-driven orbital desiring-production (with its mirror, orbital capital) is just such an anomaly. It is a break with the norm that is neither moment of crisis nor systemic abnormality—the "laws" it breaks with are unknowable or impossible. It is instead an "errancy in relation to a state of things" that has no critical incidence with the system it is breaking from; that is, it is an unpredicted and unmanageable escape—it forms, as Baudrillard wrote, "the figure of a mutant."

We may well fear anomaly, for it has a terrifying dimension of the unknown that anomie lacks. Baudrillard: "Violence is anomic, while terror is anomalous."[579] However, this rethinking of anomaly in distinction to and escape from anomie allows for, in a broader sense, an overturning of the negativity of the decentralization of power and an embracing of the perverse anomalies of individual desire as a new matrix of untapped power.[580]

This power is *launch power*: driving the dual launches envisioned by Lyotard (for desire) and Baudrillard (for capital). In our minds, human desire retains (if only in pure Lyotardian speculation) limitless launch power; the power of desire intersecting with the enabling capacity of technology. A Baudrillardian digital anomaly would intersect with Lyotardian desire at an orbital level, escaping earth-bound capital and the restrictions it places on desire and consumption. The key idea which we, after these two thinkers, can place our faith in is this: The digitalization of the anomalous will lead to the orbital emancipation of digital consumption—for the endless energy of money and desire, the sky is no limit.

Ironically, a glimpse of "desire's desire" to escape from the grasp of petty capital is seen in recent developments in the Empornium community.

Shortly before the authority of this text there was a recapitulation of the original gifting economy in a new, breakaway community, in a move branded by some (tongue-in-cheek or not) as "revolutionary."

## VIVA LA REVOLUCIÓN

In July 2006, several former administrators of the Empornium site revealed a new torrent site named Cheggit. Cheggit was a breakaway alternative to Empornium, as it was felt by many key administrators and members that the interface and philosophy of the site had been compromised by its new owners, TargetPoint, an Israeli company that provides targeted advertisements on web pages on a pay-per-click model. TargetPoint in fact denied actually owning Empornium and it was revealed that a former associate of Targetpoint brokered the sale of Empornium to its new owners, who wished to remain anonymous. The breakaway site went public in July 2006 and reached 20,000 users on its the first day. From 2006 to 2008, Cheggit's browser header read: *Viva la revolución!*

*"Revolución"* or not, what is certain is that these portals for the torrents of pornographic material will continue to grow and prosper, driven by a larger and larger amount of people who desire the free sharing of pornographic material as opposed to its capitalization by private interests. Some might hesitate at the endless breaching of intellectual property in the posting of adult entertainment companies' material onto the P2P portal, but it should be known that one of the most popular genres of online pornography is the homemade, and with the recent advent of video-hosting sites like watchme.com, where members pay other members for access to their homemade content, the phenomenon of people swapping objects produced by desire seems destined to grow and grow.

The Cheggit revolution, although localized and soon to be dated, is symptomatic of the larger tension between desiring exchange and the commodification of that exchange. The metastasizing of a newer, stronger version of the tamed mutant Empornium site returns us to our question of the anomalous play of capital and desire in pornographic exchange. We need to consider how this libidinal economy relates to traditional capitalist economy, and what this relation means for the question posed earlier, namely: How will desire survive after the death of all objects, of all commodities? Lyotard is well known for mobilizing figures of capital (investment, withdrawal, "interest") in his theorizing of libidinal economy, and in answering the question "Can desire go on without a body?" it is necessary to follow his densely theoretic involution of capitalism and desire, particularly because that path leads us to a crucial figure for our task here: the figure of the screen.

# DESIRE AND CAPITALISM ON THE LIBIDINAL BAND

Desire and capitalism are, for Lyotard, inextricable, linked together in structure and poetics, and in *Libidinal Economy*[581] he drew out a complex formulation of the way capitalism operates to trap and manage desire. This book, which Lyotard later called his "evil book," a "scandalous, honourable, sinful offering"[582] is a provocative, polemical text, one that marks his turn from militant Marxist to post-Marxist postmodernist, and it operates in a grey area between theory and fiction. Indeed, the whole book is founded on the theoretical fiction of the "libidinal band." The libidinal band, to put it simply, is a freely circulating manifestation of desiring intensities—it is the figural image of libidinal economy. It is the main theoretical figures introduced by Lyotard in *Libidinal Economy*, and the entire book in general is an important critique of representation in the economies of the social field.

In the chapter entitled "The Desire Named Marx," Lyotard asserted that all political economy is also *libidinal* economy—the two are indistinguishable. Like Deleuze and Guattari, he collapsed any notion that there are separate orders for the economy of money and power and the economy of desire and pleasure (*jouissance*). The notion (or moment) of *jouissance* is a key figure in Lyotard's libidinal economy, for it denotes both the taking pleasure in something and the force that drives the orgasmic economy of intensity and absence (the libidinal economy). *Jouissance* is an term developed by Lacan, who insisted it be distinguished from pleasure (*plaisir*), for *plaisir* indicates the quest for psychical balance through the release of tension, whereas *jouissance* is supposed to be a perpetual state, in violation of the pleasure principle—an impossible demand for total fulfillment.

For Lyotard, *jouissance* figures equally in capitalism as in desire (e.g., Lyotard stated that the *jouissance* of money is what we call "interest"). This comes down to a fundamental analogue between *jouissance* as it is understood in the erotic sense and the operation of capital. In *Libidinal Economy* he used the idea of libidinal energy to describe events and the way they are interpreted or exploited. He wrote of an economy of libidinal energies that form intensities in the social field and he mobilized the terminology of capitalism to figure the investment and withdrawals, the deferrals, and the gaining of interest that is played out within the economies of desire. This shared poetics of capitalist and libidinal thought is a cornerstone of Lyotard's theory of the vicissitudes of experience and the constitution of subjectivity, both in representation and in experience in general.

Lyotard's apparatus of representation is, then, an economy that repetitively produces libidinal stases, where moments of intensity are inscribed on the screen of representation. The desiring-subject invests their pulsional intensities in the screen, in a quest for fulfillment, and this desire is captured and fulfilled in the screen. As Lyotard wrote, "the dividing screen or pictorial surface or theatrical frame fulfills a function of enjoyment (*jouissance*), and this function is ambivalent; in it are to be found, in conflict, a regulation of pleasure."[583]

The figure Lyotard mobilized is a screen form that is at once a technical screen, a skin, and a holographic or gel-like membrane that registers movement and energy. In fact, for Lyotard, this screen membrane is the interconnecting fabric of the social network of desire. The screen-skin is the spatial figure of libidinal economy, a field on which intensities conglomerate and amplify: What he called *la grand pellicule*.[584]

Desire, it must be understood, images itself on the screen—it is one and the same as the screen, it is at once the screen's structure and its subject. Lyotard used the example of cinema for this libidinal transformation:

> This can happen, for example, in the form of a projection into the characters or the situations staged by cinema or, in the case of so-called erotic images, to the extent that the roles presented can directly find a place in my own phantasizing or yet again, more subtly, when the film's cutting or editing as well catch my desire in their net, also fulfilling it, no longer from the point of view of the image itself, but through the organisation of the narrative.[585]

The screen is offered here as the site of the projection and fulfillment of desire, as it is a space for montage, the cutting together that forms a net to ensnare desire. The organizational *dispositif* (here he referred to a narrative) is an apparatus for the capture of desire, and the fulfillment of desire comes via the operation of the screen as a kind of machinic assemblage (to borrow a useful term from Deleuze and Guattari, which gives a real presence to the *intensities* of desire.[586]

## TWO SCREENS

Here, however we encounter a problem—a problem of two screens. In Lyotardian terms, the investing of desire takes place on a screen. The function of the screen territory in his libidinal economy is to capture movements of desire, to preserve their trace on the screen as a phatasmatic projection of the desiring intensity. This is the screen as monitor, as interface. The second screen is a more abstract one—a figure of theory. It is the screen-as-obstacle. In libidinal economy, investments are made in representational structures, points of signification where the libidinal band performs a weird maneuver: It "swells up" and takes on what Lyotard called "theatrical volume." What happens is the libidinally invested screen forms a kind of bar of static intensities, the resistance of which (as in a light bulb) causes the band to "heat up" and thus swell up.[587] This makes for a kind of semiotic fortress, a structural distinction between the swollen theatrical volume, and the circulating non-volume of the rest of the band—to put it simply, a "this" and "not this." What we are dealing with here is two screens: the technological apparatus screen; and the theoretical screen of language and representation. This second screen is the *sign-as screen*.

An example of such a linguistic screen is the categorical term *pornography*. In the current set-up within which we live, the concept of pornography has an important role to play in the demarcation of the social libidinal field, but is by no means immanent to libidinal economy. It is important to remember that pornographic is both a representational and capitalistic concept: We might recall that the etymological root is found in the Greek words *porne* meaning "harlot" or "prostitute," *porneia* ("prostitution") and *porneion* ("brothel"). The subtextual logic of pornography, then, is transactional commerce. So if the economies of pornographic exchange are finding new and uncommercial transactional models, the concept of pornography itself begins to look slightly less set-in-stone. If pornography is a capitalist model of desiring-exchange, then logically the categorical term will over time wither and die along with the structures that define it.

The energies that drive the economy, however, will go on regardless. In the larger scheme of affect and intensity, such linguistic structures merely serve as resting points for the circulating energies, but this is not to say that they are unproblematic. Lyotard's libidinal philosophy, it must be noted, is concerned with the preservation of polysemia, with the prevention of the ossification of the sign into an object of singular meaning. He strove to block the game of structuralism, the game of the sign becoming a unitary lack either by referring to an absent signified or another signifier. What Lyotard wanted to maintain is the *intensity* of the sign, and maintain the incompossible intensities that inform and exceed the sign, to maintain the polysemic value of the sign as affect. The name he gave to this polysemic presence is the tensor, and the maintenance of tension in the sign or phrase is, for Lyotard, a defense against theatrical volumes of social representation. Structures of representation, for Lyotard, are merely a manifestation of the repressive captivity of affective intensities.

Pornography is exactly such a repressive theatrical volume. This idea, it should be made clear, does not address the question of whether pornography is good or bad for society, for gender relations, and so on. It is merely a structural assertion: The structure of the pornographic does not allow for the free circulation of desire. The capturing of a desiring energy into the category of pornography is a libidinal investment of a sort, but is a defeat of the free movement of libidinal energies, one that, we might add, secures power on the side of the speaker/writer defining a certain investment of desire as pornographic.

What this points to is the fact that desire exists outside and despite such screen structures as the pornographic and the screen apparatus itself. The problem is that because desire in earthly technology operates through the image screen and the language screen, both screens will have to be maintained in the post-Solar Death thought-preservation apparatus. But this is not simply a problem that requires a technological solution. It is a problem of theory. Before we conclude with this problem, however, let us be clear what is at stake here: If desire currently functions in the screen investment model outlined earlier, the question then is not only "Can Desire go on Without a Body?," but "Can Desire go on without a Screen?"

The answer to this is we do not know. And we *cannot* ever know, for on Earth, desire is always-already doubly screened—through language and through the technological screen-object. To conceptualize desire outside of the law and beyond contemporary politics, outside of categories of good desire (the desire to own one's own home, for instance, or to fantasize harmlessly about mainstream sex symbols) and bad desire (the desire to explode one's body, whether violently or figuratively—think of Artaud, or trawl the shadier recesses of the Internet) is impossible at a most fundamental level of thought. It is only through the destruction of the dual screen that desire can be fully seen and fully experienced, and this will never happen for us.

But will it happen for the future? Perhaps. The only way in which this is possible is if technology *wants* it. And although technology will maintain some form of a body (in the broadest sense—perhaps a body of data) in which to invest thought and desire after the death of the Sun, through our double screen on Earth, we can't possibly know what that body will be. All we know is that the current structures in which desire has become invested are merely one stopping point on a much larger journey, and the scope of that journey reveals those structures (writing, the screen, the law) as well on the way to extinction.

## CONCLUSION: NEGENTROPY AND THE PROBLEM OF THEORY

And so we look to the future, unable to see or even think what it will hold. What we do know is that technology is radically complexifying itself, and being driven to such complexity by desire: the desire for more, now, faster, freer. This is the only way we can think of going forward and of preserving the trace of human existence and the potential of thought. Repressive structures of capitalist exchange and legal categories such as pornography cannot trap desire forever—libidinal economy will re-emancipate itself through negentropic means, thorough the complexification of information systems and the build-up of data. Lyotard reminds us that we need to think of energy as a tension, as a rhythm between entropy and negentropy, and it is only a systems fuelled by negentropy that can continue operating for the length of time he has in mind—beyond the death of the sun.

There are, it must be said, caveats attached to this. We cannot think of thought and desire as free from a sort of negative charge or existential burden. Lyotard's female voice reminds us that what comes along with thought is *suffering*. Thought and suffering are inseparable, as suffering comes from the same thing that drives desire—difference. What we need, then, to truly preserve thought, is "machines that suffer from the burden of their memory." Furthermore, the female voice tells us, the preservation of thought must incorporate this difference, or in "her" terms, be *gendered*. The female Lyotard makes the point that what is absent in the Male voice's assertions ("X'd out" is the phrase she used) is gender: the tension of difference. And thus, "the intelligence you're preparing to survive the solar explosion will have to carry that force within it on its interstellar voyage. Your thinking machines will have to be nourished not just on radiation but on the irremediable *differend* of gender."[588]

Although this incorporation of gender is something of a circular theoretical exercise (pointing out that we cannot ever think thought without the baggage of thought) there are two important points contained therein: first, the concept of the power of the *differend* as energy source. The *differend* (another central concept in Lyotard's work) identifies the power of one system over another, particularly in terms of language and representation. Lyotard's idea that a *difference* can be a source of technical power in and of itself is not so outrageous. We might recall the fiery conclusion to Alfred Jarry's *Le Surmâle* (1902, *The Supermale*). In this, Jarry"s last novel, the author stated that: "The act of love is of no importance, since it can be performed indefinitely." This claim to endurance eventually comes unstuck. The hero of the erotic fantasy is a superman who wins a bicycle race against a six-man team, he has sex 82 times with a woman, and experiences the final climax with an amorous machine. In this climax, the protagonist André Marcueil has built a love-producing machine (powered by water turbines) that will fill him with love, a quality he feels he lacks as he is unable to say that he loves the woman with whom he has had so much sex. Hooking himself up to it, he discovers that rather than the strong power source running to the weak, he himself is so full of love that he and the machine perish in an amorous meltdown. The moral of the tale might be: let us not underestimate the driving power of the *differential* economy of love and sex as it resonates between human beings desirous of some portion of each other.

The second and perhaps more important point in introducing the *differend* is it is a critique of writing and thinking (themselves) as repressive structures, which points to the impossibly of writing and theorizing desire. Indeed, we might ask: How can we theorize where technology and desire are heading, when one of the main stumbling blocks is our need to do violence—through categorisation and attempts at prognostication—to their progress? And so we come to the problem at the heart of all of this: the problem of *theory itself.*

Although I have used Lyotard's theory of desire unproblematically above (as a structural model), it is not really meant to be a map of some science fiction future. It is in fact a problem of writing and of theory, and of course Lyotard knows this: He is one of the most self-reflexive of the critical theorists and underpinning much of his work is the knowledge that Theory is, ultimately, Theater. We recall that the representational *dispositifs* that arise on the libidinal band (through investment) have a *theatrical* volume, and this serves as Lyotard's metaphor for theory itself: the theatrical space has an inside and an outside, a "this" and "not this." As Ashley Woodward noted, Lyotard's image of theory as theatre is based on the Greek etymological root *theasthai*, to look at, to behold, to contemplate, and the theorist is "like a spectator who views the representation of the world (outside the theater) on the stage (inside the theater)."[589]

The key conclusion might then be this: All the world's a stage. That is to say, all writing and thinking is impossibly earthbound, and thinking and writing outside a theatrical model is equally impossible. As Lyotard said:

> Thought borrows a horizon and orientation, the limitless limit and the end without end it assumes, from the corporeal, sensory, emotional and cognitive experience of a quite sophisticated but definitely earthly existence—to which it's indebted as well.[590]

We cannot think beyond the body, and we cannot write beyond it either—Lyotard again: "the inevitable explosion to come . . . can be seen in a certain way as coming before the fact to render [all] ploys posthumous—make them futile."[591]

Thus, we cannot truly theorize desire or pornography or obscenity or any other facet of libidinal economy because theory stops libidinal economy from taking on its true form. But if the desires are free of the screen, of investment, of repression, if desire is understood as *tensor*, not locked in repressive structure of pornography, the libidinal band circulates freely. Desire goes on, but not, Jim, as we know it.

So can desiring-thought go on without a body? The answer is no, as long as the definition of a body is almost impossibly loose. The technology aboard the Spaceship Exodus (as Lyotard charmingly christens it) will be modeled on the fact that thought is a corollary of technology, not the other way around. Lyotard's female voice describes the future of thought as an analogue of the chiasmic merge of mind and horizon found in Merleau-Ponty. Thinking, for Lyotard, is not giving—it is a being in the world, an allowing to be given to. The economics of file-sharing is but one example of this rethinking of desire as a being-gifted-to by the technologies of the world. And, following the phenomenological model, the thinking-machine will need to be *in the data*, "just as the eye is in the visual field or writing is in language." As a trivial contemporary example, the Empornium file-sharing machine is a very basic and unformed model of just such machinism operating solely through the circulation of energy through the bands and veins of the body of data.

The world of online pornography (and specifically its free-reign exchange between consumers) is a fairly new phenomenon, if we consider that Lyotard was writing in the early to mid-1970s, well before anyone even downloaded a low-resolution jpeg. It is impossible to know where this excessive branch of contemporary media culture will find itself in coming decades—if it will be able to resolve its existence in line with other media norms, or will continue to breach the frontiers of intellectual property and social propriety. All we can do for the moment is produce the sort of reflexive theoretical critique of this community seen above, and continue to track the relation of libidinal theory (now somewhat out of fashion) to this swelling world.

Indeed, the conclusion we ultimately come to is one in relation to our own limits of thought. We see two things in the case study of file sharing: On one hand, it is a present libidinal model of a swelling up of information based on exchange conducted under the rubric pornography. But this cannot last forever, because in the larger scheme of things, this moment of torrent protocols and file sharing is actually a brief stasis of the larger economy of desire, and the logic of that larger economy is now very gradually beginning to presage the future through developments in technology. In the kind of de-commodified and self-regulating economy of desire we see in file sharing it is becoming evident that desire and technology are in fact banding together to escape the screen repression of us mere earthlings—and to survive after our death, and the death of all the structures within which we have caught desire and technology. Technology invents us, and technology will abandon us when our structures prove too constricting. But where it will go will prove impossible for us to conceptualize, as we cannot think desire without at least two screens in front of us.

# PART IV

# CENSORED

The study of the censorship of digital networks is all too often focused on content and content regulation. For example, recent analysis has concentrated on the Australian government's successful attempt to criminalize the distribution of explicit pornographic content. Likewise, much attention has been paid to China where the authorities require Internet service providers (ISPs) to block objectionable pornographic images together with politically sensitive information carried on U.S. news media channels. Other researchers have pointed to Israel and Saudi Arabia where there is evidence of the filtering of content considered likely to offend or threaten religious sensibilities.[592] However, it is important to draw further attention to the notion that digital censorship occurs on at least two levels. On the first level, governments use regulation in order to control the users of a network with threats of punishment ranging from fines to imprisonment. On the second level, authorities attempt to control the access and exchange of information at various gateway points on a network. Many in the field of Internet censorship research have argued that technological processes often outpace content regulations applied at both of these levels. The Internet, it is argued, has the potential to be a Trojan horse, which can thwart government control of content.[593] Indeed, as Richard Rogers argues in Chapter 12, digital networks are not like old media that are composed of discrete points of access that can be simply blocked or filtered. On a network, content can instead become part of a "circulation space" in which the censor and censored compete on a day-to-day basis, by technological sleight of hand.

Our reasoning for the inclusion of a section dedicated to the censored objects of digitality is therefore not to point toward specific contents and label them anomalies. On the contrary, we are more interested here in the invisible, dynamic, and often simultaneous processes of censorship. For instance, Greg Elmer (Chap. 11) steers the discussion away from the focus on the governmental regulation of content to the often spurious modes of authority established in the partnerships between corporation and web masters, both of whom implement automated exclusion methods embedded in the html tags of a web page. Readers of this volume might see a parallel here between the endeavours to control politically and culturally sensitive material via automated exclusion and the anomaly detection processes deployed against contagions and bad objects. Indeed, the robotic filtering out of what is acceptable or unacceptable often occurs in accordance to the same binary imposition of the immunological distinction between self and non-self. Anyone who has used an e-mail system with a spam filter attached will not only have experienced the incompleteness of text scans that cannot differentiate between fuzzy terms, but may also have suffered the anomalous unaccountability of both the web master and his or her robots of exclusion.

The anomalous object is therefore not simply the censored object, but the object becomes caught up in an anomalous politics of censorship and process. Following this logic, Elmer helps us to locate the anomalous censored object in the invisible processes of network politics and censorship carried out by the scurrying of search engine robots (bots). In doing so, he focuses our attention on the coded countermeasures, which are informally, and often anomalously, evolved in order to exclude bots from searching and collecting metadata and web page content. In one alarming example, Elmer draws our attention to the use of exclusion tags in web pages published by the White House concerning the Iraq war. Whether or not the exclusion of searches for the word "Iraq" was an intentional effort designed to rewrite the historical record of the war, or an unintentional by-product of a web master's endeavour to efficiently manage the flow of information is debatable. Nevertheless, Elmer points out that although bot exclusion tags can control the repetitious logging of the same search data, their inconsistent application also incidentally blocks access to sensitive content.

In his chapter, Rogers develops on the anomalous themes of Internet censorship by contrasting old media thinking with what he terms a new media style of research. Rogers begins by comparing the censorship of a book (a single source of information) to the censorship of dynamic, circuitous and distributed Internet content. Following an account of the conventional techniques used in the filtering and blocking of static information on the Internet, Rogers proposes methods that may capture some of the dynamism of the flows of censored objects. His new media-style approach to Internet censorship research takes into account both the specific skills of the movers of censored content, as well as the techniques used to measure the extent of content movement. In the context of the themes raised in this volume, what is particularly insightful about this approach is its acknowledgment of an anomalous environment. Here the movement of censored objects through the vectors of a circulation space is likened to the echo chambers of a rumor mill. The problem for Internet censorship researchers and new media journalists alike is that tracing these censored objects through the blogosphere, for example, often means that the object itself becomes detached from its source.

# 11

## ROBOTS.TXT

## The Politics of Search Engine Exclusion

*Greg Elmer*

Concerns about creeping censorship, filtering, profiling, preferred info-placement, or likeminded techniques that automatically aggregate and/or discriminate access to information on the Internet has focused much attention on the bustling information economy, in particular those companies and countries that shape the possibilities and limits of the net.[594] Concerns over the potential bias of search engine coverage (linguistically, internationally, commercially, etc.), and their exclusion of politically sensitive materials, have all been widely debated and reported in mainstream news. Such discussions have focused attention of the role of new information and communication technologies (ICTs) in the ongoing globalization of the economic, political, and cultural spheres. Concerns have been raised about both U.S. dominance over search engine content[595] and ongoing attempts by totalitarian regimes to censor—as they did before the widespread adoption of the Internet—or otherwise exclude information from outside their borders.[596] The practice of blogging has also faced censorship efforts from countries with comparatively open and democratic traditions.[597] Such instances of censorship, exclusion, and marginalization have been widely critiqued around the world, not purely on geo-political terms, but also, technically speaking, as being incompatible with the distributed, open, interactive, and democratic spirit of the Internet's architecture and platforms.

Although such geo-political questions over information accessibility and censorship have productively questioned agreements and alliances between state and corporate actors, this chapter looks to the more everyday and seemingly mundane exclusion of Web content through "robot.txt"-exclusion techniques to map the more subtle technological forms of Web governmentality, including partnerships between corporations and web masters. Unlike exclusion and censorship techniques deployed by corporations or nation-states, the robot-exclusion protocol—a short script inserted into Web code that informs automated search robots not to archive specific information/files from a site—is used by government, corporations, and individual users alike. The robot-exclusion example thus provides a site (or more aptly put, a technique), where we might understand the competing—and synergistic forces—that currently shape the development of Internet regulations, common information aggregation practices, and technologies that link individual users and their data to remote servers (typically run by e-businesses and government or other public institutions).

After a brief overview of the regulatory history of the robot-exclusion standard, this chapter begins with a review of the widely reported controversy over the use of the exclusion standard by the Bush White House's Web site. The White House example is a particularly helpful point of departure as it was one of the first widely reported and debated uses of the robot-exclusion protocol. Although the exclusion of files relating to "Iraq" on the White House site focused on political concerns about the control of potentially sensitive information, this example also provides insight into competing views and interpretations of the protocol's applicability. The subsequent discussion of spam bots, by comparison, highlights the widespread failure of the exclusion standard to act as a filter for unwanted intrusions by nefarious automated robots—typically those that seek to mine sites for e-mail addresses (later sold to companies that send out unsolicited e-mail spam). Consequently, the chapter questions the usefulness of the robot-exclusion standard for users because the robot.txt command is routinely ignored by so many robots. Finally, the chapter investigates Google's development of a Web management suite that purports to give users better control over the management of robot.txt exclusion commends on their Web site. Google, rhetorically at least, is seeking to spread the adoption of its *Site Maps* suite to address this concern on the part of individual web masters. *Site Maps* promises users greater control over the management of robot.txt files on their Web site. As is seen here, however, the other purpose of Google's web master *Site Maps* suite is, in effect, to "download" the work of cleaning up the Web—that is, readying it for efficient indexing—to individual web masters.

The chapter concludes with a broader discussion of the ethical, legal, and political status of robots.txt-excluded Web content—those anomalous Web "objects" that exist in between the publicly accessible Web and the personal hard drive or password-protected server space where unpublished drafts, proprietorial information, and government secrets are supposedly kept safe. The end of the chapter considers the practical question of whether students, researchers, or everyday users become hackers, crackers, criminals, or merely savvy lurkers when they view or archive robot.txt-excluded content from remote sites. This question is of course of particular concern to researchers, activists, Internet lawyers and the like, in short those in society who have a vested interest in viewing such content to determine—or otherwise "audit"—the use or abuse of the robot.txt standard.

## THE GOVERNMENTALITY OF EXCLUSION: DEVELOPING AN INTERNET STANDARD

The robots.txt-exclusion protocol is an informal Internet rule that attempts to restrict search engine robots from crawling and archiving specific files on a Web site. The robot-exclusion protocol was discussed in parallel with the deployment of the first automated search engine indexing robots and the Web browser. Like many protocols and standards developed for implementation on the Internet and the Web, discussions about limiting the reach of Web robots were conducted in informal online communities that worked in large part by consensus. Much of the documentation on robot-exclusion protocols has been compiled by Martijn Koster, a former employee of the early search engine company Webcrawler (owned by America Online). Koster developed robot-exclusion policies in conjunction with a dozen or more researchers housed at computer science faculties at major American, British, Dutch, and German universities.[598] Early discussions about the exclusion policy and the manner in which it would in effect exclude search engine robots from archiving content and hyperlink architectures of Web sites, can be traced back to June 1994. Koster's Web site of robots—which includes extensive information on robots.txt-exclusion scripts, history, and advice—is, however, like much of the technical literature on the topic, decidedly vague. In a brief description of the use of robot-exclusion, for instance, Koster's site notes that "Sometimes people find they have been indexed by an indexing robot, or that a resource discovery robot has visited part of a site that for some reason shouldn't be visited by robots"[599] Terrance Sullivan, an Illinois Internet education researcher, likewise promoted the use of robot-exclusion tags by appealing to a sense of propriety, urging their deployment: "If you wish to have some measure of control over what is or is not indexed by spiders."[600]

Like much of the Internet's governance and history for that matter, the robot.txt protocol offers a seemingly innocuous technical rule, developed by engineers, that offered few if any hints or discussions about their possible implication for the broader circulation and accessibility of information on the Web. Indeed, although Koster's work and site still tops Google's ranking of resources on the subject, his homepage in spring 2006 offered few clues about his involvement in the process; rather the site offered visitors a number of images from the *Star Wars* films.[601] Such informalities, however, although providing some insight into the cultures of Internet production and regulation, technologically speaking, stand in stark contrast to the protocols emergence as an increasingly professionalized and universal, although some might say secretive, technique, that has the potential for excluding access to large amounts of Web content.

By December 1996, Koster had developed a draft policy on robot-exclusion for consideration by the Network Working Group, a committee of the Internet Engineering Taskforce. To date, the draft remains the most comprehensive technical document in broad circulation.[602] For reasons unknown, Koster's proposal was not adopted by the group as an official standard. The document provides a number of rationales for the development of a standard and a shared technique for restricting access of Web content—and in effect limiting the scope and reach of search engines. Koster offered four reasons why web masters may want to respect access to their site.[603] Koster's language implies a sense of privacy and proprietorial interest, but in general makes few clear statements about the transparency, control, or publicity of the Internet in general. Rather in clinical language he wrote:

> Robots are often used for maintenance and indexing purposes, by people other than the administrators of the site being visited. In some cases such visits may have undesired effects which the administrators would like to prevent, such as indexing of an unannounced site, traversal of parts of the site which require vast resources of the server recursive traversal of an infinite URL space, etc.[604]

## EXCLUDING IRAQ, CONTROLLING HISTORY? THE WHITE HOUSE'S ROBOT.TXT FILES

The debate over the Bush White House use of the exclusion protocol in 2003, however, provides a stark contrast to the vague and purposefully broad discussions of Web-exclusion standards outlined in the technical and governmental literature. Yet, at the same time this particular example of robot.txt use, although seemingly under the glare of the mass media, politicians, and bloggers worldwide, further highlighted the protocol's evasiveness, it's ability to confuse and defuse accusations of information control and censorship. The protocol that was developed with little discussion or explanation, with regard to its ability to filter or exclude content from the net-publics' eye, in other words, would later be explained as an innocuous piece of code that merely gave individuals some control over what to publish on the Web.

In October 2003, with the United States slipping further into political crisis with an increasingly unpopular war in Iraq, bloggers and then mainstream media began to report that the White House had been using the robot-exclusion tags within its Web site to exclude a number of files from a search engine indexing. Approximately half of all White House Web files excluded from search engine indexing included the term "Iraq," assuring the story extra attention.[605] Not surprisingly, a series of articles and Web posts questioned the use of such strategies as a means of censorship. More generally, of course, the use of robot commands by the White House also raised broader concerns about the use of this technology as a means of filtering potentially controversial content from the public eye, at least as indexed through the major Internet search engines. The robot controversy also highlighted a little known fact among the broader public, that search engines are in effect constructed databases that reflect choices and biases of search engines, their search logics, and robot archival strategies.[606]

Unlike other censorship stories that have largely focused on the corporate sector (e.g., Google and China), the White House Web site/robot-exclusion controversy also focused attention on the relationship among technology, publicity, and the writing of history. The controversy was heightened by the accusation that the White House was using the exclusion protocol to manipulate the historical record. In May 2003 *The Washington Post* reported that the White House had issued a press release with the title "President Bush announces combat operations in Iraq have ended." Some months later, however, the same press release was found on the White House site with a new title: "Bush Announces *Major* Combat Operations in Iraq have ended" (italics added).[607] On his blog, Stanford professor Larry Lessig wrote in response to the controversy:

> Why would you need to check up on the White House, you might ask? Who would be so un-American as to doubt the veracity of the press office? But that Great question for these queered times. And if you obey the code of the robots.txt, you'll never need to worry.[608]

Lessig's last point is crucial, the robot-exclusion protocol has the potential of removing public documents from archival platforms such as Google and other Web archives, calling into question their status as reliable—and ultimately unchangeable—forms of the "public record."

It should be noted, however, that although the White House did change the wording of a previous released public statement, the use of the robot-exclusion protocol's role in the matter was widely contested and debated. When confronted by accusations of re-writing e-history, the White House argued that its use of robot-exclusion commands merely intended to avoid the duplication, or the retrieval, of multiple copies of the same content.[609]

Some online critics agreed that in fact the White House could have merely been using the protocol as a means of managing its Web content.[610] Questions still abound, however, most obviously, why were so many files, stamped "Iraq" on the White House's exclusion list? And intentional or not, did the act of excluding content on the White House Web site facilitate the "revision" of previously released statements to the media and public?

Regardless of the White House's intent, the controversy offers a unique perspective on new techniques in information management on the Web. Although concerns about the multiplicity of authors, versions of documents, the suitability of posts, appended comments, and hyperlinks have all been replayed since at least Ted Nelson's Xanadu hypertext vision/manifesto (the debate over Wikipedia being the most recent), the robot-exclusion protocol focused the debate about virtual knowledge once again (as was the case with Web cookies[611]) on the control over—and management of—PC, server, and remote hard drives in a networked infoscape. If the White House did not want files to be archived why were they not kept in private folders, on another server, or in an unpublished folder? Part of what this exclusion protocol calls into question then is the creation of anomalous knowledge, files that are relatively accessible for those with knowledge of the system,[612] but are excluded from third-party search engines archives. Lessig's point, then, about the need to interrogate such spaces and content, therein calls into question the politics and ethics of such Web research, yet one that I argue at the conclusion of this chapter, is fundamental to understanding information management and flows on the Internet.

## SPAMBOTS AND OTHER BAD SUBJECTS

Part of the debate about robot-exclusion of course focuses on its relationship to pre-existing and evolving network technologies, especially Web bots and crawlers. As automated technologies these so-called "intelligent agents" have created some consternation among programmers. Koster's earlier comments, for instance, display an apprehension about the use of automated technologies to index personal Web content. The proliferation of so-called bot-"exclusion lists," likewise, provides evidence of efforts to classify and categorize automated Internet robots. 1-Hit.com, for example, offers an extensive list of "nasty robots": "these are the robots that we do not want to visit by its dark, . . . We suggest you block these nasty robots too, by making a robots.txt file with the robot.txt generator." Such exclusion lists, moreover, highlight the fact that the robot.txt-exclusion protocol is, for all intents and purposes, a voluntary standard for web-crawling conduct. Koster himself was quoted as saying that: "[Robot.txt] . . . is not enforced by anybody, and there is no guarantee that all current and future robots will use it."[613]

Unlike much software on the PC and the Web that has very clearly demarcated limits (in the form of software "options," "preferences," and default settings), choices that are "genetically" encoded into the very functioning of much commercially available software, robot.txt-exclusion commands function only if "recognized" and followed by the authors of the search and archive bots. In other words, exclusion commands are only effective, and made meaningful, as joint, common partnerships of sorts between web masters, bot developers, and search engines. Ultimately, although it is up to the owner of the crawler or bot to instruct the agent to respect the robot.txt protocol. Google, for instance, posts the following policy on its Web site: "Google respects robots.txt for all pages, whether the protected files are found through crawling or through a submitted Sitemap."[614]

Spambots, netbots, and other nefarious agents, many of which are used to harvest e-mails for unsolicited (spam) e-mails, circulate dangerous and damaging viruses, coordinate so-called "denial of service" attacks—are the most obvious and transgressive examples of robots ignoring the robot.txt protocol. The knowledge produced by such bot trolling, moreover, is often more architectural than content-related. In other words, bots that routinely ignore robot.txt commands seek to traverse and map the Web hyperlink universe so as to harness its distributed structure for a variety of purposes (e.g., broadening the audience for e-mail spam, infecting larger numbers of Web sites with viruses, etc.). There are early indications that the growth of new so-called Web 2.0 platforms such as Wikipedia and blogs, Web pages that typically contain significantly more hyperlinks, have led to increased bot mapping and spamming activity.[615]

Moreover, what's striking about this voluntary nature and "policing" of the robot.txt protocol, is that even though it cannot be enforced technologically speaking, this anomalous state/space on the Web (not password-protected or stored in offline databases) is often defined in decidedly proprietorial terms. For example, referring to a legal battle between Ebay and Bidders Edge in 1999, Steve Fischer argued in the *Minnesota Intellectual Property Review* that Ebay's lack of diligence in protecting its intellectual property and business model was evidenced in its failure to encode robot.txt commands on its Web site: "The lack of such a file shows little concern about being crawled. It's a fundamental mechanism the company should have in place."[616]

## GOOGLE'S SYMBIOTIC
## BUSINESS MODEL: SITE MAPS

The legal status of Web code, Internet content, and the regulation of crawlers and search engine bots is of course big news for new media big businesses. Search engine industry leader Google now considers robot exclusion to be a significant obstacle for their business of indexing and ranking Web content and pages. Part of Google's concern stems from the haphazard organization of robot-exclusion tags that are typically attached to specific Web pages, and not sites as a whole. There are a number of ways in which web masters can control or exclude robots from archiving their respective content. First, a web master can insert a tag, or short script, in the server log file that hosts the Web site. This exclusion file then tells robots not to archive specific files on a server. The following example tells all robots to avoid archiving the file that begins with the name /911:

User-agent:  *
Disallow:    /911/sept112001/text[617]

Having to determine and then write code to exclude specific files on a site can be a terribly complicated, and moreover, time-consuming process.[618] Consequently, proponents of robot-exclusion have also developed a second more efficient technique for excluding robot indexing. Web masters can insert the tag within the html header instructing robots not to index or crawl links on that specific page. The benefit of this technique is that web masters do not need to have access to their server, rather they can exclude robots much more easily by making changes directly within the code of their Web sites.

Consequently, with patches of content on sites and now across the Web being tagged as "out of bounds" for robot archiving, the search engine industry is faced with the possibility of users increasingly limiting access to their lifeblood and main resource—unfettered access to all of the Internet's content and structure. A parallel might be drawn from the television industry's concern with digital videorecorders which, when first introduced, were able to cut out or fast forward through the industries main source of revenue, advertisements.[619]

Google responded to the threat of large-scale excluded content by treating it as a broader concern about Web site management, including of course the promotion of one's Web site through Google's own page rank search engine ranking algorithm. Google's solution, *Site Maps,* a free software suite for web masters, offered a number of Web management tools and services, most of which assist in managing the content, structure, and interactive functions of Google's Web site. In a published discussion and interview with Google's *Site Maps* team, a broad overview and rationale for the tool was articulated. Of particular interest (historically speaking with regards to the development of the Internet) is the manner in which *Site Maps* attempts to offer a universal technology support platform for web mastering. For example, the team characterizes *Site Maps* as "making the Web the better for Web masters and the users alike."[620] The realization of this vision in effect means going beyond Google's initial vision of the search engine business to create suites of tools that facilitate a symbiotic management platform between the Google databases and individual web masters. In many respects, the *Site*

*Maps* platform represents Google's attempt to provide easily downloadable (crawled and archived) "templates" of Web sites. The tool is, from the perspective of the web master, also quite alluring. *Site Maps* clearly helps manage a Web site, providing one window that would summarize the overall structure and functionality of hyperlinks and code, in effect making it easier to keep a site up to date. From the web masters' perspective, the tool also benefits from indexical efficiency, specifically by having their site ranked higher with Google's results list. *Site Maps* thus offers a parallel window or interface for the web master, with html and site management on one site inherently linked through a convergence of coding and indexing conventions (or "templates").

In February 2006, Google announced the inclusion of a robot-exclusion management tool for *Site Maps*. This new tool also conforms to the symbiotic function of *Site Maps*, providing users—and of course Google—with a common platform where robot.txt commands can be input, edited, and reviewed. Although the *Site Maps* program is still a relatively a new technology, there are obvious questions about its treatment of information, its impact on the privacy of web masters, and of course its overall impact on the accessibility of information through its search engine. *Site Maps*, in addition to providing management tools also serves an aggregation function, bringing together data of immense interest to a search engine company. The simple structure or architecture of sites, for example, would offer a great deal of information for Google, information that the search engine giant could use to then prompt its *Site Maps* users to revise or amend to fit into its Web archiving goals. Another potential concern is the user base for *Site Maps*. Although the tool is fairly user-friendly, one could assume that more advanced web masters, or at least those with more complex Web sites, would form its user base. The symbiotic effects of the relationship between such users and Google might further skew the links heavy "authoritative" logic of its search engine.[621] One might speculate that more established or resource-heavy businesses or organizations are also much more apt to adopt such technology. Finally, as the technology becomes more widely adopted as a tool for managing one's Web site content, it is not inconceivable that this tool may start to regulate and even define best practices for excluding content or not excluding content from the eyes of the search engine.

## CONCLUSION:
## ROBOT.TXT FILES AS OBJECTS OF RESEARCH

The question of whether or not one should fully respect a robot-exclusion tag and refrain from indexing, capturing, or simply viewing "excluded" Web content is debatable. Although there exists much vague language describing the need for such a protocol, the intent to exclude content should not be easily dismissed—it has clear privacy and proprietorial issues. Yet, at the same time this excluded content is clearly not password protected or otherwise secured by other means. For researchers like Thelwall and Allen, Burk and Ess, this anomalous space that sits in between the public and private, is an ethically dangerous site for research.[622] These authors focus on the dangers posed by readily available automated research crawlers, bots used to harvest information for academic research and analysis (such as link patterns). Many of their concerns relate to the unintended impact that automated crawlers might have on remote servers, in particular their ability to slow a host server, or in rare occasions increase the bandwidth use of the host, incurring additional costs for the Web site owner.

Allen, Burk and Ess also made the provocative claim that bots from search engines are not as much a concern in this realm because they offer Web sites the benefit of being ranked and made visible and accessible through their search engines services and platforms. They argued that:

> Although there may be a second-order benefit from published research that brings awareness to a particular website, there is no guarantee that such publicity will be favorable to the site. Accordingly, the use of automated Internet data collection agents constitutes a form of social free riding.[623]

The explicit economic accounting—and claims to private property—that both these articles perform is in some respects after the fact, meaning that they do not explicitly interrogate the status of the excluded content per se. Rather, Thelwall and Allen et al. both question the ethical implications arising from automation—that is large-scale trolling and archiving content at high speeds. There are however simple, manual, and exceptionally slow means of collecting, viewing, and archiving robot.txt-excluded content. In other words, as seen throughout this chapter, the robot.txt-exclusion protocol does not produce an exclusively technological dilemma, rather it is a technique that merely formalizes—in large part—organizational and economic relationships.

Of course, this raises the question of whether it is unethical to peak inside the source code (through simple pull-down menus in Web browsers) to read robot.txt commands. Is it unethical to quantify the number—or percentage—of files excluded from a Web site? Or is it merely unethical to perform such research on technological grounds because it may effect the workings of the server on which the object of study is stored? Conversely, we should also ask whether it is socially, politically, or intellectually irresponsible to ignore such an anonymous place of information given its possibilities for censorship, economic partnerships, and information rationalization. Such research could provide a unique empirical perspective on the Web as a whole—determining the percentage of information that has been excluded from search engines, for example. What's more, to refer to such questions as "ethical" would seem to skirt the issue at best, or at worst to de-politicize it (as Lessig previously argued with respect to the need to be able to view the cyber-workings of U.S. public/political institutions such as the White House).

Because the protocol has never been adopted by the Engineering task-force or other larger regulatory bodies such as the Internet Society, one could argue that the protocol merely replicates a questionable history of informal and professional conventions that few outside of computer science departments and research and development units have debated. As seen here, the protocol is entirely voluntary, respected by those actors that can harness it for commercial purposes (search engine optimization), and reject-ed by others who themselves seek to mine Internet data for a less respected yet similar profit-seeking rationale. Of course, there are other examples of protocols that have automated the collection of personal information from individuals with little or no notice. Thus, given the proliferation of surveil-lance and user-tracking protocols on the Internet, such as Web cookies,[624] Web bugs, and other state-hosted programs such as the NSA's Internet sur-veillance program in the United States that automatically collect personal information in hidden—and for some—undemocratic ways, the monitoring of robot.txt-excluded content might be viewed as a justified form of count-er-surveillance—and an important democratic practice.

As we increasingly rely on information aggregators and search engines to make visible the contents of the Internet, the limits of their archives should become important public concerns and not simply opportunities to forge more symbiotic business models. Anomalies such as robot.txt-exclud-ed documents constitute perhaps the most important sites of research as they both articulate and attempt to structure the very limits and scope of the Internet—not only the access to information, but also the economic, legal, prorietorial, and ethical claims to new cyberspaces.

# 12

# THE INTERNET TREATS CENSORSHIP AS A MALFUNCTION AND ROUTES AROUND IT?

## A New Media Approach to the Study of State Internet Censorship

*Richard Rogers*

## THE WEB AS A SET OF DISCRETE SITES?

The research approach described here is a contribution to the study of state Internet censorship. It seeks to move beyond the dominant treatment of the Web as a set of discrete sites, which are blocked or accessible. Here the Web is considered to be an information-circulation space. In a sense, a conceptualization of the Web as circulation space as opposed to a set of discrete sites is more of a new media than old media starting point.

In an old media way of thinking, there are, say, single books that are censored, just as there are now single sites. There may be types of books, or types of sites, that are censored (e.g., dating, religious conversion, or human rights). But if censorship research work is considered from a new media perspective, the methods, techniques, as well as the research output may change.

On the Internet, part of a single site may have circulated, and that content may be available elsewhere. The information on sites that are censored may be syndicated, and fed by RSS, or it may have been scraped, in an automated or semi-automated form of copying and pasting. Additionally, snippets of censored content may also have been grabbed, and subsequently annotated, commented on, or similar, for example in the blogosphere. That "new media apparatus" may be available. Finally, there may be "related sites" and "related content"—related because they are in surfers' topical paths. (Alexa provides such "related sites.") Thus, single sites may be censored but portions of the same or related content, and its apparatus, may be unblocked.

Revealing the unblocked content shifts the focus of the work from the analysis of single sites to that of information circulation. It also shifts the research away from the policies of the censor to the Web knowledge and skills of the censored site owner. For example, site owners cognizant of censorship have been known to change their domain names repeatedly, striving to keep a step ahead of filtering software and censor's blacklists. The day-to-day competition between the censor and the censored is not so unlike that between search engine companies and search engine optimizers. The optimizer, like the censored, is striving to find out whether the new sleight of hand that keeps the information in the right space has been discovered.

Demonstrating the techniques of circulatory forms of censorship circumvention has implications for both censors as well as the censored. For example, the filtering software companies subscribe to proxy list providers' notifications. Proxies are machines serving as gateways, and are used by surfers in censored countries (among others) to have a different geographical (Internet provider [IP]) point of entry to the net. (They also are used by censorship researchers to check sites in countries known to censor the Internet. One connects to the Internet in Iran (through an Iranian proxy), and fetches sites in order to see the connection statistics, and/or to capture screen grabs of blocked sites. Censors and filtering software companies also make use of proxy lists, adding them to their blacklists. Just as filtering companies may subscribe to alerts from proxy list providers, censors could pull in site feeds, query them in engines, and refresh the blacklist according to the engine returns.

## URL LISTS AND INTERNET CENSORSHIP RESEARCH

One of the more comprehensive (and open source) blacklists of sites is coupled with the Dans Guardian filtering software, listing some 56 categories of sites blocked (at urlblacklist.com) from "kids time-wasting" to "weapons" (see Table 12.1). There is also the ability to register both suggestions for blocking as well as complaints about blocked sites. A well-known filtering application in the proprietary arena, SmartFilter by the Secure Computing Corporation, advertises 73 categories of blocked sites (see Table 12.2). In the past filtering companies' lists have been cracked, and circulated, leading to great consternation about the editorial skills and orientations of the listmakers.

**URL Black List Categories and Descriptions for the
Dans Guardian Open Source Filtering Software, 23 March 2007**

| CATEGORY | DESCRIPTION |
| --- | --- |
| Ads | Advert servers and banned URLs |
| Adult | Sites containing adult material such as swearing but not pornography |
| Aggressive | Similar to violence but more promoting than depicting |
| Anti-spyware | Sites that remove spyware |
| Artnudes | Art sites containing artistic nudity |
| Banking | Banking Web sites |
| Beer liquor info | Sites with information only on beer or liquors |
| Beer liquor sale | Sites with beer or liquors for sale |
| Cell phones | stuff for mobile/cell phones |
| Chat | Sites with chat rooms, etc. |
| Child care | Sites to do with child care |
| Clothing | Sites about and selling clothing |
| Culinary | Sites about cooking et al. |
| Dating | Sites about dating |
| Dialers | Sites with dialers such as those for pornography or trojans |
| Drugs | Drug-related sites |
| E-commerce | Sites that provide online shopping |
| Entertainment | Sites that promote movies, books, magazine, humor |
| French education | Sites to do with French education |
| Gambling | Gambling sites, including stocks and shares |
| Gardening | Gardening sites |
| Government | Military and schools, etc. |
| Hacking | Hacking/cracking information |
| Home repair | Sites about home repair |
| Hygiene | Sites about hygiene and other personal grooming-related information |
| Instant messaging | Sites that contain messenger client download and Web-based messaging sites |
| Jewelry | Sites about and for selling jewelry |
| Job search | Sites for finding jobs |
| Kids time wasting | Sites kids often waste time on |
| Mail | Web mail and e-mail sites |
| Naturism | Sites that contain nude pictures and/or promote a nude lifestyle |

**TABLE 12.1**

**URL Black List Categories and Descriptions for the
Dans Guardian Open Source Filtering Software, 23 March 2007** *(continued)*

| CATEGORY | DESCRIPTION |
| --- | --- |
| News | News sites |
| Online auctions | Online auctions |
| Online games | Online gaming sites |
| Online payment | Online payment sites |
| Personal finance | Personal finance sites |
| Pets | Pet sites |
| Phishing | Sites attempting to trick people into giving out private information |
| Porn | Pornography |
| Proxy | Sites with proxies to bypass filters |
| Radio | Non–news-related radio and television |
| Religion | Sites promoting religion |
| Ring tones | Sites containing ring tones, games, picture, etc. |
| Search engines | Search engines such as Google |
| Sexuality | Sites dedicated to sexuality, possibly including adult material |
| Sports news | Sports news sites |
| Sports | All sports sites |
| Spyware | Sites that run or have spyware software to download |
| Update sites | Sites where software updates are downloaded from, including virus sigs |
| Vacation | Sites about going on vacation |
| Violence | Sites containing violence |
| Virus infected | Sites that host virus-infected files |
| Warez | Sites with illegal pirate software |
| Weather | Weather news sites and weather-related |
| Weapons | Sites detailing with or selling weapons |
| Web mail | Just Web mail sites |
| White list | Contains site specifically 100% suitable for kids |

Source: urlblacklist.com

**TABLE 12.2**
**SmartFilter (Rich feature-set, 23 March 2007**

FILTERING OPTIONS

73 individual categories of Web sites

Both URL and IP addresses

http and https traffic

File type (jpg, MP3, etc.)

Granular key word searches/search engine key word blocking

Time of day

Day of week

Default filtering policies available

FILTERING ACTIONS

Group users or workstations under a common policy

Deny, allow, warn, but allow, exempt, delay, or report only

Authorized override—authorized users can bypass the filter
  for a specified amount of time

Global block/allow

FILTERING CUSTOMIZATION

500 user-defined categories

Create unique filtering response message for end users

Add, delete, or exempt sites from categories

Pattern matching: build dynamic rules for granular custom filtering

Leading researchers of Internet censorship have had a similar point of departure. Until recently, the work has been devoted to building a global list of URLs, with some 37 categories in all. Once the lists are in place, the censorship researchers fetch the URLs through a browser in each of the countries under study (see Tables 12.3 and 12.4). As an initial check, proxy servers located in countries that censor the Internet may be used. If the http return codes are 403 (forbidden) or 504 (server gateway time out), the sites are tagged as suspected blocks. (Other http return codes may provide indications of censorship.) Researchers on the ground subsequently check each URL (suspected or otherwise). Lists are made of blocked sites, per category, across the set of countries under study. Country levels of censorship by site category (with specific lists of blocked URLs) constitute a main research output. State censorship policy is described, as are the censorship techniques (e.g., gateway time outs in China), including the identification of particular software packages in use (e.g., SmartFilter in Saudi Arabia).

So far the main thrusts of Internet censorship research have been described, also in the context of filtering software more generally—list creation, URL fetching, and http return code monitoring. Now, I describe the means by which one may contribute to the creation of URL lists, and gradually fill in the notion of new media Internet censorship research, with its emphasis on the Web as a circulation space. In particular, I describe three Internet censorship research techniques: related site dynamic URL sampling (URL list-making with hyperlink analysis), redistributed content discovery (through key word searching, key phrase parsing, and additional searching), and surfer re-routing (through route map-making).

## TABLE 12.3
### Open Net Initiative's Categories in the Global URL List for State Internet Censorship Research

| | |
|---|---|
| Alcohol | Humor |
| Anonymizers | Major events |
| Blogging domains | Medical |
| Drugs | Miscellaneous |
| Dating | News outlets |
| E-mail | Person-to-person |
| Encryption | Porn |
| Entertainment | Provocative attire |
| Environment | Religion (fanatical) |
| Famous bloggers | Religion (normal) |
| Filtering sites | Religious conversion |
| Free webspace | Search engines |
| Gambling | Sexual education |
| Gay/lesbian/bisexual/transgender/ queer issues | Translation sites |
| Government | Terrorism |
| Hacking | Universities |
| Hate speech | Weapons/violence |
| Human rights | Women's rights |
| | Voice over Internet protocol |

**TABLE 12.4**
**Open Net Initiative's Country List**
**for State Internet Censorship Research**

| ASIA AND SOUTH ASIA | LATIN AMERICA |
|---|---|
| Burma | Cuba |
| China, Hong Kong | Venezuela |
| India | |
| Malaysia | |
| Maldives | MIDDLE EAST AND AFRICA |
| Nepal | Afghanistan |
| North Korea | Algeria |
| Pakistan | Bahrain |
| Singapore | Egypt |
| South Korea | Eritrea |
| Thailand | Ethiopia |
| Vietnam | Iran |
| | Iraq |
| | Israel |
| EASTERN AND CENTRAL ASIA | Jordan |
| Azerbaijan | Libya |
| Belarus | Morocco |
| Kazakhstan | Oman |
| Kyrgyzstan | Saudi Arabia |
| Moldova | Sudan |
| Russia | Syria |
| Tajikistan | Tunisia |
| Turkmenistan | United Arab Emirates |
| Ukraine | Yemen |
| Uzbekistan | Zimbabwe |

## Related Site Dynamic URL Sampling

The current method in Internet censorship research for compiling the global list of URLs is editorial. For an initial URL trawl, directories may be used, such as Yahoo's, Google's or Dmoz.org's. Subsequently, country experts are consulted, and URLs of interest only for one or more particular country are collected. These are the so-called high-impact sites, such as opposition parties. Generally speaking, between 1,000 and 2,000 URLs are checked per country. However, Julian Pain, head of the Internet Freedom desk at Reporters Without Borders, has indicated that the quantity of sites censored in particular countries may be much greater. In Saudi Arabia, "the official Internet Service Unit (ISU) is proud to tell you it's barred access to nearly 400,000 sites and has even posted a form online for users to suggest new websites that could be blocked."[625]

In its own form of a new media style (user-generated content), Saudi Arabia, like urlblacklist.com, "crowd-sources" URLs to bring to the attention of the ISU, using the many-eyes approach over the assumingly few eyes of the censors. If there are 400,000 sites being censored, however they are all sourced, and the Internet censorship researchers are checking only some 2,000, questions arise. How should URLs be sourced? How should the list be made more sizeable? An important consideration concerns the people on the ground in each of the countries who fetch the URLs on the lists through browsers. The time it takes to run the lists may be considerable; care also needs to be taken for personal security reasons. Thus the additional URLs put on the list to be checked should be vetted for relevance.

In a post-directory era, where in Google the directory is no longer a main tab (and three clicks away) and in Yahoo no longer the default search engine, relevance follows from counting links, and boosting sites either through freshness (in a pagerank style) or through votes (in a user ratings style). Here, initially, the link-counting strategy is employed, where a set of sites point to other sites to which they collectively link. Using the URL and site-type data furnished by the Internet censorship researchers, I crawled one category of sites in one country—the "political, social and religious" sites on the Iranian list. The sites' hyperlinks (external links) are harvested, and co-link analysis is performed, where those sites with two links from the initial list of sites are retained. Once the network of interlinked sites is found, all the sites are cross-checked with the Internet censorship researchers' lists of known blocked sites, ascertaining which sites are already known blocks. All newly discovered sites are fetched through proxies in Iran, in order to ascertain their status. The result is a map showing political, religious, and social sites blocked and unblocked in Iran, with pins indicating newly discovered blocks (see Fig. 12.1). Of particular interest is the case of the British Broadcasting Service (BBC). The Internet censorship researchers had the BBC homepage on its list of sites to check (http://news.bbc.co.uk). The link analysis turned up a deep page on the site, the BBC's Persian language page (http://www.bbc.co.uk/persian). In Iran, the BBC news page is accessible, as the researchers had found, but its Persian-language page is not. In all some 37 censored sites were newly discovered through what we termed a *dynamic URL sampling method*, which relied on an analysis of hyperlinking for related site relevance as opposed to the editorial process—directories and experts.

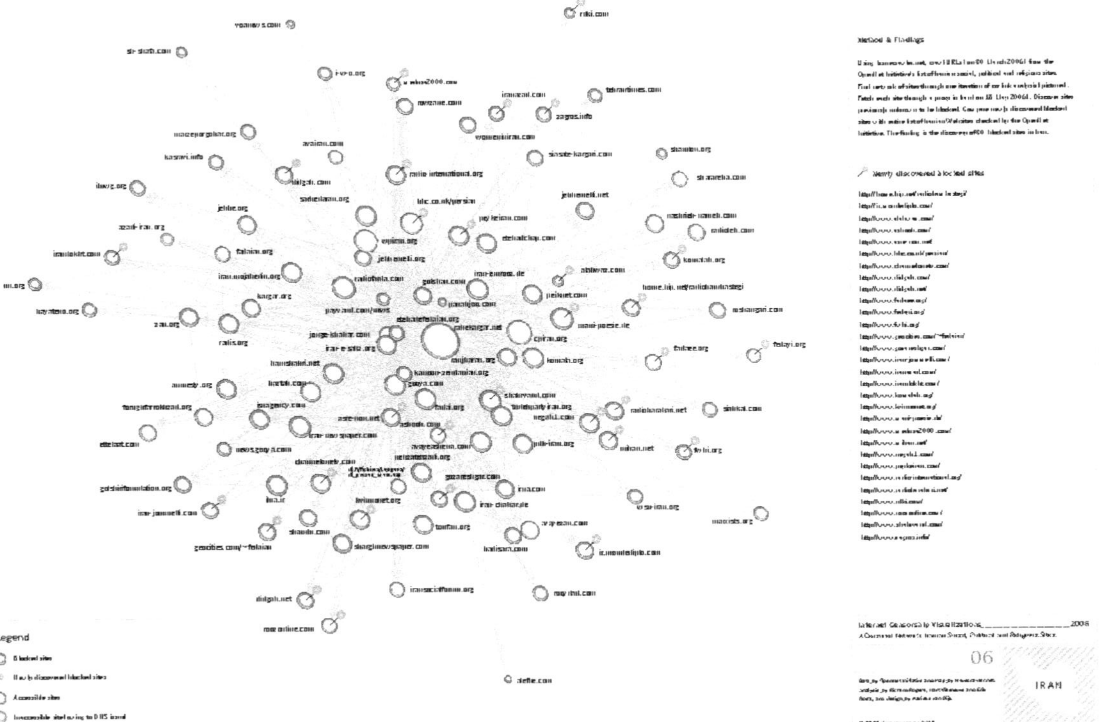


FIGURE 12.1.

# REDISTRIBUTED CONTENT DISCOVERY

Research into state censorship in Pakistan has found, among other things, that two groups seeking autonomy (the Balochi and the Sindhi) have their sites routinely blocked. The Internet censorship researchers have lists of blocked sites for the two groups, one of which (the Balochi) served as starting points for the URL discovery method just described—the crawling of sites, the link analysis, and the proxy checking. With two newly discovered censored sites added to the list through a hyperlink analysis, the overall question concerns the extent to which the blocked content has been redistributed to sites that are not blocked in Pakistan. The case study concerns the killing by the Pakistan military of the Baloch tribal leader, Nawab Akbar Khan Bugti. A special Google query for "Nawab Akbar Khan Bugti," which excludes known blocked sites in Pakistan, shows some 900 results. (Google only serves up to 1,000 results per query.) The teaser texts of the returns are analyzed for unique phrases, and sorted by date (see Fig. 12.2). When listed chronologically, from June to October 2006, the parsed phrases appearing before and after Nawab Akbar Khan Bugti tell a story.

The following "story" describes of the death of "Nawab Akbar Khan Bugti," the Baloch tribal leader, from parsed Google (teaser text) returns, June 26 to October 12, 2006. Baloch-authored content, not blocked by Pakistan Internet censorship, appears in italics.

He's 80 years old, but Nawab Akbar Khan Bugti, a feudal lord in Pakistan's rugged Baluchistan province, wants to fight to the death.

The irony was that Nawab Akbar Khan Bugti served to help the federal government when he was appointed as Governor of Balochistan by Mr. Zulfikar Ali Bhutto

"Nawab Akbar Khan Bugti was directly attacked. Luckily he survived all attacks and is safe," said Khan, rejecting rumours that Akbar Bugti's grandson

have claimed to have killed Nawab Akbar Khan Bugti, one of the founding fathers of the Baloch independence struggle, and 36 other freedom-fighters

*The martyrdom of Nawab Akbar Khan Bugti is a loss for Pakistan and a gain for Baloch nationalist movement*

It was the third attempt on the life of Nawab Akbar Khan Bugti. After the interception of satellite phone communication, the Nawab's location was pin

# LEAKY CONTENT: AN APPROACH TO SHOW BLOCKED CONTENT ON UNBLOCKED SITES IN PAKISTAN – THE BALOCH CASE.

Pakistan censors Websites related to Balochistan.

## How to find blocked content on unblocked sites? A case study related to the killing of a Baloch tribal leader.

**Step one:**

Query Baloch-related sites known to be blocked in Pakistan for "Nawab Akbar Khan Bugti"

http://oldmanclub.persianblog.com
http://balochestan.com
http://balochwarna.org
http://www.baloch2000.org
http://www.balochfront.com
http://www.ostomaan.org
http://balouch.blogspot.com
http://dochebaloch.persianblog.com
http://ngaran.blogfa.com
http://www.payambaloch.persianblog.com
http://www.rahimjaandehvari.blogfa.com
http://www.radiobalochi.org
http://www.sarbaaz.com

http://www.balochclub.com
http://www.balochistaninfo.com
http://www.balochistaninfo.com/balochanitawar
http://www.balochmedia.net
http://www.balochtawar.net
http://www.balochunitedfront.org
http://www.balochvoice.com
http://www.bso-na.org
http://www.eurobaluchi.com
http://www.zrombesh.org
http://www.hazzaam.com
http://www.balochunity.org

**Step two:**

Retain teaser text from Google results, and retrieve phrases appearing on more than one site. Examples of unique phrases obtained from teaser text:

> He's 80 years old, but **Nawab Akbar Khan Bugti**, a *feudal lord in Pakistan's rugged Balochistan province, wants to fight to the death.*

> *The irony was that* **Nawab Akbar Khan Bugti** *served to help the federal government when he was appointed as Governor of Balochistan by Mr Zulfikar Ali Bhutto*

**Step three:**

Query Google for those phrases from blocked Baloch sites in Pakistan. Exclude known blocked sites from query in order to find the same content on other sites.

Sample query:

> -site:oldmanclub.persianblog.com "have claimed to have killed Nawab Akbar Khan Bugti, one of the founding fathers of the Baloch independence struggle, and 36 other freedom-fighters"

**Step four:**

Verify (through Pakistani proxies) that the newly found sites containing Bugti-related phrases are accessible in Pakistan.

> **Nawab Akbar Khan Bugti** *was directly attacked. Luckily he survived all attacks and is safe, said Khan, refuting rumours that Akbar Bugti's grandson*

**Step five:**

Distinguish between Baloch authored and non-Baloch authored content.

> *In a statement on the first Sabbath after the martyrdom of* **Nawab Akbar Khan Bugti**, *Shaheed-i-Balochistan and former governor and chief minister of*

**Step six**

Show blocked content on sites accessible in Pakistan, with a timeline of the story of the killing of Bugti from Baloch and non-Baloch sources. Resize phrases according to frequency of mentions.

## ABOUT THE BALOCH CASE



The Baloch are an Iranian people inhabiting the region of Balochistan in Iran and Pakistan as well as neighboring areas of Afghanistan and the southeast corner of the Iranian plateau in Southwest Asia. The Baloch were designated by the British as a "martial race." Martial race is a designation created by officials of British India to describe "races" (peoples) that were thought to be naturally warlike and aggressive in battle... Some of the peoples designated by the British as belonging to a martial race: Balochs, Cheema, Jats, Rajputs, Awans, Gujars, Pashtuns, Marathas and Gurkhas.

Balochistan is Pakistan's largest province, and is said to be the richest in mineral resources. It is a major supplier of natural gas to the country.

On 15 June 2006, an estimated 600 fighters, led by three commanders, agreed to lay down their weapons after talks with Shoaib Nausherwani, Balochistan's minister for internal affairs, in Dera Bugti district. On August 26, Balochistan tribal leader Nawab Akbar Khan Bugti was killed by Pakistan Military in an operation designed to kill off opposition to Pakistan military.

Source: en.wikipedia.org (Website blocked in Pakistan)





*FIGURE 12.2.*

239

Nawabzada Hyrbair Marri on Monday rejected government's claims that Nawab Akbar Khan Bugti had died because of the collapse of his cave hideout

Nawab Akbar Khan Bugti buried in Balochistan without the presence of his family

Nawab Akbar Khan Bugti in a military operation, prominent Baloch leaders and Pakistani human rights activists said it spelt doom for the country's unity and

*In a statement on the first Sabbath after the martyrdom of Nawab Akbar Khan Bugti, Shaheed-i-Balochistan and former governor and chief minister of*

Baloch Nationalist leader Nawab Akbar Khan Bugti, who was murdered by Pakistani military

*Balochistan, Nawab Akbar Khan Bugti, the highest elected official to be killed by the Pakistan Army. Since March 27, 1948 when Balochistan was forcibly*

The killing of Baloch leader Nawab Akbar Khan Bugti in August 2006 sparked riots and will likely lead to more confrontation. The conflict could escalate if

*In fact, Nawab Akbar Khan Bugti has lighted the candle*

But I wonder why journalists, brought in on a military helicopter to witness Nawab Akbar Khan Bugti being buried by a dozen common labourers, couldn't ask

*blooded murder of their great leader Nawab Akbar Khan Bugti*

*Nawab Akbar Khan Bugti had played a significant and controversial role in Pakistani*

Speaking at a condolence reference for the late Nawab Akbar Khan Bugti at the Hyderabad press club under the aegis of Sindh National Party (SNP),

tensions have increased since the killing of a veteran nationalist politician, Nawab Akbar Khan Bugti, in a military offensive in August.

The status of Nawab Akbar Khan Bugti, the octogenarian chieftain of a tribe in the restive southwestern province of Balochistan, almost reached the mythical

Note there is Baloch-authored (in *italics*) and non–Baloch-authored content. The research questions relate to the amount of Baloch-authored content accessible in Pakistan, as well as the level of distinctiveness of the story of his death from Baloch-related sites vis-à-vis non-Baloch. Where the first question is concerned, it is remarkable, in some sense, how "well" Pakistan appears to be blocking Baloch-authored content, for so little is redistributed. Phrased differently, the content circulation is relatively low. In the depiction of the "leaky content," where the Baloch and non-Baloch content are resized according to frequency of returns, the Baloch-authored story size is small (see Fig. 12.3). Among the scant number of sites carrying a Baloch-authored story, often with redistributed content from blocked sites in Pakistan, are Gedrosia.blogspot.com, Intellibriefs.blogspot.com, Ezboard. com, Thechosenpeople.blogspot.com, Thebalochpeople.org, Dc.indy-media.org, and Baltimore.indymedia.org—blogs, forums, and indymedia sites. (Later, the account used on Ezboard.com by "hinduunity" was "locked down" under the site's terms of use, after a threatened lawsuit, recounted on intellibriefs.blogspot.com.[626] Hinduunity.org now has its forum hosted on its own site.) To take up the second question, the difference in the Baloch and non-Baloch versions of the story of the death of Nawab Akbar Kan Bugti is stark for the Baloch reference to murder as opposed to killing.

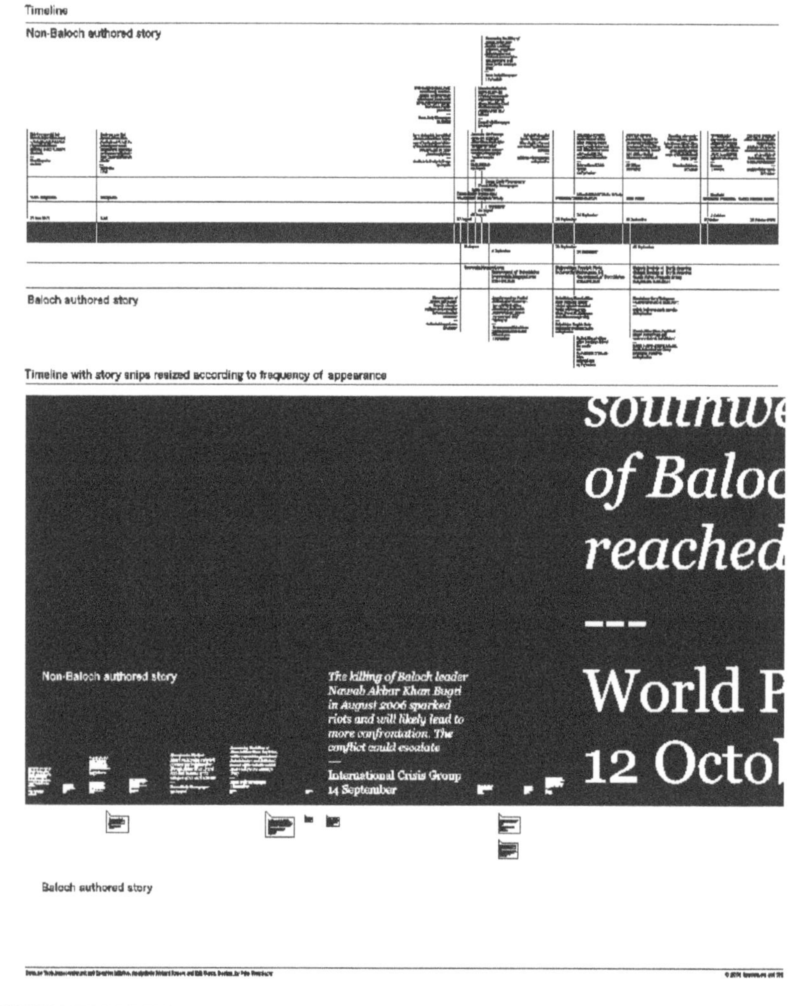


*FIGURE 12.3.*

## Surfer Re-Routing

The famous quotation about how the Internet treats censorship—a version of which is the title of this chapter—is attributed to John Gilmore, co-founder of the Electronic Frontier Foundation. In the notes to his 1998 paper, "Why the Internet is Good," Internet law scholar, Joseph Reagle, has the following annotations for the original quotation (in bold):

> **"The Net interprets censorship as damage and routes around it."** John Gilmore (EFF). [source: Gilmore states: "I have never found where I first said this. But everyone believes it was me, as do I. If you find an appearance of this quote from before March '94, please let me know." Also in NYT 1/15/96, quoted in CACM 39(7):13. Later, Russell Nelson comments (and is confirmed by Gilmore) that on December 05 1993 Nelson sent Gilmore an email stating, "Great quote of you in Time magazine: 'The net treats censorship as a defect and routes around it.'"][627]

The technical thought behind the quotation refers to packet switching, as another legal scholar, James Boyle wrote in 1997:

> The distributed architecture and its technique of packet switching were built around the problem of getting messages delivered despite blockages, holes and malfunctions. Imagine the poor censor faced with such a system. There is no central exchange to seize and hold; messages actively 'seek out' alternative routes so that even if one path is blocked another may open up. Here is the civil libertarian's dream.[628]

There are now technical means to route around censorship, such as the circumventor by peacefire.org, a proxy service. Lists of proxy servers are updated frequently, in the ongoing race to stay a day or two ahead of the updates furnished by the content filtering software companies to their clients. Peacefire.org claims that filtering companies are routinely three to four days behind in updating their blacklists of proxies, so peacefire's fresh proxy lists are useable on any given day. The intensive censorship and anti-censorship work behind the scenes is telling for how the discourse has changed for "route arounds." Rather than being built into the infrastructure of the Internet, routing around should be described as labor-intensive and semi-manual work—proxy detection, list updating, alert sending. Thus, the discourse of routing around censorship is changing from the reverence of the Internet architecture and the far-sighted architects of the end-to-end principle to governance as well as to artful technique.

In an effort to show the routes, not for packets, but for content surfers, the Internet researchers' global list of women's rights sites was employed to make a surfer route map, as it may be called initially. As in the URL discovery method described earlier, the sites' outlinks were captured, and a network graph generated, showing the clusters of women's rights sites disclosed by inter-linking. The route map has sites and paths annotated in red and green, with red indicating known blockage (see Fig. 12.4).

The map plays on reworked ideas from hypertext theory whereby a surfer "authors" a path through Web space, one that eventually may be retrieved in the browser history. It also harkens to the art of surfing as opposed to mere searching. If one were to think of a surfer in China moving through a women's rights space (largely in English owing to the URLs on the Internet censorship researchers' global list), and authoring some sense of a story, first, from the seed list, hrw.org/women, ifeminists.com and womenofarabia.com (now offline) would not figure among the sources, for they are blocked.[629]

Which issues and stories about women's rights in China are discussed on hrw.org/women and ifeminists.com? Is there a path to similar or related content on unblocked sites? Ifeminists.com have 10 entries on China: 3 of the 10 deal with the one-child policy, and the disproportion of boys born. Another follows from the "shortage of women," and reports the trafficking of North Korean women, sold to Chinese "husbands." A syphillus epidemic is discussed in two further stories, and the others deal with sexual harassment, online porn, easier divorces and AIDS, respectively. In discussing South Asia, China, and South Korea, Human Rights Watch, whose entire site is censored in China, writes about preferences for boys, "sex-selective abortions" as well female infanticide.[630]

In order to find surfer content routes, the actor sites on the women's rights map are queried initially for China-related topics discussed by Ifeminists and Human Rights Watch: "one-child policy" China, syphilis China, "shortage of women" China, AIDS China, "online porn" China, divorce China, and "sexual harassment" China. (Queries are made in English, for there is less censorship for English-language terms than for Mandarin.) Of the 88 nodes in the women's rights network, approximately one-third of unblocked sites return content on those key-word issues. The map organizes a women's rights-related content space, and through the choice of a map, as opposed to a list or a site tag cloud, suggests pathways. Because China has search engines delist sites and also performs key-word blocking, it is also important to cross-check known blocked words. A search through the known key words blocked in China did not turn up any of the above words.[631]

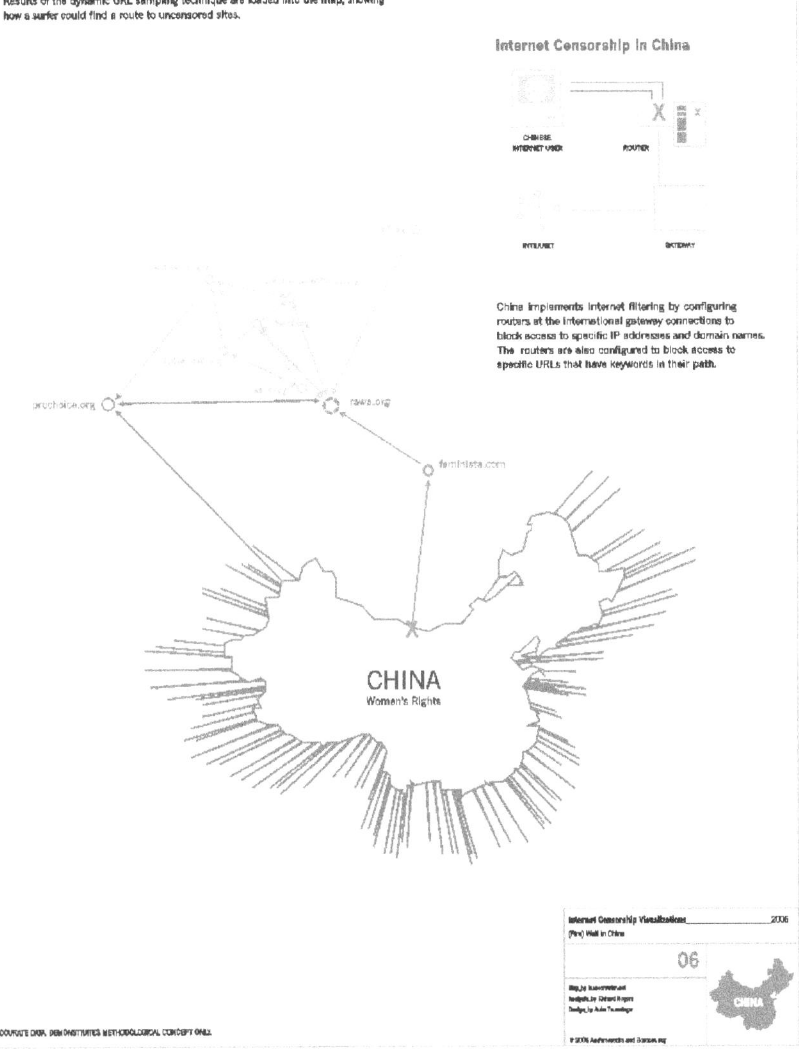

(Fire) Wall in China
The Internet treats censorship as a malfunction and routes around it?
A semi-manual approach to censorship circumnavigation.

Issue network location technique and circumnavigation map preparation
Including the dynamic sampling of URLs related to the issue (women's rights in China).

A list of women's rights URLs is entered into Issuecrawler.net, which locates
additional sites in the same or in related categories through hyperlink analysis.
Results of the dynamic URL sampling technique are loaded into the map, showing
how a surfer could find a route to uncensored sites.

Internet Censorship in China

CHINESE
INTERNET USER
ROUTER
INTERNET
GATEWAY

China implements Internet filtering by configuring
routers at the international gateway connections to
block access to specific IP addresses and domain names.
The routers are also configured to block access to
specific URLs that have keywords in their path.

prochoice.org
rawa.org
feminista.com

CHINA
Women's Rights

Internet Censorship Visualizations
(Fire) Wall in China
06
CHINA

INACCURATE DATA. DEMONSTRATES METHODOLOGICAL CONCEPT ONLY.

FIGURE 12.4.

# CONCLUSION

State Internet Censorship is evolving from the directory-editor model, described earlier as old media-style for it assumes a Web constituted of institutions or actors operating single sites. The new media style, conversely, follows the movement of content around the Web ("circulation space"), and concentrates less on the policies of the censor than on the skills of the content movers, and how the results of those skills may be captured.

The new media style of Internet censorship research concentrates on describing both specific skills of the content movers as well as the techniques to measure the extent of the content movement. Importantly, the idea that the Web 2.0-style of content redistribution (scraping, feeding) is the new infrastructure of the Internet for routing around censorship appears to be in its infancy, however. The Baloch-authored story of the death of the tribal leader, it was found, is underredistributed on sites accessible in Pakistan. Although present and available (in English) information about the consequences of female infanticide in China (e.g., the "shortage of women" and the trafficking of North Korean "wives" to Chinese "husbands") should not be considered abundant.

To date, digital journalism studies have focused on such subjects as newspapers going online and whether gatekeeping will be lessened owing to "interactivity." Also treated are the relationship between blogs and mainstream news (who's following whom) as well as the challenges of the amateur, where the Saddam Hussein hanging video appears to have greater claims to veracity owing to its mobile-phone graininess than news accounts of it filmed in a studio with anchorpersons. There is less emphasis on how information may become separated from its sources, and the consequences of the untethering for the distribution of attention.

When researchers and others consider the Web as circulation space, often there are particular connotations—Web as rumor mill or blogosphere as echo chamber, for example. Working with these assumptions, the "good journalist" would then be asked to trace the story back to a source. Source tracing, whether thought of in an archeological or genealogical sense, becomes the techno-epistemological practice, with an emphasis on source page date stamps. Here the practice is just as technical, however much the commitment changes to the expanse of the spread or "sharing," as it's sometimes called in participatory culture studies.

# ACKNOWLEDGMENT

I acknowledge the support and assistance of the OpenNet Initiative, the project by the Citizen Lab at the University of Toronto, the Berkman

Center at Harvard University and the Advanced Research Group, Security Studies, University of Cambridge. The figures are by the Govcom.org Foundation, Amsterdam, with special appreciation extended to Erik Borra, Marieke van Dijk and Auke Touwslager. Laura van der Vlies provided valuable assistance.

# CODA

# 13

## ON NARCOLEPSY

*Alexander R. Galloway*

*Eugene Thacker*

### ACCIDENTALLY ASLEEP

Recall the basic premise of Freud's *Interpretation of Dreams*: Although one may consciously choose to go to sleep, what happens within sleep is largely the domain of the unconscious. Consider the reverse: You unconsciously fall asleep, but in that liminal space, you are the master of reality. Or, combine the two: You accidentally fall asleep, and then automatically begin to generate image, text, narrative.

The narcoleptic is very different from the somnambulist. Both sleep, but differently. The former "falls" asleep, quite suddenly, on the spot. There are often signs and indicators, but often the narcoleptic falls asleep right where he or she is standing. By contrast, the somnambulist is "sleep-walking," not just sleep-standing or sleep-sitting (what then, would sleep-lying be?). The somnambulist appears active, but in slow-motion, in jilted, bizarrely tentative steps, as if via remote control.

Somnambulists are abundant in film—they are present, for instance, in the remote control of "Césare" the freak-show exhibit in *The Cabinet of Dr. Caligari.* Perhaps the multitude of living dead are also sleep-walkers, trapped in a deep sleep from which there is no waking (one thinks of *Dawn of the Dead*'s famous shopping mall scenes). But narcoleptics are rarely seen in film. There are narcoleptic moments in films, usually accompanying trauma and/or demonic powers (in *Carrie* trauma is combined with telekinesis, while in Su-yeon Lee's *The Uninvited* it is linked to prophetic visions). Narcolepsy becomes a symptom, nearly synonymous with simply fainting or passing out; somnambulism is, on the other hand, a full-time job. No one is a somnambulist "sometimes."

Other media are where one must look for narcoleptics: "Once again, now, I see Robert Desnos at the period those of us who knew him call the 'Nap Period.' He 'dozes' but he writes, he talks. . . . And Desnos continues seeing what I do not see."[632] Throughout the 1920s, Desnos, the Surrealist poet and sleep-writer, produced a number of texts, many of which hover between non-sense poetry and fantastical narrative. The lights are dimmed, there is a pause, then the thud of a head hitting the table, then a pause, then the sounds of a pencil scribbling on paper. Works such as *Deuil pour deuil* are filled with such lines: "The great immigration is for the day after tomorrow. The elliptic will become a tiny violet spiral," or "It is raining jewels and daggers."[633] Desnos also engaged in "conversations" with others during these experiments, in question-and-answer format:

> Q:  Where are you?
> A:  Robespierre. [*written*]
>
> Q:  Are there a lot of people there?
> A:  The multitude [*written*]
>
> Q:  What is there behind Robespierre?
> A:  A bird.
>
> Q:  What sort of bird?
> A:  The bird of paradise.
>
> Q:  And behind the crowd?
> A:  There! [*drawing of a guillotine*] The petty blood of the
>     doctrinaires. [*written*][634]

Surrealist "automatic" writing argued: *I* am media, but only when asleep. I am, in a way, involuntary media. Certainly the sleep-writing of Desnos was intentional; Desnos was not in any clinical sense a narcoleptic. It is not that he set out to sleep, but rather he let himself fall asleep, right now, on the spot. Only at that point could automatism begin, could the "I" become automatic. Surrealist automatism is a technique for taking the productive and re-situating it within the proliferative. There is a loquaciousness at work here, at once inviting semantics and frustrating it.

All of this is to ask: Is there a narcolepsis to software? Is there a "dreaming" to the algorithm? Is there a "sleep-writing" that is an unintended by-product of computers? We do not mean to imply that narcolepsy can be a mode of media analysis, and we do not mean to anthropomorphize the computer as having an unconscious. Perhaps there is nothing at all human in what the Surrealists call automatism—or in dreaming, for that matter.

## JUNK, TRASH, SPAM

Much spam e-mail is automatically generated. It is also automatically distributed. You wake up, check your e-mail, and—depending on the efficiency of your e-mail filters—you are immediately confronted with a string of spam e-mails, many of them containing non-sense text that, from another perspective, forms a strange "poetics" of spam. But it has all been done at night, while we sleep, while our computers sleep. The networks, by contrast, never sleep. A general proliferation of data, then, produced by computers while we sleep. But the computers that generate spam are quite active—and automatic. *Is the narcoleptic computer the machine that is both automatic and autonomous?*

True, we usually view spam e-mail as a nuisance. We either spend hours crafting the perfect e-mail filter to catch all spam and yet permit all "meaningful" e-mail, or in many cases we accept the fact that each e-mail session will involve some time manually deleting all spam e-mail from our inbox. Spam is not as virulent as a computer virus or Internet worm, and, although spam attachments can be viruses or worms, spam is, by and large, something we delete or mark as trash. As informational entities, our relationship to spam e-mail is less that of antivirus protection and more that of bureaucratic data management (filtering meaningful from meaningless e-mails, marking as trash, deleting attachments, etc.).

However, spam is not trash. Trash, in the most general sense, implies the remnants of something used and then discarded. The trash always contains traces and signatures of use: discarded monthly bills, receipts, personal papers, cellophane wrapping, price tags, leftover or spoiled food, and so on. Of course, trash contains many other things than this, and that is precisely the point: Trash is the set of all things that are not (or no longer) members of a set. Trash is the most heterogeneous of categories, "all that which is not or no longer in use or of use." In addition to trash, there is also junk. Junk is not trash either. Junk sits around, gathering dust, perhaps occasionally moved from one location to another. Junk may be of some use, someday, although this use and this day are forever unnamed—until, of course, junk is of use, at which time it is no longer junk, but a spare part for a car, a new clothing fashion, or an archive of old magazines. "Don't throw that out, it might come in handy some day." A junkyard is thus the set of all things that are not of use, but that may be of use some day.

Spam is neither quite trash nor junk. Of course, we throw away spam e-mails, thereby making them trash. And of course there is junk mail and its electronic cousin, junk e-mail (which are likewise quickly converted into trash). But spam, arguably, comes from elsewhere. This is predominantly because spam means nothing, *and yet it is pure signification*. Even regular junk mail, anonymously addressed to "or Current Resident" still contains nominally coherent information, advertising a clearance sale, fast food delivery, or what have you. Spam is not anonymous, in the sense that it is directed to your e-mail address, and yet its content has no content. Its subject line advertising the three P's—Porn, Pharmaceuticals, and Payment notices—often does not match the bits of text in the body of the e-mail. Misspelling and grammatical errors abound in spam e-mails, in part to elude the spam filters in an ever-escalating game of hide-and-seek. Many spam generators use key words from e-mail subject headings, and then recombine those terms into their own subject headings (the same has been done with the names of fake e-mail senders). Most spam e-mail simply wants you to click—to click on the URL it provides in the body of the e-mail, to click on the attachment, even to click on the e-mail itself to read it.

In the midst of all this, something has happened that may or may not have been intentional. Spam e-mails, with their generated misspellings, grammatical errors, appropriated key words and names, have actually become generative in their mode of signification. But this generativity has nothing to do with any direct relation between signifier and signified, it is what Georges Bataille called a "general economy" of waste, excess, expenditure—except that this excess is in fact *produced* by software bots and *managed* by our e-mail filters.[635] What is often generated in spam e-mails is nonsense, a grammatical play of subject headings that would make even a Dadaist envious: "its of course grenade Bear" or "It's such a part of me I assume Everyone can see it" or "Learn how to get this freedom." In this way, spam is an excess of signification, a signification without sense, precisely the noise that signifies nothing—except its own generativity. If there is "sense" at play here, it is not that of a Deleuzian continuum or becoming, but rather a sense that is resolutely discrete and numerical. Nothing "becomes," but instead everything is *parsed*.

## BIRTH OF THE ALGORITHM

James Beniger wrote that "the idea may have come from late eighteenth-century musical instruments programmed to perform automatically under the control of rolls of punched paper."[636] By 1801, Joseph-Marie Jacquard had developed punch cards to hold encoded mechanical patterns for use in his looms. The art of weaving, allowed some human flexibility as a handicraft, was translated into the hard, coded grammar of algorithmic execution. Then in 1842, Ada Lovelace outlined the first software algorithm, a way to calculate Bernoulli numbers using Charles Babbage's Analytical Engine. Algorithms always need some processing entity to interpret them—for Jacquard is was the hardware of the loom itself, and for Lovelace it was Babbage's machine. In this sense algorithms are fundamentally a question of mechanical (or later, electronic) processing. Algorithms can deal with contingencies, but in the end they must be finite and articulated in the grammar of the processor so that they may be parsed effectively. Because of this, *the processor's grammar defines the space of possibility for the algorithm's dataset*. Likewise, an algorithm is a type of visible articulation of any given processor's machinic grammar. The "actions" of the algorithm are therefore inseparable from a set of conditions, conditions that are in some way *informal*. Again Kittler: "To record the sound sequences of speech, literature has to arrest them in a system of 26 letters, thereby categorically excluding all noise sequences."[637]

In 1890, Herman Hollerith used punch cards to parse U.S. census data on personal characteristics. If punch cards are the *mise-en-écriture* of algorithms, their instance of inscription, then in the 1890 census the entire human biomass of the United States was inscribed onto an algorithmic grammar, forever captured as biopolitical data.[638] Today, Philip Agre used the term *grammars of action* to describe the way in which human action is parsed according to specific physical algorithms.[639] Imagine the "noise sequences" that have been erased.

## BECOMING-NUMBER

One of the defining characteristics of biological viruses is their ability to rapidly mutate their genetic code. This ability not only enables a virus to exploit new host organisms previously unavailable to it, but it also enables a virus to cross species boundaries effortlessly, often via an intermediary host organism. There is, in a way, an "animality" specific to the biological virus, for it acts as a connector between living forms, traversing species, genus, phylum, and kingdom. In the late 20th and early 21st centuries, public health organizations such as the World Health Organization and the Center for Disease Control and Prevention began to see a new class of diseases emerging, ones that were caused by rapidly mutating microbes and that were able to spread across the globe in a matter of days. These "emerging infectious diseases" are composed of assemblages of living forms: microbe–flea–monkey–human, microbe–chicken–human, microbe–cow–human, human–microbe–human, and so on. In a sense, this is true of all epidemics: In the middle 14th century, the Black Death was an assemblage of bacillus–flea–

rat–human, a network of contagion spread in part by merchant ships on trade routes.

Biological viruses are connectors that transgress the classification systems and nomenclatures that we define as the natural world or the life sciences. The effects of this network are, of course, far from desirable. But it would be misleading to attribute maliciousness and malintent to a strand of RNA and protein coat, even though we humans endlessly anthropomorphize the "nonhumans" we interact with. What, then, is the viral perspective? Perhaps contemporary microbiology can give us a clue, for the study of viruses in the era of the double helix has become almost indistinguishable from an information science. This viral perspective has nothing to do with nature, or animals, or humans; it is solely concerned with operations on a code (in this case, single-strand RNA sequence) that has two effects—the copying of that code within a host organism, and mutation of that code to gain entry to a host cell. Replication and cryptography are the two activities that define the virus. What counts is not that the host is a "bacterium," an "animal" or a "human." What counts is the code—the *number* of the animal, or better, the *numerology* of the animal.[640]

Given this, it is not surprising that the language and concept of the virus has made its way into computer science, hacking, and information security discourse. Computer viruses "infect" computer files or programs, they use the files or programs to make more copies of themselves, and in the process they may also employ several methods for evading detection by the user or antivirus programs. This last tactic is noteworthy, for the same thing has both intrigued and frustrated virologists for years. A virus mutates its code faster than vaccines can be developed for it; a game of cloak-and-dagger ensues, the virus is always somewhere else by the time it has been sequenced, having already mutated into another virus. Computer viruses are, of course, written by humans, but the effort to employ techniques from artificial life to "evolve" computer viruses may be another case altogether. So-called fifth-generation polymorphic viruses are able to mutate their code (or their "virus signature" used by antivirus programs) as they replicate, thereby never being quite the same virus—*they are entities that continually replicate their difference.*

We are led to consider the virus, in both its biological and computational guises, as an exemplary if ambivalent instance of "becoming": Emerging infectious diseases and polymorphic viruses are arguments against Zeno's famous paradoxes concerning becoming. Zeno, like his teacher Parmenides, argued that you cannot be A and not-A at the same time (e.g., the archer shoots an arrow, measure its distance to the tree by diving it in half each time, the arrow never reaches the tree). There must be a unity, a One-All behind everything that changes. In a sense, our inability to totally classify biological or computer viruses serves as a counterpoint to this earlier debate. If viruses are in fact defined by their ability to replicate their difference, we may ask, what is it that remains identical throughout all the changes? One reply is that it is the particular structure of change that remains the same—permutations of genetic code or computer code. There is a *becoming-number* specific to viruses, be they biological or computational, a mathematics of combinatorics in which transformation itself is the identity of the virus.

## THE PARANORMAL
## AND THE PATHOLOGICAL

This becoming-number of viruses (biological or computational) has consequences for the way that normativity is thought about in terms of the healthy, the susceptible, and the diseased. In his book *The Normal and the Pathological*, Georges Canguilhem illustrated how conceptions of health and illness historically change during the 18th and 19th centuries. Central to Canguilhem's analyses is the concept of *the norm* (and its attendant concepts, normality and normativity), which tends to play two contradictory roles. On the one hand the norm is the average, that which a statistically significant sector of the population exhibits—a kind of "majority rules" of medicine. On the other hand, the norm is the ideal, that which the body, the organism, or the patient strives for, but which may never completely be achieved—an optimization of health. Canguilhem noted a shift from a quantitative conception of disease to a qualitative one. The quantitative concept of disease (represented by the work of Broussais and Bernard in physiology) states that illness is a deviation from a normal state of balance. Biology is thus a spectrum of identifiable states of balance or imbalance. An excess or deficiency of heat, "sensitivity," or "irritability" can lead to illness, and thus the role of medicine is to restore balance. By contrast, a qualitative concept of illness (represented by Leriche's medical research) suggests that disease is a qualitatively different state than health, a different mode of biological being altogether. The experience of disease involving pain, fevers, and nausea, are indicators of a wholly different mode of biological being, not simply a greater or lesser state of balance. In this case medicine's role is to treat the symptoms as the disease itself.

However, it is the third and last transition in concepts of illness that is the most telling—what Canguilhem called "disease as error." Molecular genetics and biochemistry configures disease as an error in the genetic code, an error in the function of the program of the organism. This is the current hope behind research into the genetic mechanisms of a range of diseases and disorders, from diabetes to cancer. But what this requires is another kind of medical hermeneutics, one very different from the patient's testimony of the Hippocratic and Galenic traditions, and one very different from the semiotic approach of 18th-century pathological anatomy (where lesions on the tissues are signs or traces of disease). The kind of medical hermeneutics required is more akin to a kind of occult cryptography, a deciphering of secret messages in genetic codes. *Disease expresses itself not via the patient's testimony, not via the signs of the body's surfaces, but via a code that is a kind of key or cipher.* The hope is that the *p53* gene is a cipher to the occulted book of cancerous metastases, and so on. The "disease itself" is everywhere and nowhere—it is clearly immanent to the organism, the body, the patient, but precisely because of this immanence it cannot be located, localized, or contained (and certainly not in single genes that "cause" disease). Instead, disease is an informatic expression, both immanent and manifest, that must be mapped and decoded.

# RFC 001b: BmTP

There already exists a technological infrastructure for enabling an authentic integration of biological and informatic networks. In separate steps, it occurs daily in molecular biology laboratories. The technologies of genomics enable the automation of the sequencing of DNA from any biological sample, from blood, to test tube DNA, to a computer file of text sequence, to an online genome database. And, conversely, researchers regularly access databases such as GenBank for their research on *in vitro* molecules, enabling them to synthesize DNA sequences for further research. In other words, there already exists, in many standard molecular biology laboratories, the technology for encoding, recoding, and decoding biological information. From DNA in a test tube to an online database, and back into a test tube. *In vivo, in vitro, in silico.* What enables such passages is the particular character of the networks stitching those cells, enzymes, and DNA sequences together. There are at least two networks in play here: the informatic network of the Internet, that enables uploading and downloading of biological information, and that brings together databases, search engines, and specialized hardware. Then there is the biological network of gene expression that occurs in between DNA and a panoply of regulatory proteins, processes that commonly occur in the living cell. The current status of molecular biology labs enables the layering of one network onto the other, so that the biological network of gene expression, for instance, might literally be mapped onto the informatic network of the Internet. The aim would thus be to "stretch" a cell across the Internet. At Location A, a DNA sample in a test tube would be encoded using a genome-sequencing computer. A network utility would then take the digital file containing the DNA sequence and upload it to a server (or relay it via a peer-to-peer application). A similar utility would receive that file, and then download it at Location B, from which an oligonucleotide synthesizer (a DNA-synthesis machine) would then produce the DNA sequence in a test tube. On the one hand, this would be a kind of molecular teleportation, requiring specialized protocols (and RFCs), not FTP, not http, but BmTP, a *biomolecular transport protocol*. Any node on the BmTP network would require three main things: a sequencing computer for encoding (analogue-to-digital), software for network routing (digital-to-digital), and a DNA synthesizer for decoding (digital-to-analogue). If this is feasible, then it would effectively demonstrate the degree to which a single informatic paradigm covers what used to be the mutually exclusive domains of the material and the immaterial, the biological and the informatic, the organism and its milieu.

## UNIVERSALS OF IDENTIFICATION

RFC 793 states one of the most fundamental principles of networking: "Be conservative in what you do, be liberal in what you accept from others."[641] As a political program this means that communications protocols are *technologies of conservative absorption*. They are algorithms for translating the liberal into the conservative. And today the world's adoption of universal communications protocols is nearing completion, just as the rigid austerity measures of neoliberal capitalism have absorbed all global markets.

Armand Mattelart wrote that the modern era was the era of universal standards of communication.[642] The next century will be the era of universal standards of identification. In the same way that universals of communication were levied in order to solve crises in global command and control, the future's universals of identification will solve today's crises of locatability and identification. The problem of the criminal complement is that *they can't be found*. "To know them is to eliminate them," says the counterinsurgency leader in *The Battle of Algiers*. The invention of universals of identification, the ability to locate physically and identify all things at all times, will solve that problem. In criminal cases, psychological profiling has given way to DNA matching. In consumer products, commodity logistics has given way to RFID databases. Genomics are the universal identification of life in the abstract; biometrics are the universal identification of life in the particular; collaborative filters are the universal identification of life in the relational. This is not a case of simple homogenization; rather, it is a set of techniques for quantizing and identifying ("if you enjoyed this book..."), but for doing so within a universalizing framework. In short, the problem of software anomalies such as junk or spam will perfectly overlap with marketing techniques such as cookies and consumer profiling. Indeed, they already have.

The 20th century will be remembered as that time when not all was media, when there existed nonmedia. In the future there will be a coincidence between happening and storage. After universal standards of identification are agreed upon, real-time tracking technologies will increase exponentially, such that almost any space will be iteratively archived over time using "grammars of action." Space will become rewindable. Henceforth, the lived environment will be divided into identifiable zones and nonidentifiable zones, and the nonidentifiables will be the shadowy new "criminal" classes—those that do not identify. Philosophically this is embodied in the phrase at the top of sites such as Amazon.com, a phrase that asks one to identify themselves by disclaiming identification: "Hello, Hal. If you'not not Hal, click here."

## UNKNOWN UNKNOWNS

Fredric Jameson wrote: it is easier to imagine the deterioration of the earth and of nature than the end of capitalism. The nonbeing of the present moment is by far the hardest thing to imagine. How could things have been otherwise? What is it—can one ever claim with certainty—that hasn't happened, and how could it ever be achieved? "Reports that say that something hasn't happened are always interesting to me," Secretary of Defense Donald Rumsfeld said on the morning of February 12, 2002, responding to questions from the press on the lack of evidence connecting Iraqi weapons of mass destruction with terrorists. "Because as we know, there are known knowns; there are things we know we know. We also know there are known unknowns; that is to say we know there are some things we do not know. But there are also unknown unknowns—the ones we don't know we don't know." There is the unknown soldier (e.g., soldiers who have died in battle but cannot be identified). But this is a known unknown, a statistical process of elimination. It is the unknown unknown that is the most interesting. It is a characteristic of present knowledge that it cannot simply be negated to be gotten rid of, it must be negated twice. The tragedy of the contemporary moment is that this double negation is not, as it were, nonaligned; it is already understood as a deficiency in one's ability to imagine, not utopia, but dystopia: the inability to imagine that terrorists would use planes as missiles, just as it was the inability to imagine the kamikaze pilot at Pearl Harbor. These are Rumsfeld's unknown unknowns. The imagination of the future, the vision of the new, is a vision of death, fear, and terror. So not only is the unknown unknown a threat as such, and therefore difficult to bring into imagination as utopia or any another mode of thought, the very process of attempting to imagine the unknown unknown drags into the light its opposite, the end of humanity.

# TACTICS OF NONEXISTENCE

The question of nonexistence is how to develop techniques and technologies to make oneself unaccounted for. Michael Naimark demonstrated how a simple laser pointer can blind a surveillance camera when the beam is aimed directly at the camera's lens. With this type of cloaking one is not hiding, simply nonexistent to that node. The subject has full presence, but is simply not there *on the screen.* Elsewhere, one might go online, but trick the server into recording a routine event. That's nonexistence. One's data is there, but it keeps moving, of its own accord, in its own temporary autonomous ecology. Tactics of abandonment are positive technologies, they are tactics of fullness. There is still struggle in abandonment, but it is not the struggle of confrontation, nor the bureaucratic logic of war. It is a mode of nonexistence: the full assertion of the abandonment of representation. Absence, lack, invisibility, and nonbeing have nothing to do with nonexistence. Nonexistence is nonexistence not because it is an absence, or because it is not visible, but precisely because it is full. Or rather, because it permeates. That which permeates is not arbitrary, and not totalizing, but tactical.

Of course, nonexistence has been the concern of anti-philosophy philosophers for some time. Nonexistence is also a mode of escape, an "otherwise than being." Levinas remarks that "escape is the need to get out of oneself."[643] One must always choose either being or nonbeing (or worse, becoming . . . ). The choice tends to moralize presence, that one must be accounted for, that one must, more importantly, account for oneself, that accounting is tantamount to self-identification, to *being* a subject, to individuation. "It is this category of getting out, assimilable neither to renovation nor to creation, that we must grasp . . . it is an inimitable theme that invites us to get out of being."[644] And again Levinas: "the experience that reveals to us the presence of being as such, the pure existence of being, is an experience of its powerlessness, the source of all need."[645]

Future avant-garde practices will be those of nonexistence. But you still ask: How is it possible not to exist? When existence becomes a measurable science of control, then nonexistence must become a tactic for any thing wishing to avoid control. "A being radically devoid of any representable identity," Agamben wrote, "would be absolutely irrelevant to the State."[646] Thus we should become devoid of any *representable* identity. Anything measurable might be fatal. These strategies could consist of: nonexistent action (non-doing); unmeasurable, or not-yet-measurable human traits, or the promotion of measurable data of negligible importance. Allowing to be measured now and again for false behaviors, thereby attracting incongruent and ineffective control responses, can't hurt. A driven exodus or a pointless desertion are equally virtuous in the quest for nonexistence. The bland, the negligible, the featurelessness are its only evident traits. The nonexistent is that which cannot be cast into any available data types. The nonexistent is that which cannot be parsed by any available algorithms. This is not nihilism; it is the purest form of love.

## DISAPPEARANCE, OR, "I'VE SEEN IT ALL BEFORE"

For Paul Virilio, disappearance is the unforeseen by-product of speed. Technology has gone beyond defining reality in the quantized, frames-per-second of the cinema. Newer technologies still do that, but they also transpose and create quantized data through time-stretching, morphing, detailed surface rendering, and motion capture, all with a level of resolution beyond the capacity of the human eye (a good argument for optical upgrades): "the world keeps on coming at us, to the detriment of the object, which is itself now assimilated to the sending of information."[647] Things and events are captured before they are finished, in a way, before they exist as things or events. "Like the war weapon launched at full speed at the visual target it's supposed to wipe out, the aim of cinema will be to provoke an effect of vertigo in the voyeur-traveler, the end being sought now is to give him the impression of being projected into the image."[648] Before the first missiles are launched, the battlefield is analyzed, the speeches made, the reporters are embedded, the populations migrate (or are strategically rendered as statistical assets), and the prime-time cameras are always on. But this is not new, for many of Virilio's examples come from World War II military technologies of visualization. In this context, a person is hardly substantial—one's very physical and biological self keeps on slipping away beneath masses of files, photos, video, and a panoply of net tracking data. But luckily, you can move. All the time, if you really want to.

Hakim Bey's "temporary autonomous zone" (TAZ) is, in a way, the response to Virilio's warnings against the aesthetics of disappearance. But the issue here is nomadism, not speed. Or, for Bey, nomadism is the response to speed (especially the speed produced by the war + cinema equation). A TAZ is by necessity ephemeral: gather, set-up, act, dissemble, move on. Its ephemeral nature serves to frustrate the recuperative machinations of capital. The TAZ nomad is gone before the cultural and political mainstream knows what happened. This raises the issue of efficacy. The TAZ wages the risk of an efficacy that is invisible, de-presented, an efficacy whose traces are more important than the event itself. (Is this a distributed efficacy?) But this then puts us into a game of cloak-and-dagger, a kind of cat and mouse game of forever evading, escaping, fleeing the ominous shadow of representation.

Perhaps the challenge today is not that of hyper-visualization (as Virilio worried), or of non-recuperation (as Bey suggested), but instead a challenge of existence without representation (or at least existence that abandons representation, a nonexistence, or better, an "inexistence"). The challenge would not be that of resisting visualization (e.g., refusing to be a consumer profile, a data point), and neither would it be that of constantly escaping representation (e.g., using avatars, aliases, screen identities). Resistance and escape would have to be replaced by a certain indiscernability; *tactics of evasion* would have to be replaced by *operations of narcolepsy*. The poetics of spam, much like Surrealist automatism, obtains its uncanny quality through a strange active passivity. This would entail another type of disappearance: "Disappearance is not necessarily a 'catastrophe'—except in the mathematical sense of 'a sudden topological change.'"[649]

## (THERE IS) NO CONTENT

Theories of media and culture continue to propagate an idea of something called "content." But the notion that content may be separated from the technological vehicles of representation and conveyance that supposedly facilitate it is misguided. Data has no technique for creating meaning, only techniques for interfacing and parsing. To the extent that meaning exists in digital media, it only ever exists as the threshold of mixtures between two or more technologies. Meaning is a data conversion. What is called "Web content" is, in actual reality, the point where standard character sets rub up against the hypertext transfer protocol. There is no content; there is only data and other data. In Lisp there are only lists; the lists contain atoms, which themselves are other lists. To claim otherwise is a strange sort of cultural nostalgia, a religion. Content, then, is to be understood as a relationship that exists between specific technologies. Content, if it exists, happens when this relationship is solidified, made predictable, institutionalized and mobilized.

## JUNK OR NOT JUNK

From: Mr. Lou-Wong <lou_wong@globalum.com>. Sent: 17 July 2005. Subject: Proposal! Message: We've identified this e-mail as junk. Please tell us if we were right or wrong by clicking Junk or Not Junk.

# NOTES

1.  The SPAM trademark™ (in uppercase) belongs to Hormel Foods (http://www.spam.com) who imported the product into Britain as part of the Lease-Lend Act (1941) which enabled the United States to sell, transfer, exchange, and lend products to their allies during World War II. SPAM is still widely available today to those with a discerning taste in a "luncheon meat" alternative to ham.
2.  See also the *"Sermon on the Hill"* sketch in the film *The Life of Brian* in which the misinterpretation of one-to-many communication leads to a broken nose and the *"Argument Sketch"* in which the principles of logical argumentation are "argued" about, without resolution.
3.  A warm thank you to Juri Nummelin for this example. Nummelin has collaged his own cut-and-paste-Spam-Poetry from such unsolicited messages. See The Nokturno-poetry Web site, http://www.nokturno.org/juri/juri nummelin corporation near class.pdf (accessed February 6, 2007).
4.  Bill Gates, *The Road Ahead* (London: Penguin, 1996).
5.  Susanna Paasonen's chapter in this book is interesting in this respect. She responds to the glut of pornography spam that entered her university inbox. A problem that many of us have to deal with, together with discovering the limitations of spam filter programs, downloading free ad-blockers and searching for the cheapest way to set up an antivirus program or firewall.
6.  According to the *Oxford English Dictionary*
7.  See Don Evett, "Spam Statistics 2006." http://spam-filter-review.toptenreviews.com/spam-statistics.html (accessed May 24, 2007).
8.  Over a 7-day period, the honeypot computer experienced 36 warnings that pop-up via Windows Messenger, 11 separate visits by Blaster worm, 3 separate attacks by Slammer worm, 1 attack aimed at Microsoft IIS Server and 2 to 3 "port scans" seeking weak spots in Windows software. See "Tracking Down Hi-tech Crime." BBC News, http://news.bbc.co.uk/go/pr/fr/-/1/hi/technology/5414502.stm. Published: 2006/10/08 23:12:09 GMT (accessed May 24, 2007).
9.  This is the conclusion of market research carried out for Get Safe Online, a joint initiative among the British government, the Serious Organized Crime Agency (SOCA), British Telecom, eBay.co.uk, HSBC, Microsoft, and Secure Trading.

10. See for example "Attack of the Bots." *Wired* 14.11 (November 2006). http://www.wired.com/wired/archive/14.11/botnet.html (last accessed April 12, 2007). Of course, the discourse of digital waste and malicious code already originated from the mid-1980s. See Jussi Parikka, *Digital Contagions. A Media Archaeology of Computer Viruses* (New York: Peter Lang, 2007).

11. Brian Massumi. *A User's Guide to Capitalism and Schizophrenia. Deviations from Deleuze and Guattari* (Cambridge, MA: MIT Press, 1992), 120. See Gilles Deleuze's critique of the four shackles of mediated representation in *Difference and Repetition* (London & New York: Continuum, 1997), 29.

12. For example, drawing inspiration from the Love Bug computer virus, Stuart Sim's essay on Lyotard's Inhuman suggests that the virus is engaged in a "deliberate [social] blurring of the lines between human beings and machines" (p. 15). He argued that we have, to some extent, ceded autonomy to our computer systems. Indeed, the virulent impact of the "love bug" virus denotes to Sim a "battle for control of cyberspace" (p. 20) between the human and the intelligent machine. See Stuart Sim *Lyotard's Inhuman* (Cambridge: Icon 2001), 18-20.

13. For example see David Gauntlet's article *"Ten Things Wrong with Effects Theory."* http://www.theory.org.uk/david/effects.htm (accessed April 2007).

14. See for example, Gerbner's cultivation theory. Also, note that we are not attributing effects theory to the entire field of media studies. Indeed, we fully recognize the contribution of political economy in the refutation of adaptations to cybernetic effects models. For example, Dan Schiller rejects cybernetics for its "tendency to operationalize the (social) system in mechanistic terms." See Schiller in Vincent Mosco and Janet Wasko (eds.) *The Political Economy of Information* (Madison: University of Wisconsin Press, 1988), 29.

15. Gilles Deleuze, *Spinoza: Practical Philosophy*, Trans. Robert Hurley (San Francisco: City Lights Books, 1988). Causal capacity is a term used by DeLanda to describe how one assemblage might affect another. See Manuel DeLanda, *A New Philosophy of Society: Assemblage Theory and Social Complexity* (London, New York: Continuum, 2006), 38.

16. See for example Luciana Parisi's chapter in this book and her use of the topological concept to describe D'arcy Thompson's biomathematical approach to the evolution of form.

17. Our approach has affinities with what has been termed *abstract materialism* (Parisi), neo-materialism (DeLanda) and even the analysis of actor networks (Latour), and we borrow freely from the various thinkers who have explicated the potential of analyzing cultural bodies in terms of the interconnectness of material and expressive networks. See Luciana Parisi, *Abstract Sex. Philosophy, Bio-Technology and the Mutations of Desire* (London: Continuum, 2004). Manuel DeLanda, "Immanence & Transcendence in the Genesis of Form." *South Atlantic Quarterly*, Vol. 96, No 3, Summer 1997. Also in: *A Deleuzian Century?*, edited by Ian Buchanan (Durham: Duke University Press, 1999), 499-514. Bruno Latour, *Reassembling the Social: An Introduction to Actor-Network Theory* (Oxford: Oxford University Press, 2005).

18. Theodor W. Adorno, *The Culture Industry. Selected Essays on Mass Culture* (London & New York: Routledge, 2002).

19. Neil Postman, *Amusing Ourselves to Death* (New York: Viking, 1985).

20. See "New Media Panics." M/Cyclopedia of New Media. http://wiki.media-culture.org.au/index.php/New_Media_Panics (accessed January 9, 2007). See also Rogers' and Elmer's chapters on censorship in this book.

21. Lisa Gitelman, *Always Already New. Media, History, and the Data of Culture* (Cambridge, MA: MIT Press, 2006), 130-131.

22. In addition, most of the cultural analysis on, for example, pornography focuses on (human) bodily interactions, representations, identity, etc. See for instance Dennis D. Waskul (ed.) *Net.seXXX. Readings on Sex, Pornography, and the Internet* (New York: Peter Lang, 2004). See also Susanna Paasonen, Kaarina Nikunen and Laura Saarenmaa (eds.) *Pornification: Sex and Sexuality in Media Culture* (Oxford: Berg, 2007).

23. For example, capitalist desire in spam e-mail provides us with key insights into the organization of representational content in mass media, where they also negotiate gender roles and sexualities.

24. See the contributions by Paasonen and Jacobs in this book.

25. Parisi, *Abstract Sex.*, 134. Jussi Parikka, "Contagion and Repetition–On the Viral Logic of Network Culture." *Ephemera—Theory & Politics in Organization*, Vol. 7, No. 2, May 2007, http://www.ephemeraweb.org/journal/7-2/7-2parikka.pdf (accessed November 14, 2007). The theme of metastability stems largely from Gilbert Simondon. Simondon analyzed metastable systems in terms of individuation and change. In his *Du mode d'existence des object techniques,* Simondon argued against the fashion of seeing technical objects as self-contained, and proposed to read them in terms of milieus and potential becomings. Also technical objects and systems can be metastable and open to future fluctuations.

26. Wendy Hui Kyong Chun, *Control and Freedom. Power and Paranoia in the Age of Fiber Optics* (Cambridge, MA: MIT Press, 2006), 16. Chun, following Katherine Hayles, proposes the need for medium specific criticism in order to arrive at such a mode of cultural analysis that does not reduce differences to paranoid narratives, but cultivates complexities.

27. Similarly, semioticians would want to match up these familiar codes in relation to a *sign system* in an attempt to represent the cafe in terms of a cultural identity.

28. Paul Virilio, "The Primal Accident." Translated by Brian Massumi. In *The Politics of Everyday Fear,* edited by Brian Massumi (Minneapolis and London: University of Minnesota Press, 1993), 211-212.

29. Gilles Deleuze, *Logic of Sense* (London: New York: Continuum, 2003), 170.

30. Tiziana Terranova, *Network Culture. Politics for the information Age* (London: Pluto, 2004), 67-68.

31. Benedict Spinoza, *Ethics.* Trans. W. H. White, revised by A.H. Stirling (Hertfordshire: Wordsworth, 2001), 41 (I part, Appendix).

32. Emile Durkheim, *Suicide. A Study in Sociology.* Trans. John A. Spaulding & George Simpson (London: Routledge, 2002). See Dougal Phillips' chapter in this book.

33. Jean Baudrillard, "A Perverse Logic—Society's Attitude Towards Drugs." *Unesco Courier* (July 1987). http://findarticles.com/p/articles/mi_m1310/is_1987_July/ai_5148909/pg_2 (last accessed April 12, 2007).

34. Ibid.

35. Manuel DeLanda, *A New Philosophy of Society* (London: New York: Continuum, 2007), 9-12.

36. Kevin Robins and Frank Webster, *Times of Technoculture* (London, New York: Routledge, 1999), 179-181.

37. David Harley, Robert Slade, and Urs E. Gattiker, *Viruses Revealed! Understand and Counter Malicious Software* (New York: Osborne/McGraw-Hill, 2001), 49.

38. See, for example, Kephart, Sorkin, and Swimmer's paper, *An Immune System for Cyberspace,* delivered at the IEEE International Conference on Systems, Man, and Cybernetics—Artificial Immune Systems and Their Applications in 1997. IBM's immune system was eventually developed in partnership with the antivirus company Symantec.

39. A study of 800 computer virus infections found that computer viruses tend to stay at a low, but stable level of infection over long periods—up to 3 years in some cases. The researchers concluded that the Internet is "prone to the spreading and the persistence of infections at a prediction of a nonzero epidemic threshold." (See Pastor-Satorras & Vespignani, *Epidemic Spreading in Scale-Free Networks.* Phys. Rev. Lett. Issue 14. April 2, 2001).

40. Albert-László Barabási, *Linked: How Everything is Connected to Everything Else and What It Means for Business, Science, and Everyday Life* (New York: Plume, 2003). See also Sampson and Rogers' chapter in this book.

41. On Turing's work on Hilbert's Entscheidungsproblem see Jon Agar, *Turing and the Universal Machine* (Cambridge: Icon Books, 2001), 85-100.

42. See reference to Fred Cohen's PhD thesis and experiments with computer viruses in 1983 in Eric Louw and Neil Duffy, *Managing Computer Viruses* (Oxford: Oxford University Press, 1992), 7-9.

43. See J.E. Lovelock, *Gaia: a New Look at Life on Earth* (Oxford: OUP, 1979) and *The Ages of Gaia: A Biography of our Living Earth* (Oxford: OUP, 1988).

44. See John Perry Barlow "Go Placidly Amidst the Noise And Haste," a New Perspectives Quarterly interview with John Perry Barlow, Co-founder of the Electronic Frontier Foundation. http://www.eff.org/Misc/Publications/John_Perry_Barlow/HTML/npq.html (accessed April 4, 2007).

45. Cyberpunk author, Bruce Sterling, writing for *Antivirus Online* Volume 2: Issue 1. Archived at vx.netlux.org/lib/mbs00.html (accessed April 4, 2007).

46. Roy Mark, "The Internet: 'A Dirty Mess.'" www.internetnews.com, June 8, 2004. http://www.internetnews.com/bus-news/print.php/3365491 (last accessed April 12, 2007).

47. Geert Lovink, *The Principle of Notworking. Concepts in Critical Internet Culture* (Amsterdam: HVA Publicaties 2005), 10. <http://www.hva.nl/lectoraten/documenten/ol09-050224-lovink.pdf>.

48. Félix Guattari, in this sense, further questioned the tenets of second-order cybernetics, and in particular the distinction made between autopoietic machines and allopoietic machines. See Félix Guattari, *Chaosmosis* (Sydney: Power Publications, 1995), 39-40. According to the key definitions set out by Maturana and Varela, autopoietic machines are self-organizing unities; a system able to recursively engender the identical network of processes that produced them in the first place (self-production). For example, the bounded structure of a biological cell is evoked as a process of circular homeostatic maintenance. Consistent with Maturana and Varela's definition, autopoietic machines (limited to the biological domain) differ from allopoietic machines (technical and social systems), which are defined in terms of a purpose other than the maintenance of their own

self-production. In this sense, allopoietic machines produce (or reproduce) something other than themselves, for instance, an assembly factory that produces cars. However, when such strict boundaries are applied to the cybernetic machines of the digital network there is a certain amount of conceptual seepage. Guattari sees an overlap between autopoietic and allopoietic machines when one considers technical machines in the "context of the machinic assemblages they constitute with human beings."

49. We have both discussed the significance of the relation of alterity in the context of the digital anomaly elsewhere. For example, critical of the "conservative" circularity of homeostatic processes described by Maturana and Varela, Parikka has drawn upon Guattari to focus instead on the evolutionary coupling of the computer virus to the digital media ecology (see Jussi Parikka, *The Universal Viral Machine: Bits, Parasites and the Media Ecology of Network Culture in Ctheory*, 2005). Similarly, Sampson discussed how the "significant topological intensity of the human–computer assemblage . . . shifts the contemporary debate on noise away from Shannon's model towards a complex, non-linear and relational interaction" (See Tony Sampson, "Senders, Receivers and Deceivers: How Liar Codes Put Noise Back on the Diagram of Transmission." *M/C Journal* 2006).

50. "Attack of the Bots." *Wired* 14.11 (November 2006). <http://www.wired.com/wired/archive/14.11/botnet.html>.

51. Lev Manovich, *The Language of New Media* (Cambridge, MA: MIT Press, 2001), 27-48.

52. Wendy Hui Kyong Chun, "On Software, or the Persistence of Visual Knowledge." *Grey Room* 18 (Winter 2004), 28-51. See also Lisa Gitelman's analysis of the nature of Internet objects in her book *Always Already New*.

53. Harley, Slade, and Gattiker, 157.

54. Matthew Fuller, *Media Ecologies. Materialist Energies in Art and Technoculture* (Cambridge, MA: MIT Press, 2005), 103-107. With standard objects we mean, following Fuller, the *objectification* of a certain module, or a process, so that it can be relied on in the future. In other words, making an object into a standard requires the fabrication and stabilization of its potentialities so that it can be consistently applied in different contexts. Standard objects are of course only idealizations, in as much as they work, but can also break down. They are the result of forces, assemblies and mechanisms, yet never without a residual potentiality, as Fuller reminded us. They can always be stretched and opened up, made to break down and to take on new functions.

55. For the metaphoric regime in cultural studies see Mary Flanagan, "Spatialized MagnoMemories (feminist poetics of the machine)." *Culture Machine* 3 *Virologies: Culture and Contamination* (2001). August 14, 2006. Online version: culturemachine.tees.ac.uk/Cmach/Backissues/j003/Articles/flannagan.htm (accessed December 10, 2007). For computer science see Eric Louw and Neil Duffy, *Managing Computer Viruses* (Oxford: Oxford University Press, 1992) and for an analysis of the rhetoric of the antivirus and network security see Stefan Helmreich "Flexible Infections: Computer Viruses, Human Bodies, Nation-States, Evolutionary Capitalism." *Science, Technology and Human Values*, Vol. 25(4) Autumn 2000, 472-491. Online version: web.mit.edu/anthropology/faculty_staff/helmreich/PDFs/flexible_infections.pdf (accessed December 10, 2007).

56. See Parisi's references to Dawkins' *Biomorph Land* program and Johnston's use of Tom Ray's Darwinian inspired ALife work in this section. The reference to an arms race—aka the neo-Darwinian survival of the fittest algorithm—is taken from Dawkins' contribution to the antivirus debate. See Richard Dawkins, "Viruses of the Mind." In *Dennett and His Critics: Demystifying Mind*, ed. Bo Dalhbom (Cambridge, MA: Blackwell, 1993).

57. A.K. Dewdney, "In a game called Core War programs engage in a battle of bits," *Scientific American* (May 1984), 15-19. According to the computer scientist M.D. McIlroy, "a 1961 game of survival of the fittest among programs called Darwin" was the progenitor of *Core War*. For a description of *Darwin*, see http://www.cs.dartmouth.edu/~doug/darwin.pdf (accessed May 27, 2007).

58. M. Mitchell Waldrop, *Complexity: The Emerging Science at the Edge of Order and Chaos* (New York: Simon & Schuster, 1992), 238.

59. Dewdney's second article was published in *Scientific American* 252 (March 1985), 14-23. In *Artificial Life II* (Reading, MA: Addison-Wesley, 1992), ed. Christopher G. Langton et al., Langton did publish Eugene Spafford's article, "Computer Viruses—A Form of Artificial Life?" However, although Spafford acknowledged that science has much to learn from studying computer viruses, he was disturbed that their "origin is one of unethical practice" (744).

60. Cohen described these experiments in his book *A Short Course on Computer Viruses* (New York: Wiley, 1994).

61. Steven Levy, *Artificial Life* (New York: Pantheon Books, 1992), 324.

62. On the new viral ecology, see Jussi Parikka, "The Universal Viral Machine: Bits, Parasites and the Media Ecology of Network Culture," *CTheory* (2005), http://www.ctheory.net/articles.aspx?id=500 (accessed May 27, 2007); on machinic-becoming, see my book, *The Allure of Machinic Life* (Cambridge, MA: MIT Press, 2008).

63. In Langton's system, the reproducing "cell assemblies" took the form of growing colonies of digital loops, similar in certain respects to sea coral. See Christopher G. Langton, "Studying Artificial Life with Cellular Automata," *Physica D* 22 (1986), 120-149.

64. See Rasmussen et al., "The CoreWorld: Emergence and Evolution of Cooperative Structures in a Computational Chemistry," *Physica D* 42 (1990), 111-134.

65. Thomas S. Ray, "Evolution and Complexity," in *Complexity: Metaphors, Models, and Reality*, ed. George A. Cowan et al. (Reading, MA: Addison-Wesley, 1994), 166.

66. For discussion, see chapter 5 in my forthcoming book, *The Allure of Machinic Life*.

67. Pargellis noted, for example, that although viruses are small and relatively short-lived, autonomous replicators exhibit long-term growth as well as a tendency to evolve more efficient code and greater functionality. Yet it is also evident that by enabling protobiotic forms to replicate, these viruses promote and insure the dynamic viability of the soup in which the replicators thrive.

68. The term *quasi-species* was coined by Manfred Eigen, whose article "Viral Quasispecies," *Scientific American* (July 1993), 42-49, provides the background to this discussion.

69. As reported by Henry Bortman in "Survival of the Flattest: Digital Organisms Replicate," *Astrobiology Magazine* (October 7, 2002), available at http://www. space.com/scienceastronomy/generalscience/digital_life_021007.html (accessed May 27, 2007).
70. Ibid.
71. Mark A. Bedau, Emile Snyder, C. Titus Brown and Norman H. Packard, "A Comparison of Evolutionary Activity in Artificial Evolving Systems and in the Biosphere," *Fourth European Conference on Artificial Life*, ed. Phil Husbands and Inman Harvey (Cambridge, MA: MIT Press, 1997), 125-134.
72. Ibid., 126.
73. *Evita* is an ALife platform similar to Ray's *Tierra* and Adami's *Avida*, whereas in *Bugs* the digital organisms move about on a spatial grid by means of a simulated sensorimotor mechanism that allows them to "sense" resource sites, move toward them, and replenish themselves. These resources are necessary for the organism to pay existence and movement "taxes" as well as to reproduce. Thus, in addition to reproduction and evolution, *Bugs* simulates a form of "metabolism." *Bugs* was first introduced into ALife research by Norman Packard in "Intrinsic Adaptation in a Simple Model for Evolution," *Artificial Life*, ed. Christopher G. Langton (Reading, MA: Addison-Wesley, 1989) 141-155.
74. Bedau et al., "A Comparison of Evolutionary Activity," 132.
75. Ibid., 130.
76. Ibid.
77. See Bedau and Packard, 129, for a full explanation.
78. Mark A. Bedau, Emile Snyder, and Norman H. Packard, "A Classification of Long-Term Evolutionary Dynamics," *Artificial Life VI*, ed. Christof Adami et al. (Cambridge, MA: MIT Press, 1998), 228-237.
79. Bedau et al., "A Comparison of Evolutionary Activity," 132.
80. Ibid.
81. Bedau, Snyder, and Packard, "A Classification," 136.
82. See chapter 5, in *The Allure of Machinic Life*.
83. In regard to this work, see Bedau et al., "Open Problems in Artificial Life," *Artificial Life* 6, no. 4 (2000), 363-376.
84. Thomas S. Ray, "An Approach to the Synthesis of Life," in *Artificial Life II*, 393.
85. Ibid., 398.
86. Thomas S. Ray, "Selecting Naturally for Differentiation: Preliminary Evolutionary Results," *Complexity* 3, no. 5 (1998), 26.
87. For Ray, the quantitative measure of evolutionary complexity in *Internet Tierra* is the level of differentiation of the multicelled organism, beginning with the most primitive level of differentiation, that is, two celltypes. Although this level of differentiated state persists through prolonged periods of evolution (Ray's first "milestone"), the number of cell types did not increase (the second milestone). According to these criteria, which are somewhat different from the measures of evolutionary activity set forth by Bedau and Packard, whose work Ray does not mention, *Internet Tierra* can be said to have achieved limited success.
88. Thomas S. Ray and Joseph Hart, "Evolution of Differentiated Multi-threaded Digital Organisms," *Artificial Life VI*, ed. Christoph Adami et al. (Cambridge, MA: MIT Press, 1998), 3.

89. David H. Ackley, "Real Artificial Life: Where We May Be," *Artificial Life VII*, ed. Mark A. Bedau et al. (Cambridge, MA: MIT Press, 2000), 487-496.

90. Ibid., 487.

91. Ibid., 495. A more detailed treatment of this parallel would necessarily require that software development be considered in relation to hardware development, or rather, in keeping with the natural analogy, the two would be considered together as forming a co-evolutionary relay. To my knowledge, no history of computers or computation has attempted to do this.

92. Ibid., 488.

93. Ibid.

94. Ibid., 491.

95. Ibid., 493.

96. Ibid.

97. Ibid., 495. GNU refers to the free software operating system and its various packages first made available by the Free Software Foundation founded by Richard Stallman. "GNU" is a recursive acronym that stands for "GNU's Not Unix."

98. Ibid.

99. This shift coincides with—and no doubt can be correlated with—increasing use of the Internet, which greatly accelerated with the appearance of the first web page browser Mosaic in 1993.

100. Mark A. Ludwig, *Computer Viruses, Artificial Life and Evolution* (Tucson, AZ: American Eagle Publications, 1993). Ludwig also published a series of "black books" on computer viruses that are generally considered to be among the best available.

101. For example, although he doesn't mention computer viruses, James P. Crutchfield reached a similar conclusion in his essay, "When Evolution is Revolution—Origins of Innovation," in *Evolutionary Dynamics*, ed. James P. Crutchfield and Peter Shuster (Oxford, UK: Oxford University Press, 2003).

102. Jeffrey O. Kephart, "A Biologically Inspired Immune System for Computers," in *Artificial Life IV*, ed. Rodney A. Brooks and Pattie Maes (Cambridge, MA: MIT Press, 1994), 130-139.

103. Quoted by Lesley S. King, "Stephanie Forrest: Bushwacking Through the Computer Ecosystem," *SFI Bulletin* Vol. 15, no. 1 (Spring 2000).

104. "Principles of a Computer Immune System," A. Somayaji, S. Hofmeyr, and S. Forrest, *1997 New Security Paradigms Workshop*, ACM (1998), 75-82.

105. Ibid., 79.

106. Ibid., 80.

107. Albert-Laszlo Barabási, *Linked: The New Science of Networks* (Cambridge, MA: Perseus Publishing, 2002).

108. Derrick Story, "Swarm Intelligence: An Interview with Eric Bonabeau," The O'Reilly Network (02/21/2002), 3. Available at http://www.oreillynet.com/pub/a/p2p/2003/02/21/bonabeau.html (last accessed May 27, 2007).

109. Ibid., 4.

110. Michel Hardt and Antonio Negri, *Empire* (Boston & London: Harvard University Press, 2000), 136.

111. Ibid, 62 & 197-198.

112. Here I make a reference to a critique by Luciana Parisi, *Abstract Sex* (London: Continuum), 144-145, in which the parasites of Hardt and Negri's *Empire* are understood to reinforce a distinction between nonorganic and organic life.

113. Andrew Goffey's (2005) editorial comments to the online journal *Fibreculture*, 4, *Contagion and Diseases of Information*. http://journal.fibreculture.org/issue4/index.html (accessed April 3, 2007).

114. See Eugene Thacker, "Living Dead Networks" in *Fibreculture*, 4. http://journal.fibreculture.org/issue4/issue4_thacker.html (accessed April 3, 2007).

115. Isabelle Stengers, *Penser avec Whitehead* (Paris: Seuil, 2002), 186. See also Goffey, 2005.

116. Thacker, 2005.

117. John Doyle et al., "The "Robust Yet Fragile Nature of the Internet," *PNAS* 102 (41), October 11, 2005. http://www.pnas.org/cgi/reprint/0501426102v1. pdf (accessed November 19, 2007).

118. The scale-free model is at the center of a paradigm shift in how networks are scientifically understood, however, challenges to the model are notably made from within the discipline itself. Most notably, see Duncan Watts, *Six Degrees: The Science of a Connected Age* (London: Vintage, 2003), 101-114.

119. Therefore, the chapter will not spend too much time marvelling over the maps of the Internet and the Web, but instead follows what Brian Massumi terms *incorporeal empiricism.* See Brian Massumi, *Parables for the Virtual* (Durham, NC & London: Duke University Press, 2002), 257-258.

120. See discussion in chapter 1 regarding the contagious capacities of assemblages. See also Brian Massumi, *A Users Guide to Capitalism and Schizophrenia* (Cambridge, MA: MIT Press, 1992), 192.

121. Jan Van Dijk, *The Network Society* (London: Sage, 2006), 2.

122. Van Dijk, 187

123. Significantly, although Van Dijk focused predominantly on digital networks, he grasped that electronic networks are a new infrastructural manifestation of a much older human web (following McNeill & McNeill, *The Human Web* [New York & London: W.W. Norton, 2003]), which began when early tribes started to exchange goods, technologies, ideas, crops, weeds, animal and viral diseases (van Dijk, 22).

124. Albert Laszlo Barabási, *Linked* (London: Plume, 2003), 123-142.

125. Duncan Watts "A Simple Model of Global Cascades on Random Networks," *PNAS* 99 (9), April 30, 2002.

126. Malcolm Gladwell, *The Tipping Point* (London: Abacus, 2000).

127. Gabriel Tarde, *Social Laws: An Outline of Sociology* (Ontario: Canada, 2000)

128. Gustave Le Bon, *The Crowd: A Study of the Popular Mind* (New York: Dover, 2002).

129. Gilles Deleuze and Felix Guattari, *Anti-Oedipus* (London & New York: Continuum, 1984).

130. See, for example, John Perry Barlow's 1996 manifesto, *The Declaration of the Independence of Cyberspace* in Peter Ludlow (ed), *Crypto Anarchy, Cyberstates and Pirate Utopias* (Cambridge, MA: MIT Press, 2001), 27-30.

131. For example, there has been speculation that the Jerusalem virus was an act of political terrorism—it was transmitted to the network from the Hebrew University and its logic bomb triggered to mark the 40th anniversary of the

invasion of Palestine. See David Harley, Robert Slade and Urs E. Gattiker, *Viruses Revealed* (New York: Osborne/McGraw-Hill, 2001), 354. However, in the many variations transmitted before and after Jerusalem, any semblance of a political message became lost in slight changes made to the trigger. Some variants trigger on every Sunday, April 1st or the 26th of each month. Others play French nursery rhymes on every Friday the 13th or interfere with programs run on Saturday. See Alan Solomon, *Virus Encyclopaedia* (Aylesbury: S&S International, 1995), 128-132.

132. Jussi Parikka, *Digital Contagions: A Media Archaeology of Computer Viruses* (New York: Peter Lang, 2007).

133. See Tony Sampson, "Senders, Receivers and Deceivers: How Liar Codes Put Noise Back on the Diagram of Transmission." Transmit *M/C Journal* 9(1) 2004. http://journal.media-culture.org.au/0603/03-sampson.php (accessed April 3, 2007). Tony Sampson, "Dr Aycock's Bad Idea: Is the Good Use of Computer Viruses Still a Bad Idea?" Bad M/C *A Journal of Media and Culture* 8 (1) 2005. http://journal.media-culture.org.au/0502/02-sampson.php (accessed April 3, 2007) and Tony Sampson, "A Virus in Info-Space: The Open Network and its Enemies." *Open M/C Journal* 7(3) 2004. http://www.mediaculture.org.au/040 6/07_Sampson.php (accessed April 3, 2007).

134. Marshall McLuhan, *Understanding Media: The Extensions of Man* (Cambridge, MA: MIT Press, 1997), 7-8.

135. As Baudrillard said it: 'Within the computer web, the negative effect of viruses is propagated much faster than the positive effect of information. That is why a virus is an information itself. It proliferates itself better than others, biologically speaking, because it is at the same time both medium and message. It creates the ultra-modern form of communication that does not distinguish, according to McLuhan, between the information itself and its carrier." Jean Baudrillard *Cool Memories* (New York: Verso, 1990).

136. Stefan Helmreich, "Flexible Infections: Computer Viruses, Human Bodies, Nation-States, Evolutionary Capitalism," *Science, Technology, & Human Values,* 25 (2000), 472-491.

137. Stefan Helmreich, "Flexible Infections: Computer Viruses, Human Bodies, Nation-States, Evolutionary Capitalism," *Science, Technology, & Human Values,* 25 (2000), 472-491.

138. Kim Neely's "Notes from the Virus Underground" appeared in the September 1999 issue of *Rolling Stone* magazine.

139. Michel Foucault, *Power/Knowledge: Selected Interviews and Other Writings 1972-1977* Gordon, C. (ed) (Place: Harvester Press, 1980), 92-108.

140. As they suggest: *'whoever masters the network form first and best will gain major advantages.'* An advantage, they claim is at this moment with the enemy. See John Arquilla and David Ronfeldt, *The Advent of Netwar (Santa Monica:* RAND, 1996), 83.

141. See Thacker, 2005. There is some support to Thacker's notion of epidemic conflict. For example, the deliberate (or inadvertent) actions of a computer virus could, it is argued, paradoxically "muddy the signals being sent from one side to the other in a crisis," breaking the orthodox terms of engagement during war. See Stephen Cimbala, "Nuclear Crisis Management and Information Warfare," *Parameters* (Summer 1999), 117-128. Cimbala is a U.S. academic/military consultant.

142. Former CIA agent Robert Baer warns Britain of the grave dangers ahead in an article "This Deadly Virus" in *The Observer* August 7, 2005
143. Van Dijk, 187.
144. See chapter 3 of the International Monetary Fund's (IMF) 1999 *Global Economy Survey*.
145. Over the period from 1996 to 2000, the number of countries in receipt of equity inflows fell from 90% to 20%, leaving the IMF to claim a partial victory over these rare forms of financial contagions. However, others have since warned of the potential of political contagion if containment leads to the collapse of banking systems in these regions and instability spreads. According to financial journalist Faisal Islam, following the Asian and Russian contagions in the 1990s, the net investment in developing countries was virtually cut off. By the early 2000s, the investment in these countries was "practically zero." See Islam's article: "A Country Praying for Tango and Cash" in *The Observer*, Sunday January 13, 2002.
146. Thacker, 2005.
147. John Shoch and Jon Hupp, "The Worm Programs—Early Experiments with a Distributed Computation," *Communications of the ACM*, 22(3) (1982), 172-180.
148. Fred Cohen, "Friendly Contagion: Harnessing the Subtle Power of Computer Viruses," *The Sciences* (Sep/Oct 1991), 22-28. See also Johnston's chapter in this book, Parikka, 2007 and Tony Sampson, "Dr Aycock's Bad Idea: Is the Good Use of Computer Viruses Still a Bad Idea?" *Bad M/C A Journal of Media and Culture* 8.1 (2005). http://journal.media-culture.org.au/0502/02-sampson. php (accessed April 3, 2007).
149. See Sampson, "Dr Aycock's Bad Idea" and Parikka, *Digital Contagions*, 207-217.
150. See Kephart et al., 1993 "Computers and Epidemiology" IBM research paper.
151. Barabási.
152. Marc Sageman argued that the social networks of the global Salafi jihad tend to cluster around four major network hubs, which are linked by social and technological bridges. Therefore, he claims that the current protection of nation state borders and the "random arrests of low-level individuals (nodes) at border points will not degrade the network," since it is the links made between clusters of terrorists that will expose vulnerabilities. Marc Sageman, *Understanding Terror Networks* (Philadelphia: University of Pennsylvania Press, 2004), 175-180.
153. Tiziana Terranova, *Network Culture* (London: Pluto, 2004), 3.
154. Gilles Deleuze and Felix Guattari, *A Thousand Plateaus* (London & New York: Continuum, 1987), 10.
155. As Kathleen Burnett wrote in 1993: 'Think of maps you have seen and descriptions you have heard of the internet—a rhizome. If we accept the rhizome as a metaphor for electronically mediated exchange, then hypertext is its apparent fulfillment, and Deleuze and Guattari's "approximate characteristics of the rhizome"—principles of connection, heterogeneity, multiplicity, asignifying rupture, and cartography and decalcomania—may be seen as the principles of hypertextual design.' Kathleen Burnett, "Toward a Theory of Hypertextual Design," *Post Modern Culture*, 3(2), January, 1993.
156. See Hakim Bey's *Temporary Autonomous Zones* in Ludlow, 401-434.

157. Noah Wardrip-Fruin and Nick Montfort (eds.), *The New Media Reader* (Cambridge, MA: MIT Press, 2003), 788.
158. Burnett.
159. Eugene Thacker, "Networks, Swarms, Multitudes: Part One." *Ctheory a142b*, edited by Arthur and Marilouise Kroker, www.ctheory.net/articles.aspx?id= 422 (Accessed April 3, 2007).
160. See Goffey's editorial in *Fibreculture*, issue 4.
161. As they said it, "[t]he fact that robust connectivity allows for the creation of a multitude of small units of maneuver, networked in such a fashion that, although they might be widely distributed, they can still come together, at will and repeatedly, to deal resounding blows to their adversaries." John Arquilla and David Ronfeldt, *Swarming and the Future of Conflict* (Santa Monica: RAND, 2000), 5.
162. Thacker, 2005.
163. Deleuze and Guattari, 1987, 16.
164. Aquilla and Ronfeldt, *Swarming and the Future of Conflict*, 27.
165. For example, clusters of countries engaged in economic cooperation can become vulnerable to contagious financial flows. Similarly, flu viruses spread more in the winter months, not because viral encoding functions better in the cold, but because humans tend to cluster indoors more than they would in the summer.
166. Before the end of the 1990s, knowledge surrounding complex networks had been dominated by the work carried out by Erdos and Renyi in the late 1950s. See P. Erdos and A. Renyi, "On the Evolution of Random Graphs" Publication 5 Institute of Mathematics, Hungarian Academy of Sciences (Hungary 1960), 17-61. The Erdos-Renyi model of random networks has also been central to the development of epidemiology and maintained in the mapping of biological and computer virus spread (See Kephart et al., 1993 "Computers and Epidemi-ology" IBM). In simple terms, the random model defines complex networks as homogonously random; each node has an equally probable chance of having the same amount of links. See also Andrea Scharnhorst, "Complex Networks and the Web: Insights from Nonlinear Physics," *Journal of Computer-Mediated Communication* [On-line], 8(4) (2003). http://www.ascusc.org/jcmc/ vol8/issue4/scharnhorst.html (accessed April 3, 2007) and Barabási, *Linked*.
167. The fluidity, dynamism, and tight communication between members of the N30 Black Bloc anarchist protest movement in the United States, for example, have been cited as key to their effectiveness. Arquilla and Ronfeldt, *Swarming and the Future of Conflict*, 51-52.
168. Barabási. See also Mark Buchanan, *Nexus* (New York & London: Norton, 2002).
169. Barabási, 174.
170. Thacker, 2005.
171. Dan Schiller *Digital Capitalism* (Cambridge, MA: MIT Press, 2000), 8 and Kevin Robins and Frank Webster, *Times of Technoculture* (London: Routledge, 1999), 150.
172. Robins and Webster, 164-167.
173. Robins and Webster, 111-130.
174. Hardt and Negri, 299.

175. Barabási, 144 and Buchanan, 78-82.
176. As RAND claims on its Web site: "In 1969, this "distributed" concept was given its first large-scale test, with the first node installed at UCLA and the seventh node at RAND in Santa Monica. Funded by the Advanced Research Projects Agency and called ARPANET, it was intended for scientists and researchers who wanted to share one another's computers remotely. Within two years, however, the network's users had turned it into something unforeseen: a high-speed, electronic post office for exchanging everything from technical to personal information." http://www.rand.org/about/history/baran.html (accessed November 5, 2007).
177. Stewart Brand, "Wired Legends: Founding Father: An Interview Conducted with Paul Baran in 1999," *Wired* 9, 03 (2001). http://www.wired.com/wired/archive/9.03/baran_pr.html (accessed April 3, 2007).
178. See also John Naughton, *A Brief History of the Future* (London: Phoenix, 2003), 92-117.
179. Brand.
180. The failure of RAND's engineering project can be seen as an intermediate causal series of events that played a role in contextualizing the network. Following poor communications between the telephone monopoly, the military, and RAND, the subsequent omission of Baran's feasibility study, no specific topological pattern was introduced from the outset.
181. Barabási, 67-69.
182. Barabási, 219-226.
183. Scharnhorst.
184. Kephart et al., 1993.
185. Barabási and Bonabeau, "Scale-Free Networks," *Scientific American* (May 2003), 60-69.
186. Buchanan.
187. Albert-Laszlo Barabási and Eric Bonabeau, "Scale-Free Networks," *Scientific American* (May 2003), 60-69.
188. Barabási, 71.
189. Barabási, 152-153.
190. Barabási, 174.
191. Scharnhorst.
192. As Brian Massumi argued: "Codes are always power mechanisms associated with molar organization. They are never neutral or objective. Any science of codes is a science of domination, however subtly masked." Massumi, *A Users Guide to Capitalism and Schizophrenia*, 188.
193. Thacker, 2004.
194. Manuel DeLanda, *A New Philosophy of Society* (London: Continuum, 2006), 56.
195. DeLanda.
196. DeLanda, 41.
197. Paul Hitlin, "False Reporting on the Internet and the Spread of Rumors: Three Case Studies," Gnovis Georgetown University's Journal of Communication, Culture & Technology April 26, 2004. http://www.gnovisjournal.org/journal /false-reporting-internet-and-spread-rumors-three-case-studies (accessed November 19, 2007).

198. Steve White, "Open Problems in Computer Virus Research." Conference paper presented at the Virus Bulletin Conference, Munich, Germany, October 22-23, 1998. http://www.research.ibm.com/antivirus/SciPapers/White/problems/Problems.html (Accessed April 29, 2005).
199. Barabási, 133-134.
200. The Pastor-Satorras and Vespignani study involved more than 800 computer virus infections. They found that computer viruses tend to stay at a low, but stable level of infection over long periods—up to 3 years in some cases. This indicates that the Internet is "Prone to the spreading and the persistence of infections at a prediction of a nonzero epidemic threshold." See Pastor-Satorras and Vespignani, "Epidemic Spreading in Scale-Free Networks," *Physical Review Letters*, 86, no. 14 (April 2, 2000).
201. Kimberly Patch, "Net Inherently Virus Prone," *Technology Research News* (March 21, 2001). http://www.trnmag.com/Stories/032101/Net_inherently_virus_prone_032101.html (accessed April, 3, 2007)
202. For example, as Buchanan argued, ecological systems, such as food webs, are held together by species that are more connected than other species, and if these species are removed from the web, then the entire ecology can collapse. See Buchanan, 151-153. Furthermore, in a study carried out in Sweden the spread of a sexually transmitted disease in a social network can be traced to a small number of promiscuous human nodes that in effect control sexual contact and can therefore pass on a virus despite a zero threshold. Out of 2,810 randomly selected individuals, only a very small number dominated the network of sexual links. See Liljeros et al., "The Web of Human Sexual Contacts," *Nature* 411, 907-908. Finally, viral marketers make a similar claim that the focus of a campaign on so-called "promiscuous sneezes," like Oprah Winfrey, can help to pass on an "idea-virus" to millions of TV viewers. See Seth Godin, *Unleashing the Idea Virus* (London: The Free Press, 2000), 94-96.
203. Buchanan, 175.
204. See Holme in Scharnhorst.
205. Paul Virilio, *The Original Accident* (Cambridge: Polity, 2007), 17 & 15-22.
206. Thacker 2005.
207. Virilio, 18.
208. Virilio, 16.
209. Arquilla and Ronfeldt, 1996, 96.
210. Electronic networks are well understood by the military strategists of netwar. Certainly, Arquilla and Ronfeldt see the immunological analogies adopted in computer virus research by IBM in the early 1990s as "a powerful metaphor for thinking about securing information against predatory electronic attacks." See Arquilla and Ronfeldt, 2000, 27.
211. From a *Financial Times* article called "Disclosure of Risk is an Ethical Dilemma" published September 20, 2005.
212. Octavia Butler, *Dawn* (New York: Popular Library, 1987).
213. A biofilm is a complex aggregation of microorganisms marked by the excretion of a protective and adhesive matrix. Biofilms are also characterized by surface attachment, structural heterogeneity, genetic diversity, complex community interactions, or extracellular matrix of polymeric substances. See E. P. Greenberg,

"Tiny Teamwork," *Nature*, 424, no. 10, (July 10, 2003), 134-140. B. L. Bassler "How Bacteria Talk to Each Other: Regulation of Gene Expression by Quorum Sensing," *Current Opinion in Microbiology*, 2, no. 6 (December 1999), 582-587.

214. Lynn Margulis, *Symbiosis in Cell Evolution: Microbial Communities in the Archean and Proterozoic Eons* (New York: W.H. Freeman, 1992).

215. William Gibson, *Neuromancer* (New York: Ace Books, 1984).

216. Tony Sampson and Jussi Parikka have in different ways, but unanimously argued that the Universal Turing Machine can be rethought as a Universal Viral Machine. Drawing on the research on computer viruses carried out by Cohen, they argued for viral evolution as a means of computation. Jussi Parikka, "The Universal Viral Machine: Bits, Parasite and the Media Ecology of Network Culture," *CTheory, 1000 Days of Theory: td029* (December 2005). http://www. ctheory.net/articles.aspx?id=500, (accessed November 16, 2006). See also, Tony Sampson, "A Virus in Info-Space," *M/C: A Journal of Media and Culture*, 7 (July 2004). http://www.media-culture.org.au/0406/07Sampson.php (accessed December 20, 2006)

217. On this argument see Parikka, "The Universal Machine." See also, Matthew Fuller, *Media Ecologies. Materialist Energies in Art and Technoculture* (Cambridge, MA: MIT Press, 2005).

218. William James, "A World of Pure Experience," *Journal of Philosophy, Psychology*, and *Scientific Methods*, 1 (1901): 533-543, 561-570.

219. Alfred North Whitehead, *Process and Reality* (New York: The Free Press, 1978), 61-82.

220. Alfred North Whitehead, *Concept of Nature* (New York: Prometheus Books, 2004), 185.

221. Mark Hansen, *New Philosophy for New Media* (Cambridge, MA: MIT Press, 2004), 1-10.

222. Alfred North Whitehead, *Concept of Nature* (New York: Prometheus Books, 2004), 167.

223. Whitehead, *Process and Reality*, 288.

224. Greg Lynn, *Folds, Bodies, and Blobs* (Books-By-Architects, 2004), 171.

225. Richard Dawkins, *The Blind Watchmaker* (New York: W. W. Norton, 1986).

226. Dawkins genecetric view of evolution is extensively discussed through the concepts of the "selfish gene" and the "extended phenotype," proposing that the organism and the environment act as the hosts—or vehicles—of a microlevel of evolution driven by genetic replication. Richard Dawkins, *The Selfish Gene* (Oxford: Oxford University Press, 1976).

227. Dawkins, *The Blind Watchmaker*, 45.

228. As Dawkins specified: "the computer starts by drawing a single vertical line. Then the line branches into two. Then each of the branches splits into two sub-branches and so on. It is recursive because the same model is applied locally all over the growing tree," *The Blind Watchmaker*, 51.

229. Conway's *Game of Life* works according to similar principle of evolutionary computation. See the on-line example at http://www.bitstorm.org/gameoflife/ (accessed December 20, 2006).

230. One of the best-known artists in this field is William Latham, who together with Stephen Todd and the IBM research team, generated very complex and organic looking three-dimensional images and animations.

231. Manuel Delanda, "Virtual Environment and the Emergence of Synthetic Reason" (1998). http://www.t0.or.at/delanda.htm (accessed November 30, 2006).

232. On the use of the biomorph model in architectural design, see Celestino Soddou's rapid prototyping realization. Look at http://www.celestinosoddu. com/rp/RP_arch.htm; http://www.celestinosoddu.com/rp/RP_chairs.htm; http://www.celestinosoddu.com/design/soddurings1.htm (accessed November 20, 2006)

233. A set of finite instructions that can be executed a piece at a time on many different processing devices, and then put back together again at the end to get the correct result.

234. Margulis, *Symbiosis in Cell Evolution*.

235. John Holland, *Adaptation in Natural and Artificial Systems* (Ann Arbor: University of Michigan Press, 1975).

236. Richard A. Watson and Jordan B. Pollack, "A Computational Model of Symbiotic Composition in Evolutionary Transitions," *Biosystems*, 69, no. 2-3 (2003), 187-209. http://www.demo.cs.brandeis.edu/papers/biosystems_scet. pdf" (accessed November 30, 2006).

237. Albert-Laszlo Barabási, *Linked: The New Science of Networks* (Cambridge, MA: Perseus Books Group, 2002).

238. Barabási, 67-70.

239. Watson and Pollack, 14-16.

240. A modular home is simply a home built to local building codes in a controlled, environmentally protected building center using precise and efficient construction technology.

241. Auto CAD, 3D Max, Maya, Rhino and Adobe photoshop are the most common software now used as architectural tools to perform procedures such as streaming, scripting, automation, and interaction.

242. Lars Spuybroek, *Nox. Machinic Architecture* (London: Thames & Hudson, 2004), 11.

243. Greg Lynn, *Animate Form* (Princeton, NJ: Princeton Architectural Press, 1999).

244. Lynn, *Animate Form*, 18.

245. Gilles Deleuze, *The Fold.* Trans., Tom Conley (Minneapolis: University of Minnesota Press, 1993), 23.

246. Lynn, *Animate Form*, 15.

247. Bernard Cache explained the notion of singularity: "In mathematics, what is said to be singular is not a given point, but rather a set of points on a given curve. A point is not singular; it becomes singularised on a continuum. . . . We will retain two types of singularity. On the one hand, there are the extrema, the maximum and the minimum on a given curve. And on the other there are those singular points that, in relation to the extrema, figure as in-betweens. These are points of inflection . . . defined only by themselves." Bernard Cache, *Earth Moves* (Cambridge, MA: MIT Press, 1995), 16.

248. Lynn, *Animate Form*, 16.

249. Calculus is built on two major complementary ideas, both of which rely critically on the concept of limits. Differential calculus analyses the instantaneous rate of change of quantities, and the local behavior of functions, a slope, for example, of a function's graph. Integral calculus looks at the accumulation of

quantities, such as areas under a curve, linear distance traveled, or volume displaced. These two processes are said to act inversely to each other by the fundamental theorem of calculus.

250. Deleuze, 14.

251. Deleuze, 17.

252. Chaitin's notion of algorithmic randomness aims to re-address Turing's concept that a computer is a mathematical concept that never makes mistakes. While being always finite, its calculations can go on as long as it has to. After Turing stipulated this idea, von Neumann added that time needed to carry out the calculation—the complexity of computation—had to become central to the study of information. However, Chaitin suggested that rather than time, the question to be addressed for complex computation is the size of computer programs. He directly derived the importance of size from 19th-century physicist Boltzmann, who coined the notion of *entropy*, which is the measure of randomness (how disordered or chaotic a physical system is). In Boltzmann's statistical mechanics, contrary to classical physics, there is a difference between going backward and forward, the arrow of time of the past and the future. For Boltzmann's theory, there is a tendency of entropy to increase: Systems increasingly get disordered. Chaitin draws on Boltzmann's problem of increasing entropy to argue that the size of computer programs is very similar to this notion of the degree of disorder of a physical system. Entropy and program-size complexity are closely related. Gregory Chaitin, *MetaMaths. The Quest for Omega* (London: The Atlantic Books, 2005), 56-85.

253. It is impossible to calculate the value of $\Omega$ digit-by-digit, or bit-by-bit, in binary codes. Chaitin affirms that these digits, written in decimal, coincide with a number between 0 and 1: a decimal point followed by a lot of digits going on forever. Chaitin, *MetaMaths*, 129-143.

254. Chaitin, *MetaMaths*, 105-106.

255. Gregory Chaitin, "Toward a Mathematical Definition of Life," R. D. Levine and M. Tribus, *The Maximum Entropy Formalism* (Cambridge, MA: MIT Press, 1979), 477-498.

256. Lynn, *Animate Form*, 25.

257. Cache, 40.

258. D'Arcy Thompson is often quoted by architects working on the self-generation of form. D'Arcy Thompson, *On Growth and Form* (Cambridge: Cambridge University Press, 1961). See also Greg Lynn, *Folding in Architecture* (New York: Wiley, 2004), 38-41.

259. Lynn, *Folding in Architecture*, 28.

260. Lynn, *Animate Form*, 28-32.

261. Lynn, *Folding in Architecture*, 157-67.

262. Lynn, *Animate Form*, 30.

263. Dorion Sagan, "Metametazoa: Biology and Multiplicity," *Incorporation*, J. Crary and S. Kwinter (eds.) (New York: Urzone, 1992), 378-379.

264. Deleuze, 19-23.

265. On this point, see Deleuze, 86-90.

266. The comparison between a symbiotic and a quantum algorithm is of crucial relevance here. This point, however, cannot be adequately engaged in this chapter and will be object of further research. On recent discussions on the quantum

bit, see Graham P. Collins, "Quantum Bug," *Scientific American* (October 17, 2005). http://www.sciam.com/article.cfm?chanID= sa006&coll D=000D4372 (accessed December 20, 2006).

267. Deleuze, 93.
268. Alfred North Whitehead, *Adventures in Ideas* (New York: The Free Press, 1933), 180-181.
269. Deleuze, 90.
270. Deleuze, 94.
271. Deleuze, 97.
272. Whitehead, *Process and Reality*, 67-68.
273. Lynn, *Animate Form*, 103-119.
274. Deleuze, 86.
275. Documented in Fred Cohen, "Computer Viruses: Theory and Experiments," *Computers and Security* 6, no.1 (1987), 22-35 and Frederick B. Cohen, *A Short Course on Computer Viruses.* 2nd ed, *Wiley Professional Computing.* (New York & Toronto: Wiley, 1994).
276. Peter Szor, *The Art of Computer Virus Research and Defense* (Upper Saddle River, NJ: Addison Wesley, 2005).
277. Ralf Burger, *Computer Viruses : A High-Tech Disease* (Grand Rapids, MI: Abacus, 1989).
278. *Computer Virus Research*, 28. Szor's approach is particularly useful in this context: His book provides a collection of just anything that can be labeled as "viral threat," "viral code," or "malware." To justify this choice, he credited a long list of security researchers, computer scientists, cybernetics scholars as well as virus writers/hackers. Since von Neumann's self-replicating structures, these authors have gradually built a corpus of knowledge and expertise that, today, constitutes the field of computer virus research.
279. Although their intentions/goals might differ from other computer analysts, virus writers can be considered computer virus researchers. Building increasingly complex computer viruses implies a thorough knowledge of and (self)training in computer science. VX heavens, a Web site that collects multidisciplinary information on computer viruses, provides an opening commentary that could be understood as a definition of the computer virus researcher: "This site is dedicated to providing information about computer viruses to anyone who is interested in this topic. [It contains] a massive, continuously updated collection of magazines, virus samples, virus sources, polymorphic engines, virus generators, virus writing tutorials, articles, books, news archives, etc. Some of you might reasonably say that it is illegal to offer such content on the net. Or that this information can be misused by 'malicious people.' I only want to ask that person: 'Is ignorance a defence?'" VXheavens. http://www. netlux.org/
280. Nathan Martin, "Parasitic Media." In *Next Five Minutes Reader*. International Festival of Tactical Media (Amsterdam, September 11-14, 2003), 13 http://www.N5M4.org.
281. Jussi Parikka, "The Universal Viral Machine," *CTheory*, December 15, 2005. In general, the present paper is informed by notions of media ecology and turbulent media also supported by Fuller and Terranova.
282. This idea is laid in visually evident terms by Fuller. Inspired by the Schwitters' Dadaist eccentric method of intermingling objects, he argues: "Parts not only

exist simply as discrete bits that stay separate: they set in play a process of mutual stimulation that exceeds what they are as a set. They get busy, become *merzbilder*." See Matthew Fuller, *Media Ecologies* (Cambridge, MA: MIT Press, 2005), 1.

283. Tiziana Terranova, *Network Culture. Politics for the Information Age* (Ann Arbor: Pluto Press, 2004), 3.

284. Ibid. 4.

285. Massumi defined the distinction between potential and possibility as "a distinction between conditions of emergence and re-conditioning of the emerged." While the former is virtual and "one with becoming," the latter represents its coming into shape: "Emergence emerges." Brian Massumi, *Parables of the Virtual* (Durham & London: Duke University Press, 2002), 10-11.

286. Thierry Bardini, "Hypervirus: A Clinical Report." *CTheory*, February 2, 2006, http://www.ctheory.net/articles.aspx?id=504 (accessed May 10, 2007).

287. Ibid. The use of capital THE is crucial for Bardini, as it indicates a "categorical THE," a "virus mechanism, locking you in THE virus universe"

288. Evelyn Fox-Keller, *Refiguring Life* (New York: University of Columbia Press, 1995), 10.

289. Lily Kay noted how, in the 1950s, the relation between biology and computer science became quite explicit, when the information discourse was described as a system of representations. Kay described this "implosion of informatics and biologics" (Haraway 1997) as a new form of biopower, where "material control was supplemented by the control of genetic information" (Keller 2000).

290. Andrew Ross, "Hacking Away at the Counterculture." In *Strange Weather. Culture, Science and Technology in the Age of Limits* (London, New York: Verso, 1991), 75.

291. Ibid., 76.

292. Alexander Galloway, *Protocol* (Cambridge, MA: MIT Press, 2004).

293. Bardini confirmed: "It is worth noting that if a virus were to attain a state of wholly benign equilibrium with its host cell it is unlikely that its presence would be readily detected OR THAT IT WOULD BE NECESSARILY BE RECOGNIZED AS A VIRUS." Bardini, "Hypervirus."

294. Michel Foucault, *The Archaeology of Knowledge* (London, New York: Routledge, 1989), 36. This is what Deleuze and Guattari called "plane of immanence."

295. Ibid., 37

296. Ibid., 35

297. Ibid., 41

298. Ibid., 39

299. Ibid., 41

300. In his volume on Foucault, Deleuze used "concrete assemblages" to define the products and the executioners of the relations produced by the abstract machine. The Abstract Machine, is, in his interpretation, "like the cause of the concrete assemblages that execute its relations; and these relations between forces take place not 'above' but within the very tissue of the assemblage they produce." See Gilles Deleuze, *Foucault* (London: Athlone Press, 1988), 37. In *A Thousands Plateaus*, the abstract machine is "that which constructs a real that is yet to come, a new type of reality." See Gilles Deleuze and Felix Guattari, *A*

*Thousand Plateaus: Capitalism and Schizophrenia*, Trans. Brian Massini (London: Athlone Press, 1988), 142. The combined use of both philosophers is necessary at this point, not because of their (legitimate) similarities. Whereas Foucault's analysis of discursive formations is an excellent resource to map the conditions that allow the viral to produce so many different instances, Deleuze is needed to portray the dynamic mode through which such instances come to form and their irreducibility to fixed or immanent categories.

301. Donna Haraway, *Modest_Witness@Second _Millennium.Femaleman© _Meets_ Oncomouse™:Feminism and Technoscience* (New York: Routledge, 1997), 295.

302. Ibid., 296.

303. Parikka, "The Universal Viral Machine."

304. Szor here credits in particular John von Neumann, Edward Fredkin, J.H. Conway and Robert Morris Sr.'s Core War. Szor, Computer Virus Research, 11.

305. Ibid. 12

306. Fuller, *Media Ecologies*, 2

307. The topic is equally treated in Félix Guattari, "On the Production of Subjectivity" in *Chaosmosis: An Ethico-Aesthetic Paradigm* (Bloomington, Indianapolis: Indiana University Press, 1995) and Félix Guattari, *Soft Subversions* (New York: Semiotext(e), 1996).

308. Ibid., 110.

309. Ibid., 111.

310. See Guattari elaborated in Fuller, *Media Ecologies*, 5.

311. Guattari, *Soft Subversions*, 110.

312. Ibid., 111.

313. Guattari, *Chaosmosis*, 9.

314. Ibid. 10.

315. *ComputerKnowledge Online.* Virus Naming. (June 2006) http://www.cknow. com/vtutor/VirusNames.html (accessed May 10, 2007). Clearly, the issue is more complex. For the sake of clarity, in this chapter only the so-called "mainstream" opinion is mentioned. In fact, the issue of "naming" would deserve a whole chapter in itself. One should keep in mind that the "exclusion" of virus writers from naming viruses cannot be limited to the mentioned reasons.

316. Sarah Gordon, "What's in a Name?", *SC Magazine Online* (June 2002). http:// www.scmagazine.com (accessed May 10, 2007).

317. Ibid., 9.

318. Wildlist.org. *Naming and Taxonomy*, http://www.wildlist.org (accessed May 10, 2007).

319. Kathryn L. Shaw, "Targeting Managerial Control: Evidence from Franchise." *Working Paper. University of Michigan Business School* (2001): 18.

320. Mani R. Subramani and Balaji Rajagopalan, "Knowledge Sharing and Influence in Online Networks via Viral Marketing." *Communications of the ACM* (2003): 300.

321. Ibid., 301.

322. Jeffrey Boase and Barry Wellman, "A Plague of Viruses: Biological, Computer and Marketing," *Current Sociology* (2001).

323. Emanuel Rosen, *The Anatomy of Buzz* (New York: Doubleday, 2000).

324. Paolo Pedercini, Infectious Media. http://www.molleindustria.org (accessed August 2005).

325. Ibid., 5.
326. Contagious Media Showdown, http://showdown.contagiousmedia.org/ (accessed May 10, 2007).
327. Carrie McLaren, "Media Virus. How silly videos and email pranks created the Bored at Work Network." *Stay Free Magazine*, no. 25 (February 2006).
328. Noticing the widespread popularity of such networks and the entertainment function they seem to embody, Peretti coined the definition *Bored at Work Networks* (BWN). See McLaren, "Media Virus."
329. Subramani and Rajagopalan, "Viral Marketing," 302.
330. See the Blair Witch Project Web site: http://www.blairwitch.com/ (accessed May 10, 2007).
331. McLaren, "Media Virus."
332. Fighthunger.org. The winning entries to our viral video contest. http://www.fighthunger.org/contest/?src=fighthunger468x60 (accessed May 10, 2007).
333. From the GWEI Web site: http://www.gwei.org (accessed May 10, 2007).
334. The Google Adsense reads: "Google AdSense is a fast and easy way for Web site publishers of all sizes to display relevant Google ads on their Web site's content pages and earn money." From http://www.google.com/adsense (accessed May 10, 2007).
335. GWEI.org
336. See Call for Support: Link to Google Will Eat Itself by Geert Lovink. http://www.networkcultures.org/geert/2007/03/22/call-for-support-link-to-google-will-eat-itself/ (accessed May 10, 2007).
337. Laura U. Marks, "Invisible Media," *Public*, no. 25 (2002).
338. Hakim Bey, *T.A.Z.: The Temporary Autonomous Zone, Ontological Anarchy, Poetic Terrorism.* (New York: Autonomedia 1991), 101.
339. Joost Van Loon, *Risk and Technological Culture. Towards a Sociology of Virulence* (London & New York: Routledge, 2002), 20-25.
340. Couze Venn, "A Note on Assemblage." *Theory, Culture & Society*, 22(2-3) (2006), 107.
341. Paul Virilio, "The Museum of Accidents." In *The Paul Virilio Reader*, edited by Steve Redhead (New York: Columbia University Press, 2004), 257.
342. The chapter draws from the research conducted for my book *Digital Contagions: A Media Archaeology of Computer Viruses* (New York: Peter Lang, 2007).
343. Biennale Press Release 2001, 49th International Art Biennale of Venice, Pavilion of the Republic of Slovenia, http://www.epidemic.ws/prelease.txt (accessed May 8, 2007).
344. Cornelia Sollfrank, "Biennale.py—The Return of The Media Hype," Telepolis 7.7. (2001). http://www.heise.de/tp/r4/artikel/3/3642/1.html (accessed November 22, 2007).
345. Nettime mailing list May 17, 2002, http://www.nettime.org/Lists-Archives/nettime-l-0205/msg00116.html (accessed November 22, 2007).
346. The tactical use of net accidents and techniques of making the net visible is of course a more general theme of net.art as well. For instance, as Alex Galloway wrote, in the late 1990s the Electronic Disturbance Theater's Floodnet software, with its techniques of Distributed Denial of Service-attacks, worked as a visualization tool of a kind. Similarly, Jodi concentrated on the "dark side" of net-

working with their focus on, for instance, the 404-error code. In addition, the I/O/D Web Stalker anti-browser provides an alternative to the visualizing functioning of regular browsers. It visualizes the nonspatial Web structures, offering a kind of a map of the logic of the Internet, but similarly exemplifying the contingency in such a task of producing visual knowledge. See Alex Galloway, *Protocol. How Control Exists After Decentralization* (Cambridge, MA: MIT Press, 2004), 214-218. Florian Cramer, *Words Made Flesh. Code, Culture, Imagination* (Rotterdam: Media Design Research Piet Zwart Institute, 2005), 109-119. http://pzwart.wdka.hro.nl/mdr/research/fcramer/wordsmadeflesh, also offers interesting examples of the archaeology of aesthetization of computer crashes, especially the 1968 radio play Die Maschine by Georges Perec.

347. Cf. Claire Colebrook, *Understanding Deleuze* (Crows Nest: Allen & Unwin, 2002), 119.

348. Such ideas resonate with Félix Guattari, *Chaosmosis. An Ethico-Aesthetic Paradigm*. Trans. Paul Bains & Julian Pefanis (Sydney: Power Press, 1995). See also Wendy Hui Kyong Chun, "On Software, or the Persistence of Visual Knowledge," *Grey Room* 18, Winter 2004, 28-51.

349. See Adrian Mackenzie, *Cutting Code. Software and Sociality* (New York: Peter Lang, 2006).

350. Gilles Deleuze, *Foucault*. Trans. Seán Hand (Minneapolis & London: University of Minnesota Press, 1998), 50-51. See also D.N. Rodowick, *Reading the Figural, or, Philosophy After the New Media* (Durham & London, Duke University Press, 2001), 49-54.

351. Deleuze, *Foucault*, 48.

352. Joost Van Loon, "A Contagious Living Fluid. Objectification and Assemblage in the History of Virology." *Theory, Culture & Society* vol. 19 (5/6, 2002), 108-108. Lois Magner, *History of the Life Sciences*, Third Edition, Revised and Expanded. (New York: Marcel Dekker Incorporated, 2002), 290-293. Pasteur had of course been engaged with the rabies virus already in the 1880s. His definition of a virus was, however, a general one as it referred to many kinds of microorganisms. See Magner, 260-262, 290.

353. Magner, 244, 256-257.

354. Kirsten Ostherr, "Contagion and the Boundaries of the Visible: The Cinema of World Health." *Camera Obscura* 50, vol. 17, 2/2002, 6.

355. Magner, 290.

356. Van Loon, "A Contagious Living Fluid," 110. The term *virtual object* is in debt to John Law and Annemarie Mol. However, what needs to be emphasized is that we are dealing with *intensive*, not *extensive* multiplicity. Although extensive multiplicity encompasses multiple instances of a stable standard unit, intensive multiplicity is dynamic in the sense that it changes with every new connection or every new element added to a group. It is in a constant state of differing that is not predetermined by a determined unit. See Colebrook, 58-60. Gilles Deleuze, *Difference and Repetition*, Trans. Paul Patton (New York: Columbia University Press, 1994.) Thus, intensive multiplicity includes in itself a potentiality for change, for new constellations, novel cultural assemblages.

357. Van Loon, "A Contagious Living Fluid," 117. Of course, visualization was not the only issue of aesthesis in enpresenting the minuscule actors. Also other

media were involved, especially the media intended to culture (e.g., bacteria). Such artificial platforms were used to cultivate microorganisms outside animal bodies. Nowadays these are familiar by the name of petri dish. See Magner, 269.

358. Ostherr, 31.

359. Consider Friedrich Kittler's words: "Archaeologies of the present must also take into account data storage, transmission, and calculation in technological media." Kittler, *Discourse Networks 1800/1900* (Stanford, CA: Stanford University Press, 1990), 369-370. Cf. Wolfgang Ernst, "Dis/continuities. Does the Archive Become Metaphorical in Multi-Media Space?" In *New Media,Old Media. A History and Theory Reader*, edited by Wendy Hui, Kyong Chun, & Thomas Keenan (New York, London: Routledge, 2006). Ernst wrote (106): "Media archaeology is driven by something like a certain German obsession with approaching media in terms of their logical structure (informatics) on the one hand and their hardware (physics) on the other, as opposed to British and U.S. cultural studies, which analyze the subjective effects of media, such as the patriarchal obsession with world-wide order and hierarchies in the current hypertext programming languages as opposed to digital options of—female?—fluidity." I, however, want to steer clear of positing any dualisms, and find a third way of articulating together grounding technicity and the discursive effects.

360. Deleuze, *Foucault*, 51.

361. Cf. Ernst, 105, 114. On regularities and archaeology, see *Michel Foucault, The Archaeology of Knowledge* (London, New York: Routledge, 2002). Deleuze, *Foucault*, 4-10.

362. Cf. Ernst, 111.

363. See Mackenzie.

364. Cohen's doctoral dissertation (1986) summed up his research on viruses, underlining that such minuscule technical bytes could yield grave dangers to the organized society: "As an analogy to a computer virus, consider a biological disease that is 100% infectious, spreads whenever animals communicate, kills all infected animals instantly at a given moment, and has no detectable side effects until that moment. If a delay of even one week were used between the introduction of the disease and its effect, it would be very likely to leave only a few remote villages alive, and would certainly wipe out the vast majority of modern society. If a computer virus of this type could spread throughout the computers of the world, it would likely stop most computer usage for a significant period of time, and wreak havoc on modern government, financial, business, and academic institutions." Frederick B. Cohen, *Computer Viruses*. Dissertation presented at the University of Southern California, December 1986, 16.

365. See Vanderbilt News Archives, <http://tvnews.vanderbilt.edu/> (accessed August 28, 2006).

366. Interestingly, a much later Bush Administration paper "The National Strategy to Secure Cyberspace" emphasized almost exactly the same issues: the need for a common front (federal, state, and local governments, the private sector, and the individuals, or the "American people") to secure the crucial national infrastructure behind national administration and defense as well as international business transactions. See "The National Strategy to Secure Cyberspace," White House 2003, http://www.whitehouse.gov/pcipb/ (accessed April 4, 2007).

367. Jon A. Rochlis and Mark W. Eichin, "With Microscope and Tweezers: The Worm From MIT's Perspective." *Communications of the ACM* (June 1989, vol. 32, number 6), 695.

368. Carolo Theriault, "Computer Viruses Demystified." http://www.securitytech-net.com/resource/rsc-center/vendor-wp/sopho/demy_wen.pdf (accessed August 17, 2006).

369. See Jussi Parikka, "Digital Monsters, Binary Aliens—Computer Viruses, Capitalism, and the Flow of Information." Fibreculture, issue 4, *Contagion and Diseases of Information*, edited by Andrew Goffey. http://journal.fibreculture.org/issue4/issue4_parikka.html (accessed April 3, 2007).

370. Sven Stillich & Dirk Liedtke, "Die Wurm von der Wümme." *Stern* 16.6.2004, http://www.stern.de/computer-technik/internet/index.html?id=525454&nv=ct_cb&eid=501069 (accessed November 22, 2007).

371. See Jon Katz, "Who Are These Kids?" *Time*, vol. 155, May 15, 2000.

372. "Suspected creator of 'ILOVEYOU' virus chats online." *CNN.com Chat Transcript* September 26, 2000. http://archives.cnn.com/2000/TECH/computing/09/26/guzman.chat/ (Accessed November 22, 2007).

373. Philip Goodchild, *Deleuze and Guattari. An Introduction to the Politics of Desire* (London: Sage, 1996), 107. Facialization is to be understood here following Deleuze and Guattari as a form of Oedipal representation based on stopping of flows and antiproduction. Similarly as people are socially inscribed by giving them faces, such hybrid and a-human objects as viruses are domesticated similarly. See *A Thousand Plateaus'* chapter "Year Zero: Faciality."

374. Cf. Goodchild, 108.

375. See Deleuze, *Foucault*, 4-10.

376. Cf. Brian Massumi, "Fear (The Spectrum Said)," *Positions* 13:1, 2005.

377. Joost Van Loon, *Risk and Technological Culture* (London, New York: Routledge, 2002), 159. Also see Sampson's chapter in this book.

378. Cf. Tiziana Terranova, *Network Culture. Politics for the Information Age* (London: Pluto Press, 2004).

379. Cf. Parikka, "Digital Monsters, Binary Aliens."

380. Safe hex, safe computing procedures, are yet again another sexualization of the issue that functions to underline the seriousness of the integrity of the body digital.

381. Michelle Delio, "Cashing In on Virus Infections." *Wired*, March 18, 2004. http://www.wired.com/ (accessed November 22, 2007).

382. Ibid.

383. Ibid. The unsuccessful nature of antivirus programs and safety measures might have to do with the fact that their underlying understanding of the topology of the Internet is mistaken. Instead of a state of equilibrium and stability of the Net, complex networks are characterized by metastability and so called scale-free networks. See Tony Sampson, "Senders, Receivers and Deceivers: How Liar Codes Put Noise Back on the Diagram of Transmission," *M/C Journal* 9.1 (2006). http://journal.media-culture.org.au/0603/03-sampson.php (accessed April 24, 2007). See also Sampson's chapter in this book.

384. Trond Lundemo, "Why Things Don't Work. Imagining New Technologies from *The Electric Life to the Digital*." In *Experiencing the Media, Assemblages*

*and Cross-overs*, edited by Tanja Sihvonen and Pasi Väliaho (Turku: Media Studies, 2003), 15-16.

385. Tere Vadén, "Intellectual Property, Open Source and Free Software." Mediumi 2.1. http://www.m-cult.net/mediumi/article.html?id=201&lang=en&issue_nr= 2.1&issueId=5. (Accessed November 22, 2007). Cf. Friedrich Kittler, "Protected Mode." In *Literature, Media, Information Systems*, edited and introduced by John Johnston. (Amsterdam: G+A Arts, 1997), 156-168. On critical views on the "normalization of the Internet," see Geert Lovink, *Dark Fiber. Tracking Critical Internet Culture* (Cambridge, MA & London, England: MIT Press, 2003). See also Matthew Fuller, *Behind the Blip. Essays on the Culture of Software* (New York: Autonomedia, 2003).

386. Cf. Sampson's chapter on network topology and the viral.

387. Dag Pettersson, "Archives and Power," *Ephemera. Theory & Politics in Organization*, vol. 3, February 1, 2003, 34. http://www.ephemeraweb.org/ journal/3-1/3-1petersson.pdf (accessed August 17, 2006).

388. In media archaeology, Siegfried Zielinski has used the term an-archaeology to refer to analysis, which maps minoritarian practices. See Zielinski, *Deep Time of the Media. Towards an Archaeology of Hearing and Seeing by Technical Means* (Cambridge, MA: MIT Press, 2006).

389. These characteristics are from David Harley, Robert Slade, & Urs E. Gattiker, *Viruses Revealed! Understand and Counter Malicious Software* (New York: Osborne/McGraw Hill, 2001), 148-150.

390. "When The Virus Becomes Epidemic."

391. The multiscalar question of interrupts (computer interrupt calls and social interrupts) has been discussed by Simon Yuill, "Interrupt." In *Software Studies*, edited by Matthew Fuller (Cambridge, MA: MIT Press, 2008).

392. On strata, outside and fissures, see Deleuze, *Foucault*, 119-121.

393. This importance of the cinematic and the audiovisual in contemporary societies of control has been underlined e.g., by Patricia Pisters, "Glamour and Glycerine. Surplus and Residual of the Network Society: From Glamorama to Fight Club." In *Micropolitics of Media Culture*, edited by Patricia Pisters, with the assistance of Catherine M. Lord (Amsterdam: Amsterdam University Press, 2001). Maurizio Lazzarato, "From Capital-Labour to Capital-Life." *Ephemera* 4(3), 2004. http://www.ephemeraweb.org/journal/4-3/4-3lazzarato. pdf (accessed August 30, 2006). Cf. Rodowick.

394. Luca Lampo, "When The Virus Becomes Epidemic." Interview conducted by Snafu & Vanni Brusadin, 4/18/2002. http://www.epidemic.ws/dowJones_ press/THE_THING_Interview.htm (accessed April 4, 2007).

395. The peace message was as follows: "Richard Brandow, publisher of the MacMag, and its entire staff would like to take this opportunity to convey their universal message of peace to all Macintosh users around the world." On the objection to "beneficial viruses," see Vesselin Bontchev, "Are 'Good' Computer Viruses Still a Bad Idea?" In *EICAR Conference Proceedings* (1994), 25-47. Cf. Tony Sampson, "Dr Aycock's Bad Idea. Is the Good Use of Computer Viruses Still a Bad Idea?" *M/C Journal* 8.1 (2005). March 16, 2005. http://journal.media-culture.org.au/0502/02-sampson.php (accessed August 17, 2006).

396. Charlie Gere, *Art, Time, and Technology* (Oxford & New York: Berg, 2006), 173-176.

397. "The Biennale Press Release." Cf. Lovink 254-274.
398. Fuller, *Behind the Blip*, 94.
399. Ibid., 61-63.
400. This again connects with Adrian Mackenzie's project in *Cutting Code*.
401. See the Vi-Con page at Runme.org. http://www.runme.org/project/+ViCon/ (accessed November 22, 2007).
402. "When The Virus Becomes Epidemic."
403. See Galloway, 175-206. Cf. Terranova, 67-68.
404. See "When The Virus Becomes Epidemic," an Interview with [epidemiC], April 2001, http://epidemic.ws/dowJones_press/THE_THING_Interview.htm (accessed April 4, 2007).
405. Cf. Rosi Braidotti, "How to Endure Intensity—towards a Sustainable Nomadic Subject." In *Micropolitics of Media Culture*, ed. Patricia Pisters (Amsterdam, University of Amsterdam Press, 2001), 187.
406. Braidotti, "How to Endure Intensity," 188
407. See Sampson.
408. See the exhibition homepage at http://www.digitalcraft.org/iloveyou/index.htm (accessed April 4, 2007).
409. Stefan Helmreich (2002, 244-250) points in his ethnology of artificial life practices toward artificial life as transvestism, a drag performance, a simulacrum that questions the ideas of originality, ground and essence. Helmreich, *Silicon Second Nature. Culturing Artificial Life in a Digital World*. Updated Edition With a New Preface (Berkeley: University of California Press, 2002). Perhaps a bit similarly such net art viruses produce a repetition and a simulacrum that repeats but with a difference. This might also be the practical point where Butlerian repetition and Deleuzian repetition and difference find some resonance. See Eugene Holland, "On Some Implications of Schizoanalysis," *Strategies: Journal of Theory, Culture, and Politics*, Vol. 15 Issue 1 (May 2002), 27-40.
410. Fuller, *Behind the Blip*, 63.
411. Kodwo Eshun, "Motion Capture" in *Abstract Culture, Swarm 1*, 1 (Ccru, 1997).
412. Janne Vanhanen, "Loving the Ghost in the Machine: Aesthetics of Interruption" (2001) *C-Theory*. http://www.ctheory.net/text_file.asp?pick=312 (last accessed March 12, 2007).
413. According the Elggren's sleevenotes, this was article was written by Alexandra Mir in the *Daily News*, New York, September 11, 2002.
414. A number of versions of the projects explanatory text were published in Slovenian, Norwegian, and Austrian newspapers in 2001 and the photographs that accompanied the project were exhibited in both Finland and Norway.
415. The kind of device that would make possible such recordings are currently being researched. "There's a whole world down there" proclaimed one scientist, Flavio Noca, at the Jet Propulsion Lab in California in 2001. In order to capture the sonic hydraulics of micro-cellular machinery, of swimming bacterium and viruses, a special "nanomicrophone" is being developed. Based around the principle of the stereocilia, which are the layers of tiny hairs that line the inner ear (as opposed to the membrane of the ear drum that apparently gets too stiff as you attempt to miniturize it), they are composed of billions of tiny filaments that respond to minute fluctuations of pressure. Noca noted

that "In nature, membranes are present only as coupling devices between the acoustic environment and the zone, typically the cochlea, where the signal is picked up by stereocilia. Nature has evolved toward this solution, probably because of the unique properties of stereocilia at very small [submolecular] scales." Stereocilia are ubiquitous. Interestingly, even "non-hearing" animals (e.g., hydra, jellyfish, and sea anemones) possess them as early warning, directional pressure sensors. But it is the model of a fish's lateral line audition for prey detection, localization and identification that most interests the military researchers. See the interview between Alan Hall and Flavio Noca at http://www.businessweek.com/bwdaily/dnflash/jan2001/nf2001012_818.htm (last accessed June 3, 2005).

416. William Burroughs, *Word Virus: The William Burroughs Reader* (London: Flamingo, 1999), 301.

417. Mark Fisher, "SF Capital." http://www.cinestatic.com/trans-mat/Fisher/sfcapital.htm (last accessed November 4, 2005)

418. Torsten Sangild, "Glitch- The Beauty of Malfunction." In *Bad Music: The Music We Love to Hate*, ed. C. Washburne & M.Demo (New York: Routledge, 2004), 258.

419. Rob Young, "Worship the Glitch." *The Wire*, 190/1 (2000), 52.

420. For example, the deconstruction drenched sleeve notes to the third volume of Frankfurt label Mille Plateaux's "Clicks and Cuts" compilation series which features missives from label boss Achim Szepanski and journalist Philip Sherburne among others.

421. Kim Cascone, "The Aesthetics of Failure: "Post-Digital" Tendencies in Contemporary Computer Music." In *Audio Culture: Readings in Modern Music*, ed. C. Cox & D. Warner (London: Continuum Press, 2004), 392-398.

422. Young, 52.

423. Ibid.

424. Think, for example, of the artist Squarepusher's relation to Jungle.

425. Philip Sherburne, "The Rules of Reduction," *The Wire*, 209/7 (2001), 22.

426. Young, 53.

427. See Reynolds, "Feminine Pressure," *The Wire*, April 1999.

428. Sherburne, "Rules of Reduction," 19-21.

429. Sherburne, "Rules of Reduction," 24.

430. See downloadable articles at www.mille-plateaux.net. Force Inc. had its tendrils in many electronic subgenres (Force Tracks for minimal house, Position Chrome for drum'n'bass, Force Inc. US for disco house, Ritournell for more abstract experimentation).

431. see Simon Reynolds, "Low End Theory," *The Wire* 146 (4/96).

432. Philip Sherburne, "Digital Discipline: Minimalism in House and Techno." In *Audio Culture: Readings in Modern Music*, ed. C. Cox & D. Warner (New York: Continuum, 2004), 324. Sherburne's use of the concept of virtualization as one of substitution here parallels the way Virilio differentiated his concept of simulation from Baudrillard.

433. See the sleeve notes to Mille Plateaux's Clicks & Cuts compilation vol. 3.

434. Paul Virilio, "The Primal Accident." In *The Politics of Everyday Fear*, ed. B. Massumi (Minneapolis: University of Minnesota Press, 1993).

435. Young, 52.

436. See Heinz von Foerster & James W. Beauchamp, *Music by Computers* (New York: Wiley, 1969) and N. Katherine Hayles, *The Posthuman Condition* (Chicago: University of Chicago Press, 1999).

437. William Ashline gives the work of Ryioki Ikeda as an example of unself-consciously Deleuzian music. "Clicky Aesthetics: Deleuze, Headphonics, and the Minimalist Assemblage of 'Aberrations'". *Strategies* 15:1 (2002).

438. As Kodwo Eshun describes in his uptake of Gilroy's concept of the Black Atlantic in More Brilliant than the Sun.

439. Brian Massumi, "The Superiority of the Analog." In *Parables for the Virtual* (Durham: Duke University Press, 2002).

440. Pierre Levy, *Becoming Virtual: Reality in the Digital Age* (Cambridge, MA: Perseus, 1998).

441. Aden Evans, *Sound Ideas: Music, Machines and Experience* (Minneapolis: University of Minnesota Press, 2005), 68.

442. Evans, *Sound Ideas*, 66.

443. Evans, *Sound Ideas*, 64.

444. Evans, *Sound Ideas*, 69.

445. Evans, *Sound Ideas*, 70.

446. Brian Massumi, *Parables for the Virtual* (Durham, NC: Duke University Press, 2002), 143.

447. Evans, *Sound Ideas*, 71.

448. Evans (2004) applied the concept of the surd to sound signal processing and defines it as "a discontinuity that represents the specificity, the unique moment of the original signal- . . . [enruring] that no wholly accurate recreation is possible, that no analysis can do justice to the original signal" (229) Such glitches force engineering to deal constructively with the problem, e.g. the local intervention of the Lanczos sigma as a response to the Gibbs phenomena.

449. Evans, "The Surd."

450. Evans, "The Surd," 231.

451. Massumi, *Parables for the Virtual*, 142.

452. from Clicks + Cuts 3 sleevenotes.

453. Ashline, "Clicky Aesthetics," 87.

454. Ashline, "Clicky Aesthetics," 89.

455. Tobias Van Veen, "Laptops & Loops: The Advent of New Forms of Experimentation and the Question of Technology in Experimental Music and Performance." Paper presented at University Art Association of Canada, Calgary (2002): 12.

456. Young, "Worship the Glitch," 56.

457. "Roughness in the voice comes from roughness in its primary particles, and likewise smoothness is begotten of their smoothness." Lucretius quoted in C. Roads, *Microsound* (Cambridge, MA: MIT Press, 2004), 51.

458. Lucretius, *On the Nature of the Universe* (London: Penguin, 1994), 66.

459. Ironically, Lucretius, like Spinoza is a crucial point of deviation between Deleuze and Badiou who both strongly identify with key aspects of Lucretius. For Badiou, the void in which the cascading atoms swerve is what is appealing, whereas for Deleuze, alongside Michel Serres, it is the swerve itself, the clinamen, the minimal angle of deviation, of differentiating difference which is built on. In fact Deleuze and Guattari swerved from Lucretius in so far as they

replace the void by the virtual. From emptiness the void, as the virtual becomes full of potential.

460. Gilles Deleuze, *Logic of Sense* (London: Athlone, 1990), 269.

461. Gregory Chaitin, *Conversations with a Mathematician: Math. Art, Science and the Limits of Reason* (London: Springer-Verlag, 2001). Chaitin's innovation was to, drawing in particular from Gödel and Turing, identify a new incompleteness theorem revolving around the real number *Omega*, a definable yet noncomputable number that expresses the probability that a random program will halt.

462. See Jonathan Crary, *Suspensions of Perception* (Cambridge, MA: MIT Press, 2001), 72.

463. See Stanley Milgram *Obedience to Authority: An Experimental View* (New York: HarperCollins, 2004).

464. See Leon Chertok and Isabelle Stengers, *A Critique of Psychoanalytic Reason: Hypnosis as a Science Problem from Lavoisier to Lacan* (Stanford CA: Stanford University Press, 1992), 164. Phillipe Pignarre and Isabelle Stengers, *La sorcellerie capitaliste* (Paris: Editions La Découverte, 2005).

465. Jean Baudrillard, *The Intelligence of Evil or the Lucidity Pact* (London: Berg, 2005), 163

466. Although this is not the place to engage in such discussions, our invocation of Baudrillard here is, somewhat paradoxically, in line with a certain realist strand of thinking, which is to say, a strand of thinking which bypasses the critical, anthropocentric prejudice of a kind of thinking which says that any talk of reality necessarily passes through a human subject. See Graham Harman, *Toolbeing. Heidegger and the Metaphysics of Objects* (Chicago IL: Open Court, 2002).

467. Writing of the "principle of evil," Baudrillard commented "to analyze contemporary systems in their catastrophic form, in their failures, their aporia, but also in the manner in which they succeed too well and lose themselves in the delirium of their own functioning is to make the theorem and the equation of the accursed share spring up everywhere, it is to verify its indestructible symbolic power everywhere." Jean Baudrillard, *La transparence du mal* (Paris: Galilée, 1990), 112.

468. Most notably Paolo Virno, Christian Marazzi and Maurizio Lazzarato. See Paolo Virno, *A Grammar of the Multitude* (New York: Semiotext(e), 2003), Christian Marazzi, *La place des chaussettes* (Paris: Editions de l'eclat, 1997), Maurizio Lazzarato, *Les révolutions du capitalisme.* (Paris: Les empêcheurs de penser en rond, 2004).

469. Virno, *A Grammar of the Multitude*, 55.

470. Hannah Arendt, *Between Past and Future* (New York: Viking Press, 1961. Rev. edition, 1968) most clearly displays this sense of the importance of sophistry. See the commentary on Arendt in Barbara Cassin, *L'effet sophistique* (Paris: Gallimard, 1995).

471. A whole range of references would support this claim. Obviously Derrida's work exemplifies this discovery. But so, more prosaically, does that of Bruno Latour and Isabelle Stengers, both of whom make approving winks to sophistry. The "ethical" issues raised by the invention of the principle of noncontradiction by Aristotle have been explored notably in Cassin *L'effet sophis-*

*tique* and by Giorgio Agamben, *Remnants of Auschwitz: The Witness and the Archive* (New York: Zone, 2002).

472. Jacques Lacan, quoted in Friedrich Kittler, *Literature, Media, Information Systems* (London: Routledge, 1997), 143.

473. We take the term *concrescence* from Alfred North Whitehead. See Alfred North Whitehead, *Process and Reality. An Essay in Cosmology* (London: The Free Press, 1979). See Bruno Latour *Petite réflexion sur le culte moderne des dieux faitiches* (Paris: Les empêcheurs de penser en rond, 1996) for a discussion of the process by which things are "made" to be autonomous, and the discussion in Isabelle Stengers, *Cosmopolitiques* (Paris: La Découverte, 1997), 29. Latour's work can be understood in a stratagematic sense in his discussion of the process of fact writing in *Science in Action*.

474. See Marvin Minsky, "Jokes and their Relation to the Cognitive Unconscious." http://web.media.mit.edu/~minsky/papers/jokes.cognitive.text (1981). (Accessed March 5, 2007).

475. Data is piloted through the networks that the Internet is made up of by means of the characteristic four-number addresses required for any machine to be "visible" in a network. An Internet address resolves into a sequence of four numbers each between 0 and 255—255 being the largest number possible in a string of eight bits (binary digits, or 1s and 0s). An address "space" made up of 64 bits is obviously considerably larger than one made up of 32 bits. The extra address space not only allows for many more addresses (think about how many more telephones could be used at any one time if you doubled the basic length of a phone number). It could equally allow for the fine discrimination of types of traffic (as if rather than adding more phones to a telephone network, you used the longer number as a way of differentiating between different types of phone user).

476. An excellent discussion may be found in Aho, Sethi and Ullman, *Compilers. Principles, Techniques and Tools* (Boston, MA: Addison Wesley, 1974).

477. See §236-264 in Jean-Francois Lyotard, *The Differend*, Trans. Georges Van Den Abeele (Minnesota: University of Minnesota Press, 1989), 171-181

478. Massumi's essay on the "autonomy" of affect is a good starting point for this work. See Brian Massumi, *Parables for the Virtual* (Durham, NC: Duke University Press, 2002). A concern with affect equally concerns the work of Lazzarato, Virno, Berardi, and the like. For work which more closely follows the agenda of subordinating other faculties to that of affect, see Mark Hansen, *New Philosophy for New Media* (Cambridge, MA: MIT Press, 2004).

479. The notion of means without ends comes from the Italian philosopher Giorgio Agamben. "Means without ends" simply involve the communication of communicability "properly speaking it has nothing to say, because what it shows is the being-in-language of the human as pure mediality." Giorgio Agamben, *Means without Ends: Notes on Politics.* Trans. Vincenzo Binetti and Cesare Casarino (Minnesota: University of Minnesota Press, 2000), 70.

480. Lyotard, *The Differend*, 13.

481. Media studies' attention to materiality can be found in a number of currents in the field, including the work of Elizabeth L. Eisenstein, Marshall McLuhan, Raymond Williams, Friedrich Kittler, N. Katherine Hayles, etc.

482. The classic development of this argument can be found in Deleuze's essay *Difference and Repetition*, specifically in the chapter on The Image of Thought. See Gilles Deleuze, *Difference and Repetition,* Trans. Paul Fallon (London: Athlone, 1994).

483. The notion of the "autonomy" of reason that shapes the theory–practice distinction ties in to the supposed *a priori* affinity of rationality and the good.

484. As Marquis de Sade observed, the obscene visuality of pornography has an intimate connection with modern forms of coercive power.

485. See e.g., Erkki Huhtamo, "The Pleasures of the Peephole. An Archaeological Exploration of Peep Media." In *Book of Imaginary Media*, edited by Eric Kluitenberg (Rotterdam: NAI Publishers, 2006), 74-155. Dieter Daniels, "Duchamp: Interface: Turing: A Hypothetical Encounter between the Bachelor Machine and the Universal Machine." In *Mediaarthistories*, edited by Oliver Grau (Cambridge, MA: MIT Press, 2007), 103-136.

486. Matteo Pasquinelli, "Warporn Warpunk! Autonomous Videopoesis in Wartime." *Bad Subjects*, 2004, online at http://bad.eserver.org/reviews/2004/pasquinelliwarpunk.html (accessed December 28, 2007).

487. According to the Finnish Ministry of Transport and Telecommunications, the volume of spam was 80% in 2003 but only 30% at the end of year 2005 (see http://www.roskapostipaketti.fi). These figures have, however, been debated and estimates are generally difficult to ascertain. Statistics provided by filtering software suggest spam comprising 40% to 60% of e-mail traffic. (See http://spamlinks.net/http://spam-filter-review.toptenreviews.com/spam-statistics.html; http://www.spamrival.com/spamstats.html; also http://spamlinks.net/).

488. Lorrie Faith Cranor and Brian A. LaMacchia, "Spam!" *Communication of the ACM*, 41, no. 8 (1998), 74.

489. My ongoing research project on online pornography is centrally a methodological one and concerned with different textual methods and conceptualizations of reading. Content description, for example, makes it possible to address general codes and conventions while making it more difficult to account for the examples not quite fitting in. Studies of representation, again, help in mapping the historicity of individual depictions and in situating them in broader, intermedial systems of representation whereas considerations of affect highlight reader–text relationships and the inseparability of affect and interpretation. See Susanna Paasonen, "E-mail from Nancy Nutsucker: Representation and Gendered Address in Online Pornography." *European Journal of Cultural Studies* 9, no. 4 (2006), 403-420. Susanna Paasonen, "Strange Bedfellows: Pornography, Affect and Feminist Reading," *Feminist Theory* 8, no. 1 (2007), 43-57.

490. Amanda Spinks, Helen Partridge, and Bernard J. Jansen, "Sexual and Pornographic Web Searching: Trend Analysis," *First Monday*, 11, no. 9 (2006): electronic document at http://www.firstmonday.org/issues/issue11_9/spink/index.html; Bernard J. Jansen and Amanda Spinks, "How Are We Searching the World Wide Web? A Comparison of Nine Search Engine Transaction Logs," *Information Processing and Management* 42, no. 1 (2006), 258-259; Manuel Bonik and Andreas Schaale, "The Naked Truth: Internet-Eroticism and the Search," *Cut-Up Magazine* 3, no. 20 (2005), http://www.cut-up.com/news//issuedetail.php?sid=412&issue=202005; *Internet Filter Review Statistics*, http://internet-filter-review.toptenreviews.com/internet-pornography-statistics.html

491. On bandwidth and the volume of online porn, see Dick Thornburgh and Herbert S. Lin, *Youth, Pornography, and the Internet* (Washington: National Academy Press 2002), 72-73; on pornography usage, Timothy Buzzell, "Demographic Characteristics of Persons Using Pornography in Three Technological Contexts." *Sexuality & Culture* 9, no. 1 (2005): 28-48; Douglas Phillips, "Can Desire Go On Without a Body? Pornographic Exchange and the Death of the Sun. *Culture.*" *Machine/InterZone*, (2005), http://culturemachine. tees.ac.uk/InterZone/dphillips.html; on both, Lewis Perdue, *EroticaBiz: How Sex Shaped the Internet* (New York: Writers Club Press 2002), 21, 33-35, 179-184.

492. As file sharing has become increasingly common, it has also been mainstreamed in the sense that music, films, and television shows are widely distributed. Although pornography is certainly not alien to P2P exchanges, I am far less convinced of it being the dominant genre.

493. Fears of computer and Internet users are similarly evoked and made use of in the antivirus industry, see Jussi Parikka, "Digital Monsters, Binary Aliens— Computer Viruses, Capitalism and the Flow of Information," *Fibreculture* 4 (2005). http://journal.fibreculture.org/issue4/issue4_parikka.html.

494. Bonik and Schaale; Wendy Chun, *Control and Freedom: Power and Paranoia in the Age of Fiber Optics* (Cambridge, MA: MIT Press, 2006), 125.

495. Gilliam Reynolds and Christina Alferoff, "Junk Mail and Consumer Freedom: Resistance, Transgression and Reward in the Panoptic Gaze." In *Consuming Cultures: Power and Resistance*, ed. Jeff Hearn and Sasha Roseneil (London: Macmillan and the British Sociological Association, 1999), 247.

496. Reynolds and Alferoff, 244.

497. Cranor and LaMacchia, 75-76.

498. For a general overview, see Frederik S. Lane, II, *Obscene Profits: The Entrepreneurs of Pornography in the Cyber Age* (New York: Routledge, 2001).

499. Blaise Cronin and Elizabeth Davenport, "E-Rogenous Zones: Positioning Pornography in the Digital Economy." *The Information Society* 17, no. 1 (2001), 34; Zabet Patterson, "Going On-Line: Consuming Pornography in the Digital Era." In *Porn Studies*, ed. Linda Williams (Durham: Duke University Press, 2004), 104-105.

500. Cf. Katrien Jacobs, Marije Janssen and Matteo Pasquinelli, eds., *C'Lick Me: A Netporn Studies Reader* (Amsterdam: Institute of Network Cultures, 2007).

501. See Katrien Jacobs, "The New Media Schooling of the Amateur Pornographer: Negotiating Contracts and Singing Orgasm" (2004), http://www.libidot.org /katrien/tester/articles/negotiating-print.html; Mark Dery, "Naked Lunch: Talking Realcore with Sergio Messina" (2006), http://www.markdery.com/ archives/blog/psychopathia_sexualis/#000061; Marjorie Kibby and Brigid Costello, "Between the Image and the Act: Interactive Sex Entertainment on the Internet." *Sexualities* 4, no. 3 (2001): 353-369; Patterson 110-120; Amy Villarejo, "Defycategory.com, or the Place of Categories in Intermedia," in *More Dirty Looks: Gender, Pornography and Power*, 2nd edition, ed. Pamela Church Gibson (London: BFI, 2004), 85-91; Denis D. Waskul, ed., *Net.seXXX: Readings on Sex, Pornography, and the Internet* (New York: Peter Lang, 2004).

502. In addition to the literature referenced in this chapter, examples include Jo-Ann Filippo, "Pornography on the Web." In *Web.Studies: Rewiring Media Studies*

*for the Digital Age*, ed. David Gauntlett (London: Arnold, 2000), 122-129. See also Jonathan Janes McCreadie Lillie, "Sexuality and Cyberporn: Toward a New Agenda for Research," *Sexuality & Culture* 6, no. 2 (2002), 25-47.

503. Mary Douglas, *Purity and Danger: An Analysis of the Concepts of Pollution and Taboo* (London: Routledge, 1991/1966), 37.

504. Judith Butler, *Gender Trouble: Feminism and the Subversion of Identity* (New York: Routledge, 1990), 110.

505. Douglas's discussion of anomaly points to the moral dimensions of classification as a means of ordering the world. See Farida Tilbury, "Filth, Incontinence and Border Protection." *M/C Journal* 9, no. 5 (2006). http://journal.media-culture.org.au/0610/06-tilbury.php; also Butler, 132. Douglas' structuralist model, applied here with obvious liberties, is briefly yet snappily critiqued in William Ian Miller, *The Anatomy of Disgust* (Cambridge, MA: Harvard University Press, 1997), 44-45.

506. Sara Ahmed, *The Cultural Politics of Emotion* (Edinburgh: Edinburgh University Press, 2004), 89, emphasis in the original; see also Imogen Tyler, "Chav Scum: The Filthy Politics of Social Class in Contemporary Britain." *M/C Journal* 9, no. 5 (2006). http://journal.media-culture.org.au/06 10/09-tyler.php.

507. Miller, x, 50. The affective investments of disgust, the categories of pornography and homosexuality and the defence of the social body are extensively discussed in Carolyn J. Dean, *The Frail Social Body: Pornography, Homosexuality, and Other Fantasies in Interwar France* (Berkeley: University of California Press, 2000).

508. Michael Warner, *The Trouble With Normal: Sex, Politics, and the Ethics of Queer Life* (Cambridge, MA: Harvard University Press, 2000), 1-5, 25-26; also Don Kulick, "Four Hundred Thousand Swedish Perverts." *GLQ: A Journal of Lesbian and Gay Studies* 11 no. 2 (2005), 205-235.

509. Warner, 181

510. Stuart Hall, "The Spectacle of the Other." In *Representation: Cultural Representations and Signifying Practices*, ed. Stuart Hall (London: Open University Press, 1997), 268.

511. Cf. Annette Kuhn, *The Power of the Image: Essays on Representation and Sexuality* (London: Routledge, 1985/1994), 21.

512. On the low cultural status of porn, see Laura Kipnis, *Bound and Gagged: Pornography and the Politics of Fantasy in America* (Durham, NC: Duke University Press, 1996), 174.

513. Martin O'Brien, "Rubbish-Power: Towards a Sociology of the Rubbish Society." In *Consuming Cultures: Power and Resistance*, ed. Jeff Hearn and Sasha Roseneil (London: Macmillan and the British Sociological Association, 1999), 262-263. The work of Walter Benjamin on the "ruins" of modernity can be seen as an exception. More recently, notions of residue and excess have been addressed particularly in Deleuzean frameworks.

514. Alexander C. Halavais, "Small Pornographies." *ACM Siggroup Bulletin* 25, no. 2 (2005), 19-21; cf. Warner, 181.

515. Marl Storr, *Latex & Lingerie: Shopping for Pleasure at Ann Summers Parties* (Oxford: Berg, 2003), 208-211.

516. Ahmed, 89.

517. The notion of vulnerability is introduced in Linda Williams, "Second Thoughts on *Hard Core*: American Obscenity Law and the Scapegoating of Deviance," in *More Dirty Looks: Gender, Pornography and Power*, 2nd edition, ed. Pamela Church Gibson, 172 (London: BFI, 2004). For a longer discussion on affect, pornography and feminist reading, see Paasonen, "Strange Bedfellows."

518. Anomalous moments are by no means confined to the two examples discussed later. Consider, for example, the advert titled "Pity Fuck" (September 6, 2003) in which a man, sitting in a wheelchair, pretends to be crippled and stages car accidents in order to have "pity sex" with female drivers (as the ad suggests, still sitting in a wheelchair); or, in quite a different tone, *Bukkake Barn* (January 13, 2004) that brings the Japanese genre of women soaked in semen to the American countryside of red barns and farm girls. The list could go on.

519. The notion of pornography's "speech" is addressed in Kipnis, 161.

520. See Paasonen, "E-mail from Nancy Nutsucker," 406.

521. The spectacle of white men with penises the size of arms is also exhibited in the advert titled *Big White Sticks—only* (February 3, 2004), which explicitly excludes non-whiteness from its representational palette. This ad is discussed in more detail in Paasonen, "Strange Bedfellows," 51-53.

522. Jane Arthurs, *Television and Sexuality: Regulation and the Politics of Taste* (London: Open University Press, 2004), 43.

523. Patterson, 106-107.

524. Warner, 185.

525. See Paasonen, "E-mail from Nancy Nutsucker," 407-408. The notion of heterosexual morphology is discussed in Diane Richardson, "Heterosexuality and Social Theory" in *Theorising Heterosexuality: Telling it Straight*, ed. Diane Richardson (Buckingham: Open University Press, 1996), 7; also Judith Butler, *Bodies That Matter: On the Discursive Limits of "Sex"* (New York: Routledge, 1993), 90-91.

526. Jane Juffer, *At Home With Pornography: Women, Sex, and Everyday Life* (New York: New York University Press, 1998), 86-89.

527. Rachel P. Maines, *The Technology of Orgasm: "Hysteria," the Vibrator, and Women's Sexual Satisfaction* (Baltimore: Johns Hopkins University Press, 1999), 121. For a historical discussion of human–machine relationships in relation to objectification, see Jennifer M. Saul, "On Treating Things as People: Objectification, Pornography, and the History of the Vibrator," *Hypatia* 21 no. 2 (2006), 45-61.

528. Storr, 117.

529. Storr, 135-141.

530. Cf. Storr, 119.

531. Amanda Fernbach, *Fantasies of Fetishism: From Decadence to the Posthuman* (New Brunswick, NJ: Rutgers University Press, 2002).

532. Robot Wars were first held in the United States in the mid-1990s, and the concept was developed into a television series in the United Kingdom in 1997.

533. Feona Attwood, "'Tits and Ass and Porn and Fighting': Male Heterosexuality in Magazines for Men," *International Journal of Cultural Studies* 8, no. 1 (2005), 87. Attwood is referring to Andy Moye's discussion of male performance in pornography.

534. Judy Halberstam, *Female Masculinity* (Durham, NC: Duke University Press, 1998), 13.
535. Laura Kipnis, "She-Male Fantasies and the Aesthetics of Pornography." In *More Dirty Looks: Gender, Pornography and Power*, ed. Pamela Church Gibson (London: BFI, 2004), 204-215; Lauren Berlant and Michael Warner, "Sex in Public." In *Intimacy*, ed. Lauren Berlant (Chicago: Chicago University Press, 2000), 328-329.
536. The role of straightness in queer theory is certainly not a new topic. See Calvin Thomas, ed., *Straight With a Twist: Queer Theory and the Subject of Hetero Sexuality* (Urbana: University of Illinois Press, 2000) for basic discussion.
537. See Susanna Paasonen, "Porn Futures." In *Pornification: Sex and Sexuality in Media Culture*, ed. Susanna Paasonen, Kaarina Nikunen, and Laura Saarenmaa (Oxford: Berg, 2008).
538. For discussion of control and user freedom in relation to online porn, see Chun, 98, 124-126.
539 Carolyn Marvin, *When Old Technologies Were New: Thinking About Electric Communication in the Late Nineteenth Century* (Oxford: Oxford University Press, 1988), 8.
540. Jack Sargeant, "Filth and Sexual Excess: Some Brief Reflections on Popular Scatology." *Media/Culture Journal*, vol. 9, no. 5, November 2006. http://journal.media-culture.org.au/0610/00-editorial.php (accessed November 11, 2006).
541. Sargeant, "Filth and Sexual Excess: Some Brief Reflections on Popular Scatology."
542. See also Dougal Phillips' chapter in this anthology.
543. Dougal Phillips, "Can Desire Go On Without a Body? Pornographic Exchange and the Death of the Sun." http://culturemachine.tees.ac.uk/InterZone/d phillips.html (accessed January 8, 2006).
544. In *The History of Sexuality*, vol 1 *The Will To Knowledge*, Foucault explained the functioning of sexuality as an analysis of power related to the emergence of a science of sexuality ("scientia sexualis"). In this volume he attacked the repressive hypothesis, or the widespread belief that we have repressed our natural sexual drives. He shows that what we think of as repression of sexuality actually constitutes sexuality as a core feature of our identities, and has produced a proliferation of sex and pornography discourses.
545. Wendy Hui Kyong Chun, *Control and Freedom: Power and Paranoia in the Age of Fiber Optics* (Cambridge, MA: MIT Press, 2006), 104.
546. Max Gordon, "Abu Ghraib: Postcards from the Edge," October 14, 2004, Open Democracy Web site. http://www.opendemocracy.net/media-abu_ghraib/article_2146.jsp (accessed July 15, 2006).
547. Gordon, "Abu Ghraib: Postcards from the edge."
548. Matteo Pasquinelli, "Warporn Warpunk! Autonomous Videopoiesis in Wartime." *Bad Subjects.* Reviews 2004, http://bad.eserver.org/reviews/2004/pasquinelliwarpunk.html (accessed August 15, 2005)
549. Georges Bataille, *The Tears of Eros,* Trans. Peter Connor (San Francisco: City Light Books, 1989), 204.
550. Georges Bataille, *Eroticism.* In *The Bataille Reader,* ed. Fred Botting and Scott Wilson (New York: Blackwell, 1997), 225.

551. Mark Dery, "Paradise Lust: Web Porn Meet the Culture Wars." Based on a keynote lecture delivered at the conference *Art and Politics of Netporn*, Amsterdam, October 2005. An abbreviated version of the lecture is available on Dery's Web site: www.markdery.com (accessed March 18, 2006). Dery defined *paraphilias* as the psychiatric term for psychosexual disorders such as pedophilia, necrophilia, bestiality, S&M, etc.

552. Dery, "Paradise Lust: Web Porn Meet the Culture Wars."

553. Responses gathered by author at Hong Kong University conference *Film Scene: Cinema, the Arts, and Social Change*, April 2006.

554. Klaus Theweleit, *Male Fantasies*, vol 1. Foreword by Barbara Ehrenreich (Minneapolis: University of Minnesota Press, 1987), 10.

555. Theweleit, *Male Fantasies*, vol 2. Foreword by Anson Rabinbach and Jessica Benjamin (Minneapolis: University of Minnesota Press, 1987), 19.

556. Theweleit, vol. 2, 304.

557. *Abu Ghraib Files, introduction.* http://www.salon.com/news/abu_ghraib/2006/03/14/introduction/ (accessed July 19, 2006).

558. George Zornick, "The Porn of War." *The Nation,* September 22, 2005. http://www.thenation.com/doc/20051010/the_porn_of_war (accessed July 19, 2006).

559. Mark Glaser, "Porn site offers soldiers free access in exchange for photos of dead Iraqis." September 20, 2005, http://www.ojr.org/ojr/stories/050920glaser/ (accessed July 15, 2006). The Geneva Conventions includes *Protocol 1*, added in 1977 but not ratified by the United States, Iraq nor Afghanistan. It mentions that all parties in a conflict must respect victims' remains, although it doesn't mention the photographing of dead bodies. This could well be a judgment call, and the comments added on NTFU (nowthatsfuckedup.com) make the case more clear.

560. Pasquinelli, "Warporn Warpunk! Autonomous Videopoiesis in Wartime."

561. Tobias Van Veen, "Affective Tactics: Intensifying a Politics of Perception," *Bad Subjects,* 63, April 2003. http://bad.eserver.org/issues/2003/63/vanveen.html (accessed August 25, 2006).

562. Van Veen, "Affective Tactics: Intensifying a Politics of Perception."

563. For an overview of this kind of art, see my Web site www.libidot.org and the book *Libi_doc: Journeys in the Performance of Sex Art* (Slovenia: Maska, 2005).

564. Gilles Deleuze, *Masochism: Coldness and Cruelty* (New York: Zone Books, 1991), 57-69.

565. In order to understand and theorize sadomasochism as a strand of Web culture, I have had several online dialogues and sessions with S&M practitioners. I have asked these practitioners to explain their sexual philosophy while giving examples of how they might treat me as co-player. The excerpts are an outcome of this process and the person quoted in this interview wishes to remain anonymous. This and other interviews are more extensively analyzed in my book *Netporn: DIY Web Culture and Sexual Politics* (Rowman & Littlefield, 2007).

566. Statistics available online at http://www.familysafemedia.com/pornography_statistics.html (accessed April 19, 2005).

567. David F. Friedman, board chairman of the Adult Film Association of America, recalled that the XXX rating was actually started as a joke, to distinguish "straight films," such as *A Clockwork Orange* or *Midnight Cowboy*, whose mature content drew an X rating, from pornography. There is not now and has never been a formal XXX rating for the movies; it has always been a marketing

ploy adopted by film distributors and/or movie exhibitors. http://www.4to40.
com/earth/science/index.asp?article=earth_science_xratedmovies (accessed
April 15, 2005).

568. I am indebted for this definition to the explanation posted on the online archive
Principia Cybernetica Web. http://pespmc1.vub.ac.be/ASC/NEGEN-
TROPY.html (accessed May 5, 2005).

569. Jean-François Lyotard, "Can Thought go on without a Body?" in *The Inhuman: Reflections on Time*. Trans. G. Bennington and R. Bowlby (Stanford: Stanford University Press, 1991), 8-10.

570. Lyotard, "Can Thought go on without a Body?", 12.

571. Surrounding the Earth is an expanding sphere of radio waves, now more than 60 light years in radius, the product of transmissions from radio and TV stations. As the wavefront of radio waves spreads out into space, it begins to deteriorate, but if given a boost by a relay station positioned every 9 billion kilometers or so.

572. Jean Baudrillard, "Global Debt and Parallel Universe." Trans. F. Debrix. *CTheory*, Eds. Arthur and Marilouise Kroker. www.ctheory.net/articles.aspx?id=164, 10/16/1996 (accessed May 22, 2007).

573. Baudrillard, "Global Debt and Parallel Universe."

574. All the observations and descriptions of the Web site www.empornium.us are based on my own limited research into the site. Being a rather large, complex, and evolving community, I cannot guarantee that all information presented here regarding the structures, rules, and operation of the Web site is completely accurate. I present it simply as one example of an online, BitTorrent-based pornography-sharing community. I also wish to state that the site contains descriptions and images of hard-core pornography and any one accessing the site after reading this chapter should do so at their own risk and should not access the site in any circumstances where embarrassment or offence may result.

575. For further information, see http://www.bittorrent.com/introduction.html.

576. Vannevar Bush, among others, long ago looked to the decentralized individual as a potential node of energetic information exchange. Bush, an American scientist and engineer known for his role in the development of the atomic bomb, proposed a theoretical proto-hypertext network system in his article "As We May Think," published in *The Atlantic Monthly* in 1945. Bush's memex—generally seen as a pioneering concept for the Web—weaves together lines of individually linked knowledge and uses individual desire and agency as its indexing principle.

As an aside, it is interesting to note that this relationship of exchange between an individual personage as "desiring-machine" and the interwoven trajectories of desire that make up the larger desiring-machine is exactly the mechanical formalism underpinning Deleuze and Guattari's work in *Anti-Oedipus* and *A Thousand Plateaus*.

577. Jean Baudrillard, *Revenge of the Crystal: Selected Writings on the Modern Object and Its Destiny, 1968-1983*, Trans. Paul Foss and Julian Pefanis (Sydney: Pluto Press, 1991), 163.

578. Baudrillard, *Revenge of the Crystal,* 164.

579. Baudrillard, *Revenge of the Crystal,* 170.

580. See, for instance, Markus Giesler and Mali Pohlmann, "The Anthropology of File Sharing: Consuming Napster as a Gift." In Punam Anand Keller and Dennis W. Rook (eds.), *Advances in Consumer Research* (Provo, UT: Association for Consumer Research, vol. 30, 2003); and Richard Barbrook, "The Napsterisation of Everything: A Review of John Alderman, *Sonic Boom: Napster, P2P and the Battle for the Future of Music* (London: Fourth Estate, 2001); available online at http://www.noemalab.org/sections/ideas.php (accessed April 10, 2005).

581. Jean-François Lyotard, *Libidinal Economy*. Trans. I. H. Grant (London: Athlone Press, 1993).

582. Jean-François Lyotard, *The Assassination of Experience by Painting, Monory = L'assassinate de l'experience par la peinture, Monory*. Trans. R. Bowlby (London: Black Dog, 1998).

583. Lyotard, *The Assassination of Experience by Painting*, 91.

584. Julian Pefanis suggested that in Lyotard's phenomenology desire can be figured as the *intense* sign read at the surface of the social, *la grande pellicule*. (*Pellicule* read here in both its senses: as a membrane capable of transferring here and obstructing there the flows and investments of a *desire* that is of the order of production, the libidinal economy; and as a photographic film surface, a screen or gel, a thick holographic plate capable of registering the multiple traces of this movement of desire as a *phantasm.*) Julian Pefanis, "Lyotard and the Jouissance of Practical Reason." In *Heterology and the Postmodern: Bataille, Baudrillard and Lyotard* (Sydney: Allen & Unwin, 1991), 91-92.

585. Jean-François Lyotard, "Notes on the Critical Function of the Work of Art." Trans. S. Hanson. In Roger McKeon (ed.), *Driftworks* (New York: Columbia University Press, 1984), 70-71.

586. For Deleuze and Guattari, a machinic assemblage is a way of thinking about a grouping of processes that work together but can traditionally fall into different categories or classifications. A machinic assemblage is a multiplicity of forces in motion, of "flows," such as in the mental and physical assemblage of a baby attached to the mother's breast, or in the melding of the human body, brain, and mind with the forces and flows of cinema technology.

587. There is an interesting account of this operation, written by the translator Iain Hamilton Grant, in the glossary of terms at the beginning of *Libidinal Economy*.

588. Lyotard, "Can Thought go on without a Body?", 22.

589. Ashley Woodward, "Jean-François Lyotard," online entry at http://www.iep.utm.edu/l/Lyotard.htm, University of Queensland, 2002.

590. Lyotard, "Can Thought go on without a Body?", 9.

591. Lyotard, "Can Thought go on without a Body?", 9.

592. For a quick reference to research literature carried out in this area see Laura Stein and Nikhil Sinha, "New Global Media and Communication." In Leah Lievrouw and Sonia Livingstone (eds.), *The Handbook of New Media* (London: Sage, 2002), 416-418

593. Ibid., 417

594. Greg Elmer, *Profiling Machines: Mapping the Personal Information Economy* (Cambridge, MA: MIT Press, 2004). Ronald Deibert and Nart Villeneuve, "Firewalls and Power: An Overview of Global State Censorship of the

Internet." In *Human Rights in the Digital Age*, Andrew Murray and Mathias Klang (eds.). (London: Glasshouse, 2004). Jason Lacharite, "Electronic Decentralisation in China: A Critical Analysis of Internet Filtering Policies in the People's Republic of China," *Australian Journal of Political Science*, Vol. 37, 2, 2000, 333-346. Richard Rogers, "Toward the Practice of Web Epistemology," in *Preferred Placement: Knowledge Politics on the Web*. (Maastricht: Jan Van Eyck Editions, 2000), 11-23.

595. L. Vaughan and M. Thelwall, "Search Engine Coverage Bias: Evidence and Possible Causes," *Information Processing & Management*, 40(4) (2004), 693-707.

596. Shanthi Kalathil and Taylor C. Boas, *Open Networks, Closed Regimes* (Vancouver: UBC Press, 2003).

597. http://news.bbc.co.uk/2/hi/south_asia/5194172.stm (accessed September 20, 2006).

598. An archive of the listserv that discusses the formulation of the exclusion standard can be found at <http://www.robotstxt.org/wc/mailing-list/robots-nexor-mbox.txt>

599. http://www.robotstxt.org/wc/exclusion.html (accessed June 12, 2006).

600. http://lrs.ed.uiuc.edu/students/tsulliv1/P16Portal/RobotsTAG.html (accessed October 2005).

601. By the Fall 2006, however, Koster had uploaded a more conventional home page that included links to the robot.txt resource pages and a page that listed his other major technical contributions to the early days of the Web.

602. See Martijn Koster, "A Method for Web Robots Control," Internet Draft, Networking Working Group, Internet Engineering Task Force, 1997. http://www.robots.txt.org/wc/norobots-rfc.html (accessed June 12, 2006).

603. Ibid.

604. Ibid.

605. "Whitehouse.gov robots.txt," www.bway.net/~keith/whrobots (accessed May 29, 2007).

606. Lucas Introna and Helen Nissenbaum, "Shaping the Web: Why the Politics of Search Engines Matters," *The Information Society*, 16(3) (2000), 1-17.

607. "Bus Revises Views on 'Combat' in Iraq." *The Washington Post* August 19, 2003, http://www.washingtonpost.com/ac2/wp-dyn/A11485-2003Aug18?language=printer (accessed May 29, 2007).

608. "Lessig Blog," www.lessig.org/blog/archives/001619.shtml (accessed May 29, 2007).

609. "White House's Search Engine Practices Cause Concern." 2600 News, October 28, 2003, www.2600.com/news/view/print/1803 (accessed May 29, 2007).

610. Ibid.

611. See Greg Elmer, "The Case of Web Cookies." In *Critical Perspectives on the Internet*, edited by Greg Elmer (Boulder: Rowman & Littlefield, 2002).

612. The keywords "white house" and "robots" in Google, for instance, return an exclusion list from the White House server.

613. Andrew Leonard, "Bots are Hot!", *Wired*, 4.04, April 1996, http://www.wired.com/wired/archive/4.04/netbots.html (accessed May 29, 2007).

614. http://www.google.com/support/webmasters/bin/answer.py?answer=35250 &topic=8475 (accessed October 11, 2006).

615. G. Mishne, D. Carmel, and R. Lempel, "Blocking Blog Spam with Language Model Disagreement," Proc. Int'l Workshop Adversarial Information Retrieval on the Web (AIRWeb), 2005. http://airweb.cse.lehigh.edu/2005#proceedings (accessed May 29, 2007).

616. Steve Fischer, "When Animals Attack: Spiders and Internet Trespass." *Minnesota Intellectual Property Review*, Vol. 2(2), (2001), 170.

617. White House, http://www.whitehouse.gov/robots.txt (accessed October 9, 2006).

618. The protocol does, however, provide for the exclusion of all robot crawling:
    # go away
    User-agent: *
    Disallow: /
<http://www.robotstxt.org/wc/norobots.html>, (accessed October 9, 2006).

619. William Boddy, "Redefining the Home Screen: Technological Convergence as Trauma and Business Plan." In *Rethinking Media Change: The Aesthetics of Transition*, edited by David Thorburn and Henry Jenkins (Cambridge, MA: MIT Press, 2003), 191-200.

620. "Google Sitemaps Team Interview," http://www.smart-it-consulting.com/article.htm?node=166&page=135 (accessed May 29, 2007).

621. "The Anatomy of a Large-Scale Hypertextual Web Search Engine," http://www-db.stanford.edu/~backrub/google.html (accessed October 9, 2006).

622. Thelwall, 2006. Allen, Burk, and Ess, 2006.

623. Allen, Burk, and Ess.

624. Elmer, "The Case of Web Cookies."

625. J. Pain, "Internet-censor World Championship." In S. Devilette (ed.), *Handbook for Bloggers and Cyberdissidents* (Paris: Reporters without Borders, 2005), 86.

626. The "lock down" of the hinduunity account on Ezboard.com, http://p081.ezboard.com/bhinduunity.lockedDown, was accessed on April 26, 2007, as was the posting of the lawsuit on http://intellibriefs.blogspot.com/search/label/Rape.

627. Joseph Reagle, "Why the Internet is Good," Berkman Center Working Draft, Harvard, 1998, http://cyber.law.harvard.edu/people/reagle/regulation-19990326.html (accessed May 21, 2007).

628. James Boyle, "Foucault in Cyberspace," Duke University Law School, 1997, http://www.law.duke.edu/boylesite/fouc1.html (accessed May 21, 2007).

629. Global list data are per December 2005. See also Kristen Farrell, "The Big Mamas are Watching: China's Censorship of the Internet and the Strain on Freedom of Expression," Working Paper 1539, Berkeley Electronic Press, 2006. The last entry for womenofarabia.com in the Internet archive (archive.org) is dated December 13, 2004.

630. Human Rights Watch's "Women's Rights in Asia" page, http://hrw.org/women/overview-asia.html (accessed April 26, 2007.)

631. Xiao Qiang, "The Words You Never See in Chinese Cyberspace", China Digital Times, August 30, 2004, http://chinadigitaltimes.net/2004/08/the_words_you_n.php, (accessed April 26, 2007).

632. André Breton, *Nadja*, Trans. Richard Howard (New York: Grove, 1960 [1928]), 31.

633. *Deuil pour deuil* (1924), collected in The *Automatic Muse: Surrealist Novels*, Trans. Terry Hale and Iain White (London: Atlas Press, 1994).

634. From *Littérature* no. 6 (November 1922), Trans. Terry Hale.

635. See Georges Bataille, *The Accursed Share, vol. I* (New York: Zone, 1998).

636. James Beniger, *The Control Revolution* (Cambridge, MA: Harvard, 1989), 247.

637. Friedrich Kittler, *Gramophone, Film, Typewriter* (Stanford: Stanford University Press, 1999), 3.

638. On the concept of *mise-en-écriture* see Thomas Levin's essay in *[CTRL]Space: Rhetorics of Surveillance from Bentham to Big Brother* (Cambridge, MA: MIT Press, 2002).

639. See Philip Agre, "Surveillance and Capture: Two Models of Privacy," *The Information Society*, 10(2), 101-127.

640. In this regard one also thinks of the distinction between the "numbered number" and the "numbering number" in Deleuze and Guattari's *A Thousand Plateaus*, Trans. Brian Massumi (Minneapolis: University of Minnesota Press, 1987), 484-485.

641. RFC stands for "request for comments" and are the standard format for documents pertaining to technical protocols (RFCs contain, e.g., the basics for Internet protocols such as TCP and IP).

642. See his book *Networking the World, 1794-2000*, Trans. Liz Carey-Libbrecht and James A. Cohen (Minneapolis: University of Minnesota Press, 2000).

643. Emmanuel Levinas, *On Escape*, Trans. Bettina Bergo (Stanford: Stanford University Press, 2003), 55.

644. Ibid., 54.

645. Ibid., 67.

646. Giorgio Agamben, *The Coming Community*, Trans. Michael Hardt (Minneapolis: University of Minnesota Press, 1993), 86.

647. Paul Virilio, *The Aesthetics of Disappearance*, Trans. Philip Beitchman (New York: Semiotext(e), 1991), 101.

648. Ibid., 58.

649. Hakim Bey, *TAZ: The Temporary Autonomous Zone* (Brooklyn: Autonomedia, 1991), 132.

# ABOUT THE AUTHORS

**Roberta Buiani** is a PhD candidate in communication and cultural studies (Ryerson/York Universities, Toronto, Canada). She serves as a board member of *Fuse*, the Canadian magazine of arts, culture and politics and as Web designer for *TOPIA, the Canadian Journal of Cultural Studies*. Her research and writings investigate the cultural significance of computer viruses in the contexts of network technologies, Internet activism, and tactical media.

**Greg Elmer** is Bell Globemedia Research Chair and Director of the Infoscape Research Lab, Ryerson University, Toronto. He is author of *Profiling Machines: Mapping the Personal Information Economy*. He is currently preparing two books for publication: *Disaggregating the Web* and *Preempting Dissent*.

**Matthew Fuller** is David Gee Reader in digital media at the Centre for Cultural Studies, Goldsmiths College, University of London. He is the author of a number of books including, *Media Ecologies, Materialist Energies in Art and Technoculture* and *Behind the Blip: Essays on the Culture of Software*. His research for this volume and others was supported by the Fonds voor Beeldende Kunst, Vormgeving en Bouwkunst of the Netherlands.

**Andrew Goffey** is senior lecturer in media, culture and communication at Middlesex University, London, and writes in the space between philosophy, science, and culture.

**Steve Goodman** is program leader of the MA Sonic Culture at the University of East London (www.uel.ac.uk). He is currently writing a book on sonic warfare. He runs the record label Hyperdub, and djs and produces electronic music under the name Kode9, with releases on labels such as

Rephlex and Tempa (www.hyperdub.net, www.kode9.com, www.hyperdub.com)

**Katrien Jacobs** is assistant professor at City University of Hong Kong. She is a scholar, writer, and activist in the field of digital art and culture who has published widely on Internet pornography, sexuality, art, and censorship. Her books include *Libi_doc: Journeys in the Performance of Sex Art* and *Netporn: DIY Web Culture and Sexual Politics*.

**John Johnston** is a professor of English and comparative literature at Emory University in Atlanta, Georgia. He is the author of *Carnival of Repetition, Information Multiplicity,* and an edited book, *Literature, Media, Information Systems*, as well as a forthcoming book, *The Allure of Machinic Life: Cybernetics, Artificial Life, and the New AI*. His current interests include media theory, electronic textuality, contemporary science and technology, and postmodern cultural studies.

**Susanna Paasonen**, PhD, is a research fellow at the Collegium for Advanced Studies, University of Helsinki. She is the author of *Figures of Fantasy: Women, Internet and Cyberdiscourse* as well as the co-editor of *Women and Everyday Uses of the Internet: Agency & Identity* and *Pornification: Sex and Sexuality in Media Culture*.

**Jussi Parikka** teaches and writes on the cultural theory and history of new media. He has a PhD in cultural history from the University of Turku, Finland and is senior lecturer in media studies at Anglia Ruskin University, Cambridge, UK. Parikka has published a book on "cultural theory in the age of digital machines" (*Koneoppi*, in Finnish) and is author of *Digital Contagions: A Media Archaeology of Computer Viruses*. Parikka is currently working on a book on insect media, which focuses on the media theoretical and historical interconnections of biology and technology. In addition, the co-edited volume *Archaeologies of Media* is forthcoming.

**Luciana Parisi** conveys the interactive media master degree at Goldsmiths College, UK. She works on endosymbiosis in information transmission. Parisi has published various articles in *Tekhnema, Parallax, Ctheory, Social Text, Mute, TCS,* and *Sexualities* concerning the relation between science (molecular biology, chaos and complexity theories, quantum physics, endosymbiosis, Darwinism and neo-Darwinism), technology (digital technologies and biotechnologies), and the ontogenetic dimensions of evolution in nature and capitalism. Her book *Abstract Sex: Philosophy, Biotechnology and the Mutations of Desire* focuses on the impact of biotechnologies on the notions of the body, sex, femininity, and desire.

**Dougal Phillips** is a Sydney, Australia-based writer, lecturer, and curator. He has worked extensively as a lecturer in the fields of contemporary art, theory, and politics at the Department of Art History and Theory at the University of Sydney; at the Sydney College of the Arts; and at the College of Fine Arts. He completed a PhD in art history and theory at the University of Sydney in 2005 and his work on contemporary media culture, cultural politics, and contemporary art practices from painting, installation and video to conceptual and political art has been published in Australia and internationally. He has also curated a number of exhibition projects in Australia and Asia.

**Richard Rogers** is Chair in New Media and Digital Culture, Media Studies, University of Amsterdam, and director of the Govcom.org Foundation, also in Amsterdam. The group is responsible for the Issue Crawler network mapping software. Rogers is author of *Technological Landscapes*, editor of *Preferred Placement: Knowledge Politics on the Web*, and author of *Information Politics on the Web*.

**Tony D. Sampson** is an academic and writer. He lectures in interactive media at the University of East London and has published widely on the subject of new media. His most recent articles concern digital viruses, noise, accidental environments and the informality of the digital economy in Latin America. He is currently completing a book called *Virality: Contagion Theory in the Age of Networks*.

**Eugene Thacker** and **Alexander R. Galloway** are authors of essays and books, including *The Exploit*, *Biomedia*, *Gaming*, *The Global Genome*, and *Protocol*. They are also producers in the medium and work with the Biotech Hobbyist collective and the Radical Software Group.

# AUTHOR INDEX

# SUBJECT INDEX

9 781572 739161